D1124498

RENEWALS 458-4574

DATE DUE

SEP			
AUG – 8			
JUN			
JUN			
			PRINTED IN U.S.A

Demco, Inc   GAYLORD

Vibrational Spectroscopy of Molecules on Surfaces

METHODS OF SURFACE CHARACTERIZATION

Series Editors:

Cedric J. Powell, *National Bureau of Standards, Gaithersburg, Maryland*
Alvin W. Czanderna, *Solar Energy Research Institute, Golden, Colorado*
David M. Hercules, *University of Pittsburgh, Pittsburgh, Pennsylvania*
Theodore E. Madey, *National Bureau of Standards, Gaithersburg, Maryland*
John T. Yates, Jr., *University of Pittsburgh, Pittsburgh, Pennsylvania*

Volume 1   VIBRATIONAL SPECTROSCOPY OF
MOLECULES ON SURFACES
Edited by John T. Yates, Jr., and Theodore E. Madey

Vibrational Spectroscopy of Molecules on Surfaces

Edited by

John T. Yates, Jr.
University of Pittsburgh
Pittsburgh, Pennsylvania

and

Theodore E. Madey
National Bureau of Standards
Gaithersburg, Maryland

PLENUM PRESS • NEW YORK AND LONDON

Library of Congress Cataloging in Publication Data

Vibrational spectroscopy of molecules on surfaces.

(Methods of surface characterization; v. 1)
Includes bibliographies and index.
1. Vibrational spectra. 2. Surfaces (Physics) — Optical properties. 3. Surface chemistry. I.
Yates, John T., 1935–   . II. Madey, Theodore E. III. Series.
QC454.V5V55   1987   530.4'1   87-14115
ISBN 0-306-42505-X

© 1987 Plenum Press, New York
A Division of Plenum Publishing Corporation
233 Spring Street, New York, N.Y. 10013

Printed in the United States of America

Contributors

Neil R. Avery, CSIRO Division of Materials Science, University of Melbourne, Parkville, Victoria 3052, Australia

Alexis T. Bell, Department of Chemical Engineering, University of California, Berkeley, California 94720

Alan Campion, Department of Chemistry, University of Texas at Austin, Austin, Texas 78712

R. R. Cavanagh, National Bureau of Standards, Gaithersburg, Maryland 20899

J. W. Gadzuk, National Bureau of Standards, Gaithersburg, Maryland 20899

Paul K. Hansma, Department of Physics, University of California, Santa Barbara, California 93106

Brian E. Hayden, Department of Chemistry, University of Bath, Claverton Down, Bath BA2 7AY, England

R. D. Kelley, National Bureau of Standards, Gaithersburg, Maryland 20899

P. L. Richards, Department of Physics, University of California, Berkeley, California 94720

N. V. Richardson, Donnan Laboratories, University of Liverpool, Liverpool L69 3BX, England

J. J. Rush, National Bureau of Standards, Gaithersburg, Maryland 20899

N. Sheppard, School of Chemical Sciences, University of East Anglia, Norwich NR4 7TJ, England

R. G. Tobin, Department of Physics, University of California, Berkeley, California 94720. *Present address:* AT&T Bell Laboratories, Murray Hill, New Jersey 07974-2070

Preface to the Series

A large variety of techniques are now being used to characterize many different surface properties. While many of these techniques are relatively simple in concept, their successful utilization involves rather complex instrumentation, avoiding many problems, discerning artifacts, and careful analysis of the data. Different methods are required for handling, preparing, and processing different types of specimen materials. Many scientists develop surface characterization methods, and there are extensive developments in techniques reported each year.

We have designed this series to assist newcomers to the field of surface characterization, although we hope that the series will also be of value to more experienced workers. The approach is pedagogical or tutorial. Our main objective is to describe the principles, techniques, and methods that are considered important for surface characterization, with emphasis on how important surface characterization measurements are made and how to ensure that the measurements and interpretations are satisfactory, to the greatest extent possible. At this time, we have planned four volumes, but others may follow.

This first volume brings together a description of methods for vibrational spectroscopy of molecules on surfaces. Most of the techniques are currently under active development; commercial instrumentation is not yet available for some techniques, but this situation could change in the next few years. The current state of the art of each technique is described, as are its relative capabilities. An important component of this volume is the summary of the relevant theory.

Two volumes are in preparation which will contain descriptions of the techniques and methods of electron and ion spectroscopies which are in widespread use for surface analysis. These volumes are largely concerned with techniques for which commercial instrumentation is available. The books will fill the gap between a manufacturer's handbook and review articles which highlight the latest scientific developments.

A fourth volume will give descriptions of techniques for specimen handling and depth profiling. It will provide a compilation of methods that have proven useful for specimen handling and treatment, and it will also address the common artifacts and problems associated with the bombardment of solid surfaces by electrons and ions. Finally, a description will be given of methods for depth profiling.

Surface characterization measurements are being used increasingly in diverse areas of science and technology. We hope that this series will be useful in ensuring that these measurements can be made as efficiently and reliably as possible. Comments on the series are welcomed, as are suggestions for volumes on additional topics.

C. J. Powell
Gaithersburg, Maryland

A. W. Czanderna
Golden, Colorado

D. M. Hercules
Pittsburgh, Pennsylvania

T. E. Madey
Gaithersburg, Maryland

J. T. Yates, Jr.
Pittsburgh, Pennsylvania

Preface

The observation of the vibrational spectra of adsorbed species provides one of the most incisive methods for understanding chemical and physical phenomena on surfaces. At the present time, many approaches may be applied to studies of molecular vibrations on surfaces. Some of these are used on high-area solids of technological importance (e.g., heterogeneous catalysts) while others are applied to single-crystal substrates to gain better understanding under conditions of controlled surface structure.

This book has attempted to bring together in one place a discussion of the major methods used to measure vibrational spectra of surface species. The emphasis is on *basic concepts* and *experimental methods* rather than a current survey of the extensive literature in this field.

Two introductory chapters describe the basic theoretical aspects of vibrational spectroscopy on surfaces, dealing with normal modes and excitation mechanisms in vibrational spectroscopy. The remaining seven chapters deal with various methods employed to observe surface vibrations. These are arranged in an order that first treats the use of various methods on surfaces that are not of the single-crystal type. It is in this area that the field first got started in the late 1940s with pioneering work by Terenin and others in the Soviet Union, and by Eischens and others in the United States in the 1950s. The last four chapters deal with relatively recent methods that permit vibrational studies to be made on single-crystal substrates.

The basic philosophy of *Vibrational Spectroscopy of Molecules on Surfaces* has been to present information of a fundamental and practical type that can be used by students just beginning to enter the field. In addition, the authors have often included rather recent developments to lend a timely quality to each of the chapters.

The editors wish to extend their thanks to all of the authors whose work made this book possible.

John T. Yates, Jr.
Theodore E. Madey

Pittsburgh and Gaithersburg

Contents

Normal Modes at Surfaces

N. V. Richardson and N. Sheppard

1. Introduction

Vibrational spectroscopic techniques have played a major role in extending our understanding of structure, bonding, and reactivity in all phases of matter. Only relatively recently has it become feasible to apply these powerful experimental methods to the study of surfaces and species adsorbed at those surfaces. Vibrational spectroscopy has the great advantage over many other surface-sensitive spectroscopies that one has available a vast body of data, already collected and understood, for gas phase, liquid phase, and solid systems. The concept of group frequency is of great importance. Similarly, our knowledge of spectroscopic activity in the gas phase and in three-dimensional crystalline arrays is well developed and amenable to the powerful methods of group theory in its interpretation.

An explosion of interest has occurred in the application of vibrational techniques to the surface environment. Understandably, the greatest body of data comes from adsorption on finely divided solids, using transmission infrared absorption spectroscopy or the inelastic scattering of thermal neutrons.[1-4] More recently Raman, reflection-absorption infrared, and electron energy loss (EEL) spectroscopies have been applied to adsorption on single-crystal surfaces.[5-10] These investigations can be supported by low-energy electron diffraction (LEED)

N. V. Richardson • Donnan Laboratories, University of Liverpool, P.O. Box 147, Liverpool L69 3BX, U.K.   N. Sheppard • School of Chemical Sciences, University of East Anglia, Norwich NR4 7TJ, U.K.

measurements, which can identify ordered arrays of adsorbed species.[11] The latter, in turn, puts greater emphasis on the symmetry properties, both of isolated complexes, which might occur at low adsorbate coverages, and of regular arrays occurring at well-defined coverages. Application of spectroscopic selection rules, based on the symmetry properties of the system under investigation, allows a more rigorous and penetrating analysis of the spectroscopic data. This chapter seeks, therefore, to provide a timely systematic discussion of the symmetry properties of surfaces and surface-absorbed species and then to build on this a description of the vibrational properties of the system. The discussion is supplemented and clarified by inclusion of some examples for which the existing literature exhibits a number of misunderstandings.

Some of this work has previously appeared in an article by Sheppard and Erkelens.[12] In addition, Smith and Eckstrom[13] had previously discussed some particular cases of singly adsorbed molecules and ordered arrays. Nichols and Hexter have discussed, in group theoretical terms, some spectroscopic selection rules in relation to site symmetries of adsorbed molecules and overall symmetries of the combined adsorbate/adsorbent lattices.[14,15] Richardson and Sass[16] and Hexter with Albrecht[17] and with Nichols[18] had earlier discussed the more strictly specified Raman activity of vibrations of individual adsorbed molecules on metal single-crystal surfaces. In these cases, the screening of charges by metal surfaces must be taken into account. This is also true for ir and EELS leading to the "metal-surface selection rules." An introduction to some aspects of the symmetry properties of surfaces and their influence on experimental observations has been given by Richardson and Bradshaw.[19] A very good background to many of the topics covered in this chapter can be found in the textbook on EELS by Ibach and Mills.[9]

In Section 2, we discuss the connection between gas phase degrees of freedom and the vibrational degrees of freedom for the corresponding absorbed species, together with the mixing of adsorbate- and adsorbent-derived vibrations. In Section 3, we briefly review the selection rules governing the various spectroscopic techniques available for surface investigations but concentrate on those for EELS, ir, and Raman.

Section 4 introduces the symmetry consequences of an interface. We examine the symmetry properties of clean surfaces and of particular sites on those surfaces. Finally, in this section, we consider the symmetry reductions experienced by species on adsorption and the combined symmetry of adsorbate and adsorbent site. Section 4 is supported by specific examples covered in Section 5.

Sections 6 and 7 present a similar assessment of the behavior of ordered arrays of adsorbates, including a discussion of the delocalized vibrations or phonons of the adlayer. Section 7 contains several examples discussed in detail. Section 8 concentrates on the phonon modes of the adsorbent with particular attention to those modes whose spectroscopic activity is influenced by adsorption or reconstruction.

A brief summary and forward look are covered in the last section, Section 9.

2. The Vibrational Motions of Adsorbates at Surfaces

A species in the gas phase has three degrees of kinetic freedom conferred on it by each constituent atom: $3N$ degrees of freedom for an N-atom species. For a nonlinear species, three of these are translations, three are rotations, and $3N - 6$ are vibrations. For linear species, there are only two degrees of rotational freedom and hence $3N - 5$ vibrations. A species of such low symmetry that vibrational degeneracies are not possible will have a normal mode corresponding to each of the vibrational degrees of freedom, in which all the atoms vibrate so as to pass through the equilibrium position together, i.e., without change of phase with the same characteristic frequency. Where degeneracies are possible, because of higher symmetries, some normal modes will only differ in the directions of their overall vibrational displacements and the number of different frequencies will be reduced from $3N - 6$ (or $3N - 5$) by coincidences in a manner that can be specified if the symmetry is known.

When an N-atom species is adsorbed on a solid it contributes an extra $3N$ degrees of kinetic freedom. The classification and description of those degrees of freedom, and their relationships to modes of the separated species and the solid adsorbent, depends critically on the type and in particular the strength of the adsorption. In all cases, the motion of the adsorbing species towards the surface *must* be converted to a vibration of the species against the surface, i.e., to a vibrational motion in the potential energy well defining the adsorption itself. The frequency corresponding to translation motion parallel to the surface (two degrees of freedom) may be very low if the adsorbed species feels only a weak corrugation in the adsorption potential. This is most likely for physisorption on "smooth" surfaces such as the close-packed (111) surfaces of fcc metals or (110) of bcc metals. If the frequency is low compared to kT then diffusion across the surface becomes possible and the parallel motion may then be best considered as a hindered translation. Conversely, if the lateral potential "felt" by the adsorbed species is high, a barrier to diffusion exists and the motion of the species is well described as a vibration parallel to the surface. These arguments apply independently to the two degrees of freedom parallel to the surface if the surface symmetry is low enough to ensure they are not degenerate. For example, adsorption on an fcc (110) surface may exhibit hindered translation along the close-packed atomic rows of the $\langle \bar{1}10 \rangle$ direction, while simultaneously having a well-defined vibration in the perpendicular $\langle 001 \rangle$ direction where the corrugation of the potential is likely to be much greater.

Similar arguments apply to the rotations of the gas phase species which become hindered on the surface. If the frequency of the motion is high compared to kT then it is well described as a new vibrational mode. On the other hand, if the hindrance is only slight it may be more valuable to consider the motion as pseudorotation. The latter situation is most likely to apply to rotation around the surface normal. Of course, there are no hindered rotations for single adsorbed atoms, only three vibrations, which derive from the three translational degrees of freedom of a free atom.

The above very qualitative description of normal modes on adsorption ignores two important aspects. In the case of molecular physisorption, the $3N - 6$ "internal" vibrations are likely to bear a simple, direct relationship to their gas phase counterparts with little perturbation of the associated frequencies. Chemisorption, on the other hand, may involve major rearrangement of the bonding pattern and hence a significant change in some or all of the "internal" vibrational frequencies. It is precisely this aspect that makes the study of adsorbate vibrations extremely valuable when one is attempting to understand adsorbate/substrate bonding. Secondly, as yet we have ignored any mixing of the vibrations of the adsorbed species (internal or otherwise) with the phonon modes of the substrate. In an extreme case of strong coupling, all that can be said is that the new system, as a whole, has an additional $3N$ vibrational degrees of freedom. An example of such a strong coupling case are the vibrations of an OH group bonded to a Si atom at the surface of a silica crystal. The chemisorbed OH group brings six additional vibrational degrees of freedom, as indicated in Figure 1. Three of these can be described as vibrations in which the hydrogen atom has the highest amplitude; OH stretching (ν_1), in-plane SiOH angle deformation (ν_2), and a torsional mode associated with out-of-plane motions of the hydrogen atom. Motions associated with the additional oxygen atom include in-plane (ν_5) and out-of-plane (ν_6) motions that are parallel to the surface and a motion of the oxygen atom involving stretching of the Si–O bond perpendicular to the surface (ν_4). There will, of course, be some degree of coupling between the group vibrations involving

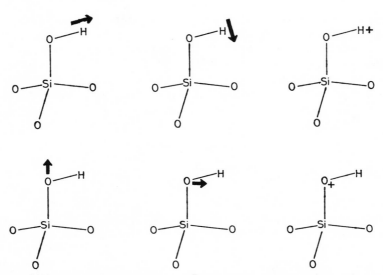

Figure 1. The schematic normal modes of vibration of the surface OH group attached to a silica lattice framework. The upper three modes are dominated by H atom motion, the lower three by O atom motion. The lower SiO_3 unit is part of the silica lattice.

principally motion of the H atom (v_1 to v_3) with those of the O atom, most likely between the Si–O bond stretching (v_4) and the in-plane OH deformation mode (v_2). The other hydrogenic modes have frequencies much different from those involving oxygen atoms. In addition, the vibrations involving oxygen motions of the OH group will be strongly coupled to modes of the bulk SiO_2 lattice. As the oxygen atom of the OH group is linked to the surface by only a single Si–O bond, the motions parallel to the surface (v_5 and v_6) involve mostly angle bending and are expected to have lower frequencies than the lattice optical modes, all of which involve some degree of Si–O stretching. However, by contrast, the surface Si–O stretch itself would have a frequency closely similar to, and probably within, the range of lattice vibrations involving coupled vibrations of Si–O bonds. The additional Si–O bond is probably best regarded as an extension of the lattice with strong vibrational participation with the lattice modes of correct symmetry. The association of a single frequency with Si–OH bond-stretching is not to be expected.

In the case of weak coupling between the lattice vibrations and those associated with the new chemisorbed species, it is meaningful to talk of $3N$ modes of vibration associated with the adsorbate complex. Such a weak coupling requires that the vibrations of the lattice be in a notably different frequency range from those of the adsorbate complex. This is achieved when adsorbates with light atoms are adsorbed on the surface of adsorbents with heavy atoms, e.g., H, CO, or N_2, etc. adsorbed on transition metal surfaces. Apart from the influence of the mass of the constituent atoms, this system is also favorable for weak vibrational coupling because the adsorbent has only relatively low-frequency acoustic modes rather than optical modes of higher frequency encountered in SiO_2. It is important to note that the symmetry considerations discussed in this chapter apply to either the weak or strong coupling cases.

3. Spectroscopic Selection Rule

A primary motivation for using a symmetry classification for the vibrational modes of adsorbate or adsorbate complex, whether isolated or in regular arrays, is to enable spectroscopic selection rules to be applied. In turn, this assists in the assignment of spectroscopic features and provides a framework for any discussion of band intensities and vibrational coupling between adsorbate molecules or between adsorbate and adsorbent.

3.1. Infrared and Raman Spectroscopy

Selection rules derived for infrared and Raman spectroscopy[20] are expected to apply in the usual manner to the surface species with an important additional qualification for metal surfaces. The high electron mobility of metals has an important influence because the electrons in metals are able to screen centers of charge in electric fields, and are able to respond fully to variations in their magnitude and position, at least at the frequencies with which we are concerned, viz., those of interatomic vibrations that are well below metal plasmon frequencies. In the case of infrared spectroscopy this is sufficiently pronounced to create a "metal-surface selection rule," which demands that only those vibrational modes with a component of dynamic dipole moment perpendicular to the surface can be observed.[21,22] Quite generally, these modes belong to the totally symmetric representation of the relevant point group.

Analogous metal-surface selection rules are expected to apply to Raman spectroscopy.[16-18] Screening of the induced dynamic dipoles parallel to the surface and screening of the parallel components of the electric field of the incident, monochromatic radiation by the metal electrons suggests that scattering will mainly arise from the α_{zz} vibrational polarizability component. Once again, only the totally symmetric modes are expected to appear in the spectrum. However, at the visible frequencies used in Raman spectroscopy, metals are appreciably less than perfect conductors and one might expect the surface selection rule to be less rigorous than for the infrared region.

3.2. Vibrational Electron Energy Loss Spectroscopy (EELS)

A primary mechanism by which electrons, of a few electron volts energy, lose some of this energy to the vibrational modes of an adsorbate is through a long-range interaction between the electric field of the incoming electron and the dynamic dipole of the adsorbate.[2,23] Electrons are forward-scattered in this inelastic collision and eventually appear, therefore, close to the specular direction. The dipole nature of this interaction imposes the same metal-surface selection rule that applies to infrared spectroscopy; only those vibrational modes with a component of their associated dynamic dipole moment perpendicular to the surface will be observed in EELS.

While these stringent selection rules would seem to severely limit the appearance of spectral features, and hence create difficulties in identifying the surface-absorbed species and in describing its bonding, the corollary is that, if the observed bands can be assigned, information is immediately available on the orientation of the adsorbed species on the surface. Such arguments have been widely applied to the structure of surface complexes.[24-27]

The dipole mechanism of EELS, thought of most importance, is not the only mechanism of inelastic scattering. Features, usually relatively weak, may appear in an EELS spectrum as a result of "impact" scattering, so called because it is a shorter-range interaction.[28-31] The scattering is not confined to the forward direction and therefore associated losses are best observed away from the specular direction so that the stronger dipole loss features are suppressed. This angular behavior of the loss intensity also provides the means of distinguishing the loss mechanism. However, a word of caution is appropriate in that disorder in the adsorbate overlayer could lead to the dipole loss feature being weakly scattered away from the specular direction.

Certain selection rules apply even under the impact scattering mechanism though they are less restrictive than those appropriate to dipole scattering.[29-31] If a plane of symmetry of the adsorbate/adsorbent system is chosen as the scattering plane, i.e., incident and detected electron beams lie in this plane, then a vibration antisymmetric to reflection in that plane will be forbidden. It could, however be observed in directions away from the incident plane. Additionally, for specular reflection impact scattering, selection rules reduce to the dipole selection rules, if the system has a twofold axis of rotation perpendicular to the surface or a symmetry plane perpendicular to the plane of incidence.

Although, in principle, appropriate off-specular EELS measurements made possible by impact scattering should permit all vibrational modes to be observed, one expects the intensity of the features to depend on the amplitude of the atomic displacements in the vibration. In practice impact scattering has, as yet, been observed only from those modes with significant hydrogen (or deuterium) character. It is to be expected that those modes involving analogous atomic displacements of the adsorbates in the same direction will give higher intensities than those with displacements in opposite directions. Finally, impact scattering is expected to be relatively more important for high-energy, short-wavelength electron beams.[29] In principle, therefore, EELS can provide more information than optical spectroscopies about the vibrational frequencies of modes involving atoms of the adsorbate or surface of the adsorbent. At present, the practical usefulness of the relaxed selection rules of impact scattering is limited by the relatively poor resolution of EEL spectra. Indeed, the resolution often must be further degraded to give adequate signal-to-noise ratios in off-specular measurements.

3.3. Other Vibrational Spectroscopies

Inelastic electron tunneling spectroscopy (IETS) has been used for obtaining vibrational spectra from adsorbed species.[32] The technique involves electron tunneling between two metal electrodes separated usually by a thin oxide-based intermediate layer which constitutes the sample. Adsorption on the high-area oxide itself or on metal particles supported on the oxide can give rise to loss processes in the tunneling electrons, corresponding to vibrational quanta. It seems that both infrared and Raman active modes of vibration can be observed by this method, and there are indications that some intensity effects are related to the metal-surface selection rule, discussed above, for these more traditional spectroscopies. The major disadvantage of IETS is that the tunneling phenomena can be efficiently measured only with the sample at liquid helium temperatures.[32]

Inelastic neutron scattering (INS) has also been used for obtaining vibrational spectra from species adsorbed on high-area samples.[4] The selection rules for this impact scattering technique are again relaxed compared to the optical spectroscopies. The disadvantage arises from the weak cross sections for inelastic neutron scattering, which, therefore, demands high-area samples. Vibrations involving hydrogen, because of their high amplitudes and the higher cross section for incoherent inelastic scattering by hydrogen, give rise to the strongest features, though the motions of heavier atoms that "carry" hydrogen during their vibrations can also be observed.

Inelastic scattering of thermal atomic beams such as helium has been used to follow the dispersion of phonon bands at the surface of materials such as LiF[33] and more recently Au[34] and Ag[35] single crystals. While having the advantage of being strongly interacting particles compared to neutrons and even electrons so that the interactions are with only the top layer of atoms, there are severe technical difficulties associated with inelastic atom scattering, and as yet no attempt seems to have been made to study adsorbed species.

4. Site Symmetries of the Adsorbent

Before embarking on a detailed discussion of symmetry properties of adsorbate/adsorbent systems and considering specific examples, it is necessary to make some general remarks about symmetry elements at an interface. The very definition of an interface implies a severe restriction on the existence of symmetry elements. Only mirror planes perpendicular to the interface are permitted and only rotation axes normal to the interface have any validity. Improper rotation axes and a center of inversion are unable to exist. Strictly, this means that only the point groups C_n and C_{nv} have any significance at a surface. Any treatment of the surface as other than a flat, uniform, unstructured layer imposes a further limitation since C_5, C_7, etc. rotation axes are impossible in extended structures. One is left with ten point groups that are compatible with the vacuum–surface interface as follows[36,37]:

$$C_1, C_s, C_2, C_3, C_4, C_6, C_{2v}, C_{3v}, C_{4v}, C_{6v}.$$

We begin this section with a discussion of bare site symmetries, by which we mean the symmetry of a potential adsorption site on the surface in the absence of an adsorbate. The symmetry elements associated with the bare site determine the highest symmetry that is possible for the site after adsorption, assuming that no adsorption-induced surface re-construction takes place.

Figure 2 illustrates the atomic arrangements to be found at the clean (100), (111), and (110) surface planes of an fcc metal. These are seen to belong to the square, hexagonal, and primitive rectangular two-dimensional lattice systems. There are five Bravais lattices possible in two dimensions, which include oblique and centered-rectangular in addition to those mentioned above. By reference to the *International Tables for*

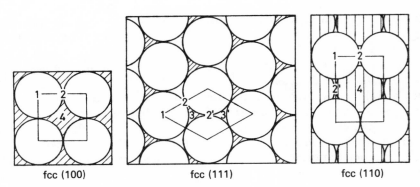

| fcc (100) | fcc (111) | fcc (110) |

Figure 2. The arrangements of metal atoms on the (100), (111), and (110) faces of a face-centered cubic (fcc) metal crystal. Second layer atoms are hatched. High-symmetry, potential adsorbate sites are labelled 1, 2, 3, or 4 according to the nearest-neighbor metal atom coordination number in the site.

X-Ray Crystallography (I.T.)[37] and taking account of the symmetry restrictions imposed by lattice layers below the surface, the arrangements of Figure 2 are found to be classified according to the $4m$, $3m$, and m point groups, respectively, and more specifically to the $p4m$, $p3m1$, and pmm space groups, respectively. To use the Schoenfliess symbols more familiar to spectroscopists, the point groups are C_{4v}, C_{3v}, and C_{2v} and the space groups C_{4v}^1, C_{3v}^1, and C_{2v}^1, respectively. Table 1 shows the relationship between crystal structures, Bravais lattices, and point groups that apply in two dimensions.

Within the unit meshes with the overall symmetry properties described above, there are general and special bare sites each with its own symmetry or point group. In Figure 2 we also indicate for later

Table 1. The Crystal Structures, Lattices, and Corresponding Point Groups That Are Possible in Two Dimensions

Crystal structure	Bravais lattice	Point group
Square	Square	C_{4v}, C_{2v}, C_4, C_2, C_s, C_1
Hexagonal	Hexagonal	C_{6v}, C_{3v}, C_{2v}, C_6, C_3, C_2, C_s, C_1
Rectangular	Primitive rectangular	
	Centered rectangular	C_{2v}, C_2, C_s, C_1
Oblique	Oblique	C_2, C_1

discussion the special, more highly symmetrical sites within each of the (100), (111), and (110) unit meshes of the fcc lattice. For convenience, the numbers used to indicate a given site correspond to the number of *equivalent* atoms in the surface layer with which an atom or molecule can be expected to interact when adsorbed on that site. For the (100) surface the sites 1 (on-top), 2 (bridge), and 4 (fourfold hollow) are seen to have C_{4v}, C_{2v}, and C_{4v} bare site symmetries, respectively; for the (111) surface the sites 1, 2, 3, and 3' (threefold hollow) have bare site symmetries C_{3v}, C_s, C_{3v}, and C_{3v}; and for the (110) surface the sites 1, 2, 2', and 4 all have bare site symmetries of C_{2v}. In each case there are also other bare sites where the symmetry is reduced to C_s, as for example at any point along the line joining sites 1 and 4, 1 and 2, or 2 and 4 of the (100) surface, along the line 1–3 of the (111) surface and finally along the lines 1–2, 1–2', 2–4, and 2'–4 of the (110) surface. All other sites have no associated symmetry elements, that is, they belong to the C_1 point group.

In the case of the (100) and (110) faces it is important to note that the correct, complete symmetry behavior can be deduced from a consideration of the top layer of atoms. Although the second layer of atoms are not directly below the top layer, the locations and numbers of all symmetry elements are the same as for the top layer alone. However, the (111) case is instructive in that this is not the situation. The top layer of atoms by itself has the higher symmetry of a close-packed array, viz., point group 6m (C_{6v}) and space group 6m (C_{6v}^1). The indicated sites 1, 2, 3, and 3' then have C_{6v}, C_{2v}, and C_{3v} symmetry, respectively. The relationship of the second layer to the first precludes the presence of true sixfold and twofold rotation axes, with the result that the correct point and space groups for the surface and the point groups for the sites are the less symmetric axes listed earlier. This example illustrates the point that, although the overall symmetry will correspond to one of the two-dimensional space groups, the correct choice depends on taking into account sufficient layers of the adsorbent. In other words, a complete three-dimensional unit cell must be used when identifying the rotation axes and mirror planes perpendicular to the surface. For this reason, it is necessary to start the discussion from a knowledge of these properties of the three-dimensional crystal lattice.

Finally, it should be noted that the same local or "chemical" type of adsorption complex on different surfaces can have different symmetry properties because of differing bare site symmetries. Thus, the bare sites on the (100), (111), and (110) planes, where in each case an adsorbed species could interact "on top" with a single metal atom (type 1 sites of Figure 2) have the symmetries C_{4v}, C_{3v}, and C_{2v}, respectively. In consequence, the lateral hindered translations of an adsorbed species bonded in these sites would remain degenerate in the first two cases but not for the third. Another example of this effect is provided by the bridge sites labeled 2 or 2′ in Figure 2. For the (100) and (110) surfaces these sites have C_{2v} symmetry, but for the (111) surface C_s is strictly appropriate because of the atoms of the second adsorbent layer.

It is perhaps worth stressing at this stage that the above discussing has been directed towards the rigorously correct description of site symmetries. It is always the case that, in practice, a strictly local symmetry may be adequate for an interpretation of the experimental results, e.g., that C_{2v} symmetry may usefully be assumed to apply to all bridge sites, or that one can use C_{6v} symmetry to discuss on-top (site 1) adsorption on the (111) fcc surface. Indeed an assumed symmetry of $C_{\infty v}$ would be adequate for all on-top sites if the system is insensitive to all but the nearest surface adsorbent atom. In fact this argument can be made somewhat less qualitative. A band observed in a vibrational spectroscopic investigation, which arises because of some reduction in symmetry, will reflect in its intensity the strength of the interaction responsible for reducing the symmetry. For example, in a site of formally C_s symmetry such as the bridge site of the (111) fcc surface, the hindered translation of an adsorbed species parallel to the surface and in the mirror plane formally belongs to the totally symmetric $A′$ representation and would therefore be a dipole-allowed transition. However, the intensity of the band in an infrared or specular EEL spectrum would directly reflect the strength of the interaction between the adsorbed species and the second layer of adsorbent atoms, which are responsible for C_s rather than C_{2v} symmetry. Under C_{2v} symmetry, this hindered translation is not a member of the totally symmetric representation A_1 and would be dipole forbidden. There is always some value in beginning any spectral analysis at some relevant high-symmetry point group and moving in stages of reducing symmetry to that point group that is formally appropriate. In that way, one can assess the importance of a particular interaction or at least the sensitivity of the experimental technique under consideration to that interaction. In addition, in vibrational spectroscopies one should remember that certain normal modes might be more sensitive than others to a perturbation to lower symmetry. For example, internal modes of an adsorbate are likely to be less sensitive to the arrangement of the surface atoms than hindered parallel translations of the complete adsorbate species. Such arguments should be considered to be relevant to all discussions about formally correct symmetry properties and resultant spectroscopic activity throughout the remainder of this chapter, and indeed to other chapters discussing spectroscopic selection rules.

The above discussion has been centered around the symmetry

properties of the bare surface sites. We now turn to a consideration of the symmetry properties of the adsorbing species and the surface complex. By "surface complex" is meant the appropriate bare site, with sufficient adsorbent atoms to define the symmetry of the bare site correctly, together with the adsorbed species. In the case of an atom or cylindrically symmetrical molecule (such as CO) oriented perpendicular to the surface, the symmetry of the surface complex is the same as that of the bare site, i.e., the bare site symmetry is the lowest common symmetry description. In other cases, the symmetry of the surface complex is determined from some more complicated compatibility between the symmetries of the isolated adsorbate and the bare surface site. An example makes this clearer: A methoxy species (CH_3O) adsorbed in an on-top site of a Cu(100) surface has a surface complex symmetry determined from the compatibility of the C_{3v} symmetry of the isolated methoxy species and the C_{4v} symmetry of the bare surface site. The result is a reduction to C_s symmetry for the surface complex as shown in Figure 3.

One cannot assume that the surface complex has the highest possible symmetry compatible with that of the bare site and the isolated adsorbing molecule. The orientation of the adsorbed species on the surface may cause a further reduction in symmetry. A cylindrically symmetrical molecule adsorbed in a bare site of C_{2v} symmetry, but with the molecular axis tilted away from the surface normal, would give rise to a surface complex of C_s symmetry if the tilting were in a mirror plane of the adsorbent and C_1 symmetry otherwise.

It is worth emphasizing again that the symmetry elements of the

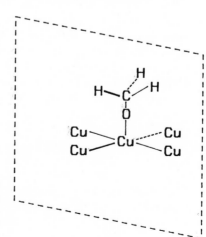

Figure 3. The structure of a methoxy species, CH_3O, adsorbed in an "on-top site" on Cu(100) showing the only remaining mirror plane, symmetry element, σ (point group C_s).

isolated adsorbate that are incompatible with the two-dimensional structured environment of the surface are necessarily lost. It may, however, be relevant to consider this loss in two stages, for the purpose of employing the stepwise symmetry reduction arguments discussed earlier. The strongest interaction of an adsorbate with a surface is likely to be the one that holds it to the surface, i.e., in the direction normal to the surface. This is the interaction that demands a reduction to C_n or C_{nv} symmetry. It is the lateral variation of this potential, dependent on the structured arrangement of surface stoms, which may cause further reduction of symmetry such as removing C_5, and C_n ($n \geqslant 7$) axes. The adsorption of benzene on a surface demands a symmetry reduction from D_{6h} to C_{6v}. Further consideration of the adsorption site causes a further reduction to C_{3v}, C_{2v}, C_3, or C_1. The two stages are best considered separately.

The above discussion has been limited to isolated adsorbate species on a surface and as such is appropriate for low surface coverages where island growth is not expected. At higher coverages, or when island growth is important, lateral interactions are expected to become important, perhaps leading to ordered arrays of adsorbates. The symmetry properties of a site then also depend on the local arrangements of other adsorbate species. This is discussed in Section 6.

5. The Vibrational Modes of Isolated Surface Adsorption Complexes

In this section, we discuss some specific examples of adsorption at low coverage where the rules developed in Section 4 can be applied. In each case, we begin by considering the full surface complex symmetry and describing the resultant vibrational modes. In addition, where appropriate, we indicate where consideration of a different, more local and higher symmetry may be useful. An attempt is made to choose adsorbate/adsorbent systems that have been studied by or are at least closely related to experimental investigations.

5.1. Hydrogen on the (100) Face of a Cubic Metal

The following discussion is appropriate also to other adsorbed atoms whose atomic weight is significantly less than that of the adsorbent atoms. The argument applies equally well to simple, face-centered or body-centered cubic metals. We consider, in turn, adsorption on the three highest symmetry sites, designated 1, 2, and 4 in Figure 2. In each case, three vibrational degrees of freedom replace the three translational degrees of freedom of a gaseous atom. In all cases, the Z axis is chosen perpendicular to the surface.

5.1.1. Site 1

In this "on top" site, the H-atom is adjacent to a single metal atom and the surface adsorption complex has C_{4v} symmetry (Figure 4a). The vibration of the hydrogen against the surface is an M–H bond stretching mode, (M = metal atom) of A_1 symmetry. The hindered translations

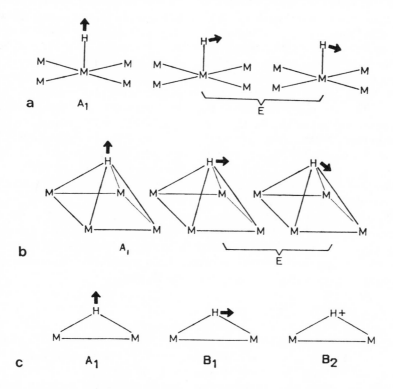

Figure 4. The three vibrations of a H atom bonded to the (100) face of a cubic metal lattice. The metal atoms (M) are assumed to be infinitely massive relative to the H atom. (a) An "on-top" site of C_{4v} symmetry. (b) A fourfold (μ_4) hollow site of C_{4v} symmetry. (c) A twofold (μ_2) bridge site of C_{2v} symmetry. A + denotes motion perpendicular to the MHM plane.

parallel to the surface form a degenerate mode of E symmetry of angle-bending character (Figure 4a). The metal-surface selection rule (Section 3) allows only the A_1, metal-surface stretch mode to be active in infrared, Raman, or specular EEL spectroscopy. The E mode would be active in off-specular EELS. An insulating adsorbent would allow all bands to be observed in all the spectroscopies.

5.1.2. Site 2

A hydrogen atom bridging two surface metal atoms gives rise to an adsorption complex of C_{2v} symmetry. In inorganic chemistry this is described as a μ_2-bridged hydride complex. The vibrational modes of the system are depicted in Figure 4c. A description of these modes in terms of bond stretching and angle bending depends crucially on the MHM angle. Since such a description is closely related to the vibrational frequencies associated with the modes, we consider this in some greater detail.

At high MHM angles, the B_1 mode is largely an antisymmetric M–H stretch.[38] It would be expected at high frequency relative to the A_1 mode, which would have more bending character. On the simplest possible model, when the MHM bond angle is 90°, the A_1 and B_1 modes would have the same frequency while at angles <90° the A_1 mode would be at higher frequency having the more bond-stretching character. The B_2 out-of-plane deformation mode is expected to always occur at the lowest of the three frequencies. The expected frequencies are a sensitive measure of the MHM bond angle.

When the metal surface selection rule is applicable, only the A_1 mode is active in infrared, Raman, and specular EELS. In off-specular EELS the B_1 and B_2 modes become allowed, though a scattering geometry coplanar with the MHM plane would permit only B_1 modes and an orthogonal scattering geometry only the B_2 modes.

5.1.3. Site 4

The adsorption complex again has C_{4v} symmetry (Figure 4b) but the μ_4-bridged complex has the hydrogen atom bonded equally to four surface metal atoms. The hydrogenic vibrations have symmetries A_1 and E as for sites of type 1. For an H-atom well above the plane of metal atoms, the A_1 mode is predominantly a stretching mode.[39] When the H atom is coplanar with the surface atoms, the vibration has out-of-plane deformation character, but this is likely to be influenced by second layer atoms, which in the case of fcc and bcc metals lie directly below the fourfold site, and hence retain some stretching character.

Since the symmetry properties are the same as site 1, the same spectroscopic selection rules apply.

5.2. CO Adsorbed Perpendicular to the (100) Face of a Cubic Metal[1,5]

In this case, the adsorbed species contributes six degrees of vibrational freedom.[19,40] Three of these derive from the translation of the CO center of mass and as such are exactly analogous to the hydrogen modes discussed above, in terms of symmetry and spectroscopic activity. It would be expected that the frequencies would be lower than for adsorbed H atoms because of the larger mass, and the greater size of the CO molecules ensures that the molecule sits further above the metal surface plane than hydrogen in sites of type 2 or 4. As a result the hindered parallel translations are expected to be of low enough frequency to fall into the frequency range of adsorbent phonon bands.[40] The higher-frequency A_1 mode, in which the molecule moves perpendicularly to the surface, should remain higher in frequency than the top of the metal phonon bands.

Of the remaining three modes, one is the internal, CO stretching mode, which belongs to the totally symmetric, A_1 representation in all the sites. It should always be active in infrared, Raman, and specular EEL spectroscopies. Finally, there are two modes that derive from rotations of the gas-phase molecule, sometimes referred to as "frustrated rotations."[40] In the C_{4v} symmetry appropriate to on-top and fourfold hollow site adsorption, they form a second degenerate pair of vibrations, of E symmetry. The degeneracy is lifted on bridge sites, and modes of B_1 and B_2 symmetry result. The frustrated rotations are active only in off-specular EELS, having the same activity as the hindered parallel translations.

The six vibrations are shown in Figure 5, for the on-top and bridge sites.

5.3. N$_2$ Adsorbed Parallel to the Surface in the Bridge Site of a (100) Cubic Face

If the four atoms (two nitrogen and the two metal atoms of the bridge) are coplanar, the adsorption complex has C_{2v} symmetry. The six degrees of vibrational freedom associated with the "light" adsorbate are illustrated qualitatively in Figure 6. Here δ denotes an in-plane deformation mode and γ an out-of-plane mode, while v refers to stretching modes.

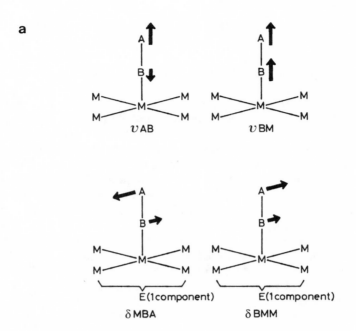

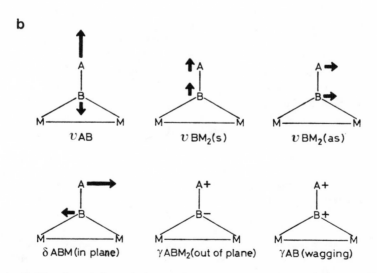

Figure 5. The six variations of a CO molecule bonded to the (100) face of a cubic metal lattice. The metal atoms (M) are assumed to be infinitely massive relative to atoms A and B. (a) An "on-top" site of C_{4v} symmetry. (b) A twofold (μ_2) bridge site of C_{2v} symmetry.

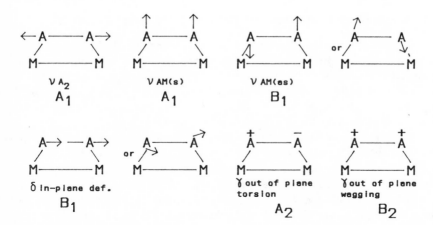

Figure 6. The six idealized vibrations of an N_2 molecule bridging two metal atoms on a cubic (100) face in a site of C_{2v} symmetry.

On a metal surface only the two A_1 modes will be active in infrared, Raman, or specular EEL spectra. It is important to note, however, that included in these active modes is the N≡N stretch, which of course, is infrared inactive in the gas-phase species. The activity derives directly from the loss of $D_{\infty h}$ symmetry on adsorption and, therefore, in line with the arguments of Section 4, the intensity of this feature in the specular EEL or infrared spectrum of parallel adsorbed N_2 molecules, should reflect the strength of chemisorption. An alternative but equivalent way of envisaging how the perpendicular dynamic dipole arises, is to consider the behavior during the vibration of the electrons forming the adsorbate/metal bond. Bonding between the nitrogen molecule and the surface involves electron donation in one direction or the other. As the NN bond stretches or contracts the degree of donation is modulated, leading to an oscillatory dipole moment perpendicular to the surface.

The B_1 and B_2 modes will be active in off-specular EELS if the scattering plane is appropriately chosen as discussed above for the H adsorption case. The A_2 mode becomes active in off-specular EELS only for a nonplanar scattering geometry, or when the scattering plane does not coincide with either of the symmetry planes of the C_{2v} adsorption complex.

On an insulating surface all but the A_2 mode are active in specular EELS and infrared. For off-specular EELS the same arguments apply as for the metal surface. All bands are active in Raman spectroscopy on the insulating surface.

5.4. Ethylene on Bridge Sites of (100) or (111) Metal Faces

This case may be considered as an extension of the N_2 example discussed above. In gas-phase, the planar ethylene molecule has D_{2h} symmetry. Including the translational and rotational modes, the 18 degrees of freedom can be classified as follows: A_g3; A_u1; $B_{1g}3$; $B_{1u}2$; B_{2g}; $B_{2u}3$; $B_{3g}1$; $B_{3u}3$, where the z axis is chosen perpendicular to the molecular plane and the x axis parallel to the C–C bond.[20,41]

On adsorption this symmetry must at least be reduced to C_{2v}. We shall consider the situation when a σ-diadsorbed species is formed by reducing the C–C bond order to 1 and forming two new M–C bonds as shown in Figure 7a. On the bridge site of an fcc (100) surface the adsorption complex has C_{2v} symmetry, but on a (111) surface the second layer of metal atoms demands a further reduction to C_s symmetry. The remaining plane of symmetry bisects and is perpendicular to the C–C bond.

A second possibility, which we consider, is that the MCCM atoms are not coplanar. This could arise from steric crowding associated with a short M–M distance. The heavy atom skeleton is shown in Figure 7b. For the bridge site on the (100) surface the twofold rotation axis indicated is retained and the point group is C_2. Again a C_2 axis is not compatible with a (111) surface because of second layer atoms and the point group is strictly C_1 (no symmetry elements).

Correlation tables are readily available linking the symmetry species for the fully symmetrical gas-phase molecule with those for the reduced symmetries introduced above. The appropriate correlation table linking D_{2h}, C_{2v}, C_2, C_s, and C_1 is provided in Table 2.

With regard to spectroscopic selection rules, on a metal surface, the completely symmetrical vibrations of A_1 A' or A symmetry are active in infrared, Raman, or the specular EELS. In principle the number of active modes, therefore, provides a criterion for the determination of the symmetry of the complex. The number of active modes, in specular EELS, for ethylene on C_{2v} sites would be 5 but on C_2 sites would rise to

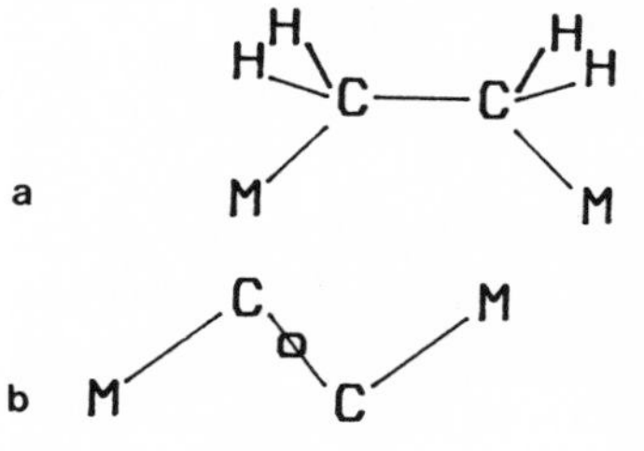

Figure 7. The bridging adsorption of ethylene on a metal surface in the so-called di-σ bonding arrangement. (a) MCCM skeleton planar; (b) MCCM skeleton nonplanar (O, twofold axis).

Table 2. The Correlation between the Symmetry Species of an Isolated D_{2h} Molecule (e.g., Ethylene) and Those That May Apply Following Adsorption

Gas-phase symmetry D_{2h}	Site symmetry				
	C_{2v}	C_2	C_s	C_s^1	C_1
A_g	A_1	A	A'	A'	A
A_u	A_2	A	A''	A''	A
B_{1g}	A_2	A	A''	A''	A
B_{1u}	A_1	A	A'	A'	A
G_{2g}	B_1	B	A'	A''	A
B_{2u}	B_2	B	A''	A'	A
B_{3g}	B_2	B	A''	A'	A
B_{2u}	B_1	B	A'	A''	A

9. The intensity of the "extra" four bands would be related to the degree of twist of the C–C axis out of the earlier MCCM plane (see Figure 7). The number of active modes for C_s or C_1 sites would be 12 or 18, respectively, though $C_{2v} \rightarrow C_s$ and $C_2 \rightarrow C_1$ would probably produce very weak features because it relies on the interaction from the second metal layers.

Ibach and colleagues have discussed similar analyses for ethylene adsorbed on the different sites of a (110) bcc surface.[42]

6. Two-Dimensional Adsorbate Arrays at Surfaces

We turn now to a discussion of those, perhaps more common, adsorbate regimes where adsorbate/adsorbate interactions are important. This will occur at higher coverages or at lower coverages when attractive adsorbate/adsorbate interactions make island growth favorable. In many systems, particularly for small or highly symmetrical molecules, ordered two-dimensional arrays of adsorbate species are found at the surface. Very often the ordering is strongly influenced by the adsorbent, in which case the periodicity of the adsorbate array has a simple relationship to that of the underlying surface. For larger molecules interacting with many surface atoms, or for those adsorbate systems in which adsorbate lateral interactions are strong relative to the corrugation of the surface potential felt by the adsorbate molecule, periodicity with respect to the lattice may be long range or absent.[11] The local arrangement of the adsorbate species may in such cases show apparent high symmetry. Within the long-range periodicity that reflects the poor registry with the adsorbent lattice, different adsorbates will have different detailed interaction with the surface and the strong features in the spectrum are likely to show greater breadth.

The experimental evidence for the size and symmetry properties of the unit cells in regular arrays of adsorbed species is usually obtained from low-energy electron diffraction (LEED).[11] The penetration depth of electrons at energies (30–300 eV) appropriate to LEED ensure also that information on the registry with the adsorbent lattice is also provided. In many respects, LEED and vibrational spectroscopic techniques are complementary. The former provides information about the arrangement and spacing of occupied adsorption sites and the latter greater details about the local site itself.

As mentioned earlier, the two-dimensional (2D) unit cell corresponds to one of the five two-dimensional Bravias lattices.[36,37] These are shown in Figure 8. There are four associated 2D crystal structures: square, rectangular, hexagonal, and oblique. Screw-axes cannot occur in two dimensions and glide-planes (glide lines) are limited to those perpendicular to the 2D lattice. Translation symmetry linked with the ten point group symmetries in two dimensions (see Section 4) result in 17 2D space groups.[36,37] These are described in greater detail in Vol. I of the *International Tables of X-Ray Crystallography* (p. 46).[37]

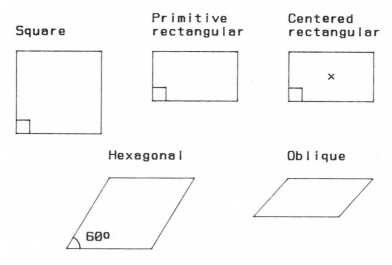

Figure 8. The five Bravais lattices that can exist in two dimensions.

6.1. The Site-Symmetry of a Surface Complex within a Regular Array

The adsorbate array itself defines a point group symmetry at the adsorbate site. Formally, the correct symmetry is the lowest common resultant of the gas-phase, bare site, and adsorbate array symmetry. The 2D array symmetry may be higher than, the same as, or lower than the symmetry dictated by the relevant bare surface site. A higher symmetry is likely only to occur when lateral interactions are dominant and adsorbate hexagonal close-packing is the result. In this case, surface adsorption site and its symmetry become meaningless concepts since a whole range of local adsorbate/surface atom arrangements are the result. Formally, the symmetry is reduced to C_1, but in practice the C_{6v} symmetry of the adsorbate array is probably appropriate for spectroscopic analysis.

The simplest situation arises when the adsorbate atoms define the same symmetry as the bare surface site. Examples are (1×1), $p(2 \times 2)$, $c(2 \times 2)$, indeed $(n \times n)$ overlayers on (100) cubic faces $(n \times n)$ on (111) cubic faces and $(m \times n)$ overlayers on (110) cubic faces. If the periodicity is greater than (1×1) then domains exist on the surface. Crossing the boundaries of these displacive domains causes simply a phase shift in the adsorbate periodicity. There are mn distinct domains in an $(m \times n)$ overlayer. However, the existence of these should not greatly affect the vibrational spectrum unless the domains are small, and the domain wall density consequently high, such that significant numbers of adsorbate species lie in reduced symmetry locations of a domain wall.

When the symmetry imposed by the adsorbate unit cell is lower than that of the bare site a further complication can arise because of rotational domains. An example is the $p(2 \times 1)$ overlayer sometimes found on (111) surfaces. The unit cell of the adsorbate dictates a C_2 symmetry, whereas the bare surface threefold hollow site has C_{3v} symmetry. Six rotational domains of the (2×1) overlayer can exist, as shown in Figure 9, related to each other by a C_3 axis and a mirror plane of the bulk

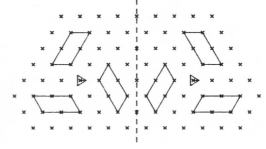

Figure 9. The six rotational domains for a $p(2 \times 1)$ overlayer on the (111) face of a cubic metal. $\triangleright$, Threefold axis; – – –, mirror plane.

lattice, which are incompatible with the adsorbate array symmetry. The implications of the existence of rotational domains for vibrational spectroscopy are that the spectroscopic behavior for each domain type must be considered separately in relation to experimental geometry, e.g., the direction of the plane of incidence, and the results added together.

7. The Vibrations of Adsorbate Arrays

When adsorbate species are at sufficiently high coverage that ordered arrays are produced, the vibrational motion of adsorbate species in different unit cells may be coupled sufficiently strongly that a description of the atomic motions in terms of phonon dispersion becomes useful. The treatment of such systems is well established in three-dimensional systems.[43-46] A knowledge of the two-dimensional, *primitive* unit cell of the system permits a determination of the *reciprocal lattice* or *surface Brillouin zone* (SBZ). The different, coupled vibrational modes can be characterized, in real space, by the phase relationship of displacements in adjacent unit cells. Alternatively, they can be characterized by a wave-vector k ($=2\pi/\lambda$, where λ is the wavelength of the associated phonon). $k = 0$ corresponds to the center (point) of the SBZ and to in-phase vibration of all unit cells, i.e., $\lambda = \infty$. $k = \pi/a_x$, where a_x is the surface unit cell dimension, is the largest meaningful wave-vector and corresponds to a point at one edge of the SBZ where neighboring unit cells, in the x direction, oscillate 180° out-of-phase. The symmetry properties of the SBZ are the same as those of the real space lattice, and individual points in the SBZ have associated point group symmetries, which have a one-to-one correspondence to the point group appropriate to sites in the real-space unit cell. Even in the absence of interunit cell coupling and phonon dispersion, it is important to consider the symmetry properties of the complete unit cell in order to describe the vibrations and their spectroscopic activity.

The degree of dispersion, the variation of vibrational frequency with phase relationship between adjacent unit cells (i.e., with phonon wave vector), depends on the extent of coupling between the vibrations of adjacent unit cells. For the internal vibrations of an adsorbate, the dispersion rarely exceeds a few percent of the vibrational frequency. It may, however, be substantially greater for the external modes of hindered rotation and translation. The surface modes of the adsorbent itself often provide examples of strong coupling between adjacent cells.[47,48]

Special attention needs to be directed at the symmetry properties of the unit cell for several reasons. Firstly, it defines the relevant factor group for analysis of all the vibrations in the absence of dispersion. Secondly, it provides the symmetry properties of the SBZ center, and all other points in the SBZ can be related to the center by simple correlation tables. Finally, and arguably most importantly, many important spectroscopic techniques are limited to probing the modes at the SBZ center, i.e., they can only excite $k = 0$ phonons. This latter limitation arises with long wavelength probes such as infrared or visible radiation for which the photons have insufficient momentum to cause significant displacement from $k = 0$.

If the wavelength of the incident probe is comparable to the dimensions of the unit cell, then phonons with $k \neq 0$ can be excited. Slow neutron beams fall into this category as do the electron beams typically used in EELS. The wavelength of an electron beam, of energy V electron volts, is given numerically in Å by

$$\lambda = (150/V)^{1/2}$$

A typical EELS experiment usually involves an electron beam of 2–10 eV corresponding to wavelengths of ca. 9–4 Å. More recently, Ibach has measured phonon dispersion using 100-eV electrons that have wavelengths of ca. 1.2 Å. Transition metals have nearest neighbor distances of 2.45–2.8 Å. In an EELS experiment, the wave vector of the excited phonon can be determined from the *change* in parallel momentum of the scattered electron since the total parallel momentum is conserved. Phonon dispersion can hence be followed by measuring the EEL spectrum as a function of off-specular scattering angle,[49,50] with each off-specular angle within the plane of incidence for a given angle of incidence corresponding to an appropriate point in the surface Brillouin zone. Specular EELS ($\Delta k_{\parallel} = 0$) probes in the SBZ center in the same manner as the optical spectroscopies. In EELS it is the impact mechanisms that allow such off-specular measurements to be made. For the optical spectroscopies the reflection is sharply confined to the specular direction because of the low momentum of phonons.

For general points in the SBZ the symmetry elements are lost, but along special lines and at special points symmetry elements may be preserved. For points away from the SBZ center and edges, the highest possible symmetry, C_s, occurs along lines connecting the center to an apex or edge midpoint. These lines are probed by EELS measurements in a scattering plane coincident with one or other mirror plane of the adsorbate/adsorbent system. At points on the SBZ edges additional reflection planes and rotation axes may be present. Figure 10 shows the SBZ for a simple square lattice with the principal points and lines

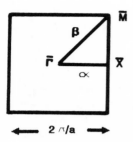

Figure 10. The surface Brillouin zone (SBZ) for a simple
square lattice, showing the principal points and directions. a is
the lattice spacing in the real space unit cell.

conventionally designated. The point groups appropriate to the points
$\bar{\Gamma}$, $\bar{X}$, $\bar{M}$, α, β are C_{4v}, C_{2v}, C_{4v}, C_s, and C_s, respectively.

Having determined the primitive unit cell of the adsorbate/adsorbent
array, the consideration of its symmetry elements leads to the identification of the *two-dimensional space group,* for which Hermann–Maugin or
Schoenfliess symbols can be used. The latter clearly identifies the
isomorphous point group (usually in this case referred to as the *factor
group*)[18] used for determining the symmetry species and spectroscopic
activities of the vibrations of the unit cell as a whole. The factor group is
derived from the space group by elimination of all translational symmetry
operations and reduction of glide lines to simple reflection planes such
that all remaining symmetry elements pass through one point.

Where a unit cell contains a single adsorbate species, the site group
analysis of spectral activity also applies to the unit cell as a whole, i.e.,
the site group is the same as the factor group. The situation is more
complex when more than one adsorbate molecule exists in the unit cell.
In such cases, of which there are many examples, each adsorbate species
contributes a full set of internal and external vibrations to those of the
unit cell. The unit cell has a total of $3NM$ adsorbate vibrations when
there are M adsorbate molecules of N atoms in the primitive unit cell.
There are $3NM$ phonon modes at each point in the SBZ. Two extreme
possibilities can still be distinguished. In the first of these, the M
adsorbate species in the unit cell occupy sites unrelated to each other by
any symmetry element of the unit cell. The spectroscopic selection rules
appropriate to each site can then be assessed independently, although
coupling between modes of the same symmetry in different adsorbate
molecules will affect the frequencies observed. The extent of such
coupling depends on the similarity of the frequencies of the modes
involved and the extent of intermolecular vibrational interaction. The
second possibility is that a subset of the M adsorbate molecules in the
unit cell are symmetry related. The molecules are then necessarily in
equivalent sites, whereas in the first possibility molecules are in inequiv-

alent sites. *Resonance* coupling is then to be expected between the vibrations of the set of symmetry-related adsorbate molecules, and a set of coupled vibrational modes occur involving for two such molecules in-phase and out-of-phase displacements of the symmetry-related molecules. These modes belong to different symmetry representations in the factor group and different frequencies will occur shifted from that of the molecule on an isolated site by an amount depending only on the strength of the resonant coupling. For an analysis of the spectroscopic activity, the factor group must be used. Of course, more complex intermediate examples can exist where the unit cell contains one set of symmetry-related molecules in one type of site and a second set of symmetry-related molecules in a different site. Though more complex, this situation is amenable to the same type of analysis. A simple one-dimensional example can make these ideas much clearer.

In Figure 11a, the adsorbate atom, A, forms an array with twice the lattice dimension of the metal adsorbent, M. There is one adsorbate atom and two metal atoms in the new unit cell with the adsorbate occupying a bridge site. The corresponding one-dimensional Brillouin zone is shown in Figure 11d. The symmetry-based analysis of the substrates of the adsorbate in the unit cell is simply the same as is derived from the point group of the adsorbate site. If we consider the motion of the adsorbate atoms perpendicular to the metal atom chain, (A–M stretch) then phonon modes can be described. Each point in the Brillouin zone has a single phonon arising from A–M stretching motion. The adatoms move in-phase at $k = 0$, the BZ center, and out-of-phase at $k = \pi/2m$, the BZ boundary, where m is the metal dimension. Consider now Figure 11b. The adsorbate-imposed unit cell now has a dimension $3m$. There are two adsorbate atoms in the unit cell but they are not related to each other by any symmetry elements. One atom occupies a bridge site and one an on-top site There are now two A–M stretches in the unit cell corresponding to motions of A atoms perpendicular to the MM chain, one characteristic of each type of site, though some nonresonant vibrational coupling between the two modes may lead to small frequency shifts relative to the values expected for the isolated adsorbates. Both the modes are symmetric with respect to the mirror planes perpendicular to the *mm* direction (Figure 11b) which characterizes the unit cell symmetry. Correspondingly, each point of the SBZ will also have two phonon modes associated with A–M stretching vibrations.

Finally, Figure 11c shows the case of two adsorbate atoms per unit cell. The atoms are in equivalent sites and are symmetry related. The unit cell again has a length $3m$. The two A–M stretches associated with the unit cell now partake in resonance coupling forming new modes, one with

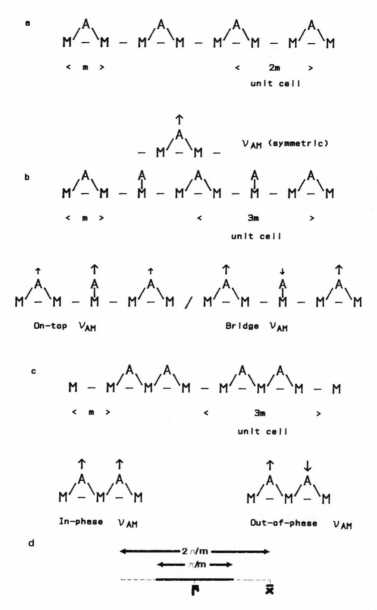

Figure 11. Some possible bonding arrangements of an adsorbate atom (A) on a chain of metal atoms (M), and the vibrations that arise from the motion of the adsorbate perpendicular to the chain. (a) One adsorbate atom per unit cell; (b) two inequivalent adsorbate atoms per unit cell; (c) two equivalent adsorbate atoms per unit cell; (d) Brillouin zone corresponding to (a) (full line) and metal lattice (dashed line).

in-phase displacements on the two atoms and the other with 180°
out-of-phase displacements. Each adsorbate atoms within the unit cell
contributes equally to the two modes. More importantly, from the view
of symmetry and spectroscopic selection rules, the two modes have
different symmetries relative to the mirror plane of the unit cell, i.e., one
belongs to the symmetric representation of the factor group while the
other belongs to the antisymmetric representation. In a two-dimensional
array on a metal surface, only the in-phase symmetry component could
be active in the infrared or specular EELS experiments. The two
adsorbate atoms of the unit cell contribute equally to each of the two
phonon modes of A–M stretching character, which occur at each point of
the SBZ. The two modes would have different symmetries at the zone
center and zone edge, but this is not the case at a general point in the
SBZ where coupling can occur under the reduced symmetry.

When symmetry-related species are present in a unit cell, as described
above, correlations can be made between the occupied site symmetry (S)
representations and those of the unit cell as a whole, the factor group
(F).

The observation of more than one frequency associated with each
mode of vibration that would be associated with a single site is termed
factor group splitting or *correlation field splitting*. We shall prefer the
former description.

We further illustrate some of the above ideas with examples chosen
to demonstrate different aspects of the vibrational analyses of regular
adsorbate/adsorbent arrays. The first two examples are situations in
which the symmetry of the array is closely similar to that of the clean
surface; one illustrates the simplest possibility with only a single
adsorption complex in the primitive unit cell, whereas the second
involves the presence of two symmetry-related adsorbates in the unit cell.
The third and fourth examples involve primitive adsorbate-determined
unit cells which have different shapes and symmetries from that of the
bare surface. Both examples have two molecules per unit cell. In one
case, the two molecules are symmetry related; in the other they are not.
Our final example involves a molecule with a low site symmetry but
where a higher factor group symmetry is appropriate to the unit cell. In
the following examples we shall confine our discussion of spectroscopic
activities to the expectations for adsorption on metal surfaces.

7.1. $c(2 \times 2)$ Array of Adatoms in the Fourfold Hollow Site of the (100) Face of a Cubic Metal

This is a frequently encountered surface structural arrangement.
Examples are O/Ni(100),[51] S/Ni(100),[51] N/W(100),[52] and Br/

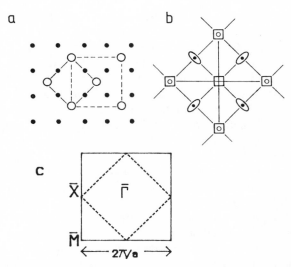

Figure 12. The atomic positions (a) and symmetry elements (b) for a ($\sqrt{2} \times \sqrt{2}$) R45°—$c(2 \times 2)$ arrangement of adsorbate atoms (O) and metal atoms (•) on a (100) face. O···O, outline of centered unit cell; O—●—O, outline of the primitive unit cell. □, fourfold rotation axis; ⊘ or ⊖, twofold rotation axis; ——, reflection planes. (c) Brillouin zones for the adsorbent (full lines) and adsorbate (dashed lines) lattices.

Cu(100).[53] Figure 12 shows the structure and the $c(2 \times 2)$ unit cell is indicated by dashed lines. This is not, however, the primitive unit cell. The latter is designated by the bold lines and contains only a single adsorbate atom. A better description of the array is $p(\sqrt{2} \times \sqrt{2})R45°$, which places greater emphasis on the primitive unit cell. There are two metal atoms per unit cell and the fractional coverage describing this structure is $\theta = \frac{1}{2}$. Figure 12 shows the symmetry elements associated with this array. The space group is $p4m$ or C_{4v}^I (International Tables, p. 92).[32] The symmetry of the adsorbate site is C_{4v}, and in this simple example this point group also describes the symmetry elements of the unit cell; i.e., C_{4v} is appropriate also for the factor group. There is hence a trivial one-to-one correlation between the site group (S) representations and the factor group (F) representations for the unit cell as a whole shown in Table 3. The interchange of B_1 and B_2 is a formality reflecting the 45° reorientation of the unit cell relative to the adsorption site. Also in Table 3 are given the number of adsorbate modes of each symmetry at the SBZ center and the symmetry of the dipole (x, y, and z) and polarizability components (x^2, y^2, z^2, xy, xz, yz), respectively. In this example, there are no modes of A_2, B_1, or B_2 symmetry, and the A_1 and E vibrations correspond exactly to those of the isolated atom on the

Table 3. The Symmetry Species, Numbers of Vibrations, and Spectroscopic Activities (x, y, z, Infrared; xx, xy, etc., Raman) for an Atomic Adsorbate Giving Rise to the Unit Cell Illustrated in Figure 12, e.g., $O/Ni(100)$—$c(2 \times 2)$

Site symmetry $S = C_{4v}$	Factor group symmetry $F = C_{4v}$	Number of modes/activity
A_1 ———————— A_1		$1\ z$; xx, yy, zz
A_2 ———————— A_2		$0\ xy$
$B_1{}^a$ $\diagdown$ $\diagup$ B_1		0 —
B_2 $\diagup$ $\diagdown$ B_2		0 —
E ———————— E		$1\ x$, y; xz, yz

^{a} This inversion occurs because the diagonal reflection planes of the site become the principal planes of the unit cell and vice versa.

fourfold site discussed in Section 5.1. The A_1 mode involves atomic motion perpendicular to the surface and is infrared, Raman, and specular EELS active. The E mode is the doubly degenerate parallel vibration, which is inactive at the SBZ center, i.e., for specular EELS.

Away from the SZB center, i.e., for off-specular measurements, the symmetry is reduced to C_s along the $\bar{\Gamma}\bar{X}$ line and $\bar{\Gamma}\bar{M}$ line. The E mode splits into A' and A'' modes, which are, respectively, symmetric and antisymmetric with respect to the remaining mirror plane of symmetry. The A_1 mode now also takes on A' symmetry and can in principle mix with the parallel mode of the same symmetry, creating new modes involving both parallel and perpendicular displacements. Both are expected to be active in the off-specular EELS experiment. The antisymmetric mode, A'', is a transverse phonon mode which remains inactive for measurements in the chosen plane of incidence. At $\bar{X}$ the symmetry is higher again at C_{2v} and three phonon modes are expected. The mode perpendicular to the surface has A_1 symmetry. The other symmetric mode along $\bar{\Gamma}\bar{X}$ belongs to the B_1 representation at $\bar{X}$, i.e., it is once more a longitudinal mode with no perpendicular component. Both are EELS active at the appropriate off-specular angle. The third mode has B_2 symmetry and is again spectroscopically inactive within the plane of incidence. Finally, at $\bar{M}$ the appropriate point group is C_{4v} again. The parallel modes are again degenerate modes of E symmetry. The perpendicular A_1 mode and the component of the E mode involving displacement in the scattering plane are active in the off-specular EELS.

In specular EELS measurements the A_1 feature has been identified at 314 cm^{-1} for O/Ni(100) and at 355 cm^{-1} for S/Ni(100)[51]. Detailed measurements of the dispersion along $\bar{\Gamma}\bar{X}$ and $\bar{\Gamma}\bar{M}$ have not yet been made, but a phonon has been observed at 450 cm^{-1}.[54]

7.2. Bridge Site Adatom Occupancy of the (100) Surface of a Cubic Metal

An example of this structure is the saturation coverage of hydrogen on W(100), whose vibrational properties have been discussed in some detail.[55,56] The structure is shown in Figure 13. The occupancy of all bridge sites on a (100) face leads to a coverage $\theta = 2$, i.e., two hydrogen atoms per surface metal atom. The site symmetry for an adsorbed hydrogen atom is clearly C_{2v}. However, the unit cell, which is the same as that of the clean surface (see Figure 13) has the associated factor group symmetry C_{4v}, derived from the two-dimensional space group $p4m$ on C_{4v}^{I}. The unit cell vibrations *must* be analyzed in terms of the factor group C_{4v} and not the local C_{2v} site group. The correlation is shown in Table 4. Two adsorbate atoms per unit cell ensure six adsorbate derived degrees of freedom. Each point in the SBZ has six new phonon modes. For an isolated atom, the C_{2v} site group demands that the two parallel motions of the hydrogen with respect to the surface have different frequencies (Section 5.1). These were designated ν_{HM_2} (as), for the essentially bond stretching mode in the HM_2 plane, and γ_{HM_2} for the out-of-plane bending mode. In the C_{4v} factor group analysis for the SBZ center, these modes each form a doubly degenerate mode of E symmetry as shown in Figure 14. The perpendicular motions of the symmetry related H atoms couple in a more interesting way to produce a mode of A_1 symmetry and one of B_1 symmetry. Four fundamental frequencies are, therefore, expected at $k = 0$, the SBZ center. Only the A_1 mode, however, is expected to be dipole active. It has been observed in specular EELS at 1038 cm^{-1}.[55,56] Away from the specular direction, along $\bar{\Gamma}\bar{X}$,

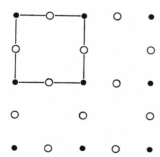

Figure 13. The atomic position of H atoms (○) and tungsten atoms (●) in the (1×1) structure found for H/W(100) at $\theta = 2$. The unit cell is outlined by ——.

Table 4. The Symmetry Species, Numbers of Vibrations, and Spectroscopic Activities for Atomic Adsorbates Giving Rise to the Unit Cell of Figure 13, e.g., H/W(100)—(1 × 1)

Site symmetry $S = C_{2v}$	Factor group symmetry $F = C_{4v}$	Number of modes	Activity
A_1	A_1	1	z; xx, yy, zz
A_2	A_2	0	xy
	B_1	1	—
	B_2	0	—
B_1	E		
B_2		2	x, y; xz, yz

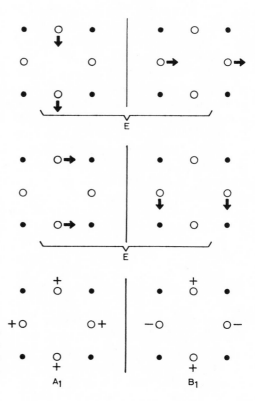

Figure 14. The six vibrational degrees of freedom contributed by the H atoms (○) for the (1 × 1) structure of Figure 13. The metal atoms (●) are assumed to be of infinite mass.

Table 5. The Point Group Symmetries Appropriate to High-Symmetry Points of the SBZ for H/W(100)—(1 × 1), together with the Symmetry of the Hydrogen Vibrations[a]

Vibration	C_{4v} $\bar{\Gamma}$	C_s $\bar{\Gamma}\bar{X}$	C_{2v} $\bar{X}$	C_s $\bar{\Gamma}\bar{M}$	C_{4v} $\bar{M}$
Perpendicular	A_1 B_1	A' A'	A_1 B_1	A' A''	E
Out-of-plane	E	A' A''	A_1 B_2	A' A''	A_1 B_2
In-plane	E	A' A''	B_1 A_2	A' A''	B_1 A_2
Number of active modes	1	4	4	3	3

[a] The number of modes active in EELS is also indicated.

the C_s symmetry splits the E modes into A' and A'' components. The symmetric components, A', should be active in EELS and, indeed, have been observed at 644 cm^{-1} and 1248 cm^{-1} derived from γ_{HM_2} and $\nu_{HM_2}(as)$, respectively.[55,56]

The two modes involving perpendicular motion with respect to the surface have A' symmetry and should be active in the off-specular EELS experiment. They can interact with each other and A', parallel, longitudinal modes. Only a single band has been observed which originates in the A_1 mode at $\bar{\Gamma}$ and shows no dispersion. Along $\bar{\Gamma}\bar{M}$ the type of analysis is the same as along $\bar{\Gamma}\bar{X}$, with the exception that the mode deriving from the B_1 phonon at $\bar{\Gamma}$ has A'' symmetry and is therefore EELS inactive. At $\bar{M}$ the perpendicular modes form a doubly degenerate pair of E symmetry in the relevant C_{4v} point group. The symmetry classification of the modes in the SBZ together with their spectroscopic activity is given in Table 5.

7.3. c(4 × 2) Structure for CO/Pt(111)

The EEL spectrum for this system has been discussed by Ibach and colleagues,[57,58] who show from the internal frequencies that the ordered array contains both on-top or linearly bonded CO molecules and bridging CO molecules. The CO axis is perpendicular to the surface for all molecules and the coverage is $\theta = \frac{1}{2}$.

The proposed structure is shown in Figure 15 with the c(4 × 2) unit cell, containing two adsorbate molecules of each type, outlined by dashed lines. For our purposes, we choose the alternative, primitive unit cell,

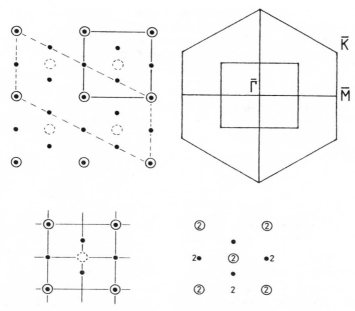

Figure 15. The proposed structure for $c(4 \times 2)$—CO/Pt(111) showing the "on-top" and bridging CO molecules (○ and ⊙, respectively) and metal atoms (•). The $c(4 \times 2)$ unit cell is outlined by $\cdots$ and the primitive unit cell by ———. The symmetry elements and SBZ are also shown.

which contains a single molecule of each type. The symmetry elements of this unit cell and the surface Brillouin zone are also shown in Figure 15. The space group is pmm (C_{2v}^{1}) and factor group and site group is C_{2v}. As discussed in Section 6, the lower symmetry of the adsorbate array relative to the bare surface gives rise to rotational domains which may influence the vibrational spectroscopic data. The trivial correlation table linking site group and factor group is shown in Table 6, together with the number of vibrational modes of the unit cell in each symmetry representation and the symmetry of the dipole and polarizability components. The displacements of the atoms associated with CO vibrations on linear and bridge sites have already been described in Section 5.2. The only change is that the E modes, arising from the frustrated rotations and frustrated translations, are split into B_1 and B_2 components by the lower symmetry imposed by the other CO molecules.

The allowed modes in infrared, Raman, and specular EELS are the A_1 modes and these have been detected at $2100 \, \text{cm}^{-1}$ (ν_{CO} stretch) and $470 \, \text{cm}^{-1}$ (ν_{CPt} stretch) for the linear CO molecules and at $1850 \, \text{cm}^{-1}$

Table 6. The Symmetry Species, Number of Vibrations, and Spectroscopic Activities for Two Diatomic Adsorbates Adsorbed in the Linear and Bridging Sites to Form the Unit Cell Illustrated in Figure 15, as for CO/Pt(111)—$c(4 \times 2)$

Site symmetry $S = C_{2v}$	Factor group symmetry[a] $F = C_{2v}$	Number of modes	Activity
A_1	A_1	4^b	z; xx, yy, zz
A_2	A_2	0	xy
B_1	B_1	4^b	x; xz
B_2	B_2	4^b	y; yz

[a] The x and y axes for the unit cell have been chosen with the same orientation as for the C_{2v} site.
[b] Two in each symmetry representation, for each of the two different molecules, linear and bridged in the unit cell.

(ν_{CO} stretch) and $380 \, cm^{-1}$ (ν_{CPt} stretch) for the bridging species.[57,58] In addition, weak overtones in the EEL spectrum have identified some of the modes involving parallel displacements. A combination frequency between the linear and bridging CO stretches suggests appreciable coupling between the two sites.[58]

Away from the $\bar{\Gamma}$ point but along a mirror plane of symmetry, either the B_1 or B_2 modes, depending on the scattering plane, become allowed (A' modes) in off-specular EELS. The existence of rotational domains, however, leads us to expect both B_1 and B_2 modes in all scattering planes. Such modes have not yet been detected owing to the very low amplitudes of the displacements.

7.4. $c(4 \times 2)$ Structure for CO adsorbed on Ni(111)

Although the LEED pattern is very similar for $c(4 \times 2)$ structures of CO/Pt(111) and CO/Ni(111), the vibrational spectrum is quite different.[59] The spectrum of CO/Ni(111) is indicative of bridge site occupancy only. Figure 16 shows the proposed structure, the $c(4 \times 2)$ unit cell and the more appropriate primitive unit cell. The two adsorbed CO molecules per unit cell are chemically equivalent since they are in identical sites related to each other by symmetry elements of the array. The space group is pgg (C_{2v}^{II}), giving rise to a C_{2v} factor group, but the site symmetry is only C_2. The correlation table is shown in Table 7.

Some representative vibrations of the unit cell (phonon modes at $\bar{\Gamma}$)

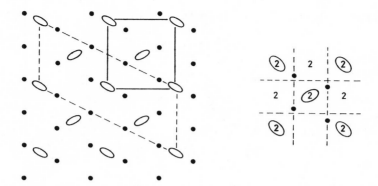

Figure 16. The proposed structure for $c(4 \times 2)$—CO/Ni(111) showing the adsorbate ⬭ and metal atom (•) positions. The $c(4 \times 2)$ unit cell is outlined by $\cdots$ and the primitive unit cell by ——. The symmetry elements of the primitive unit cell are also shown. 2 denotes a twofold rotation axis and − − − − a glide plane.

are shown in Figure 17. Only the two A_1 modes are expected to be spectroscopically active on specular, and they have been observed in EELS at 1900 and 400 cm^{-1} for the ν_{CO} stretch and ν_{CNi_2} stretch, respectively.[59] In off-specular EELS away from the SBZ center, and taking into account three rotational domains, all bands should be active. However, once again, the parallel modes have not been observed. It is the higher symmetry of the unit cell and the chemical equivalence of the CO molecules that leads to the simpler spectrum than for CO/Pt(111) at the same coverage.

Table 7. The Symmetry Species, Number of Adsorbate Vibrations, and Their Spectroscopic Activities for Two Diatomic Molecules Adsorbed in the Bridging Mode, so as to Give Rise to the Unit Cell of Figure 16, as for CO/Ni(111)—$c(4 \times 2)$

Site symmetry $S = C_2$	Factor group symmetry $F = C_{2v}$	Number of modes (two molecules) and activity
A	A_1	2 s; xx, yy, zz
	A_2	2 xy
B	B_1	4 x; xz
	B_2	4 y; yz

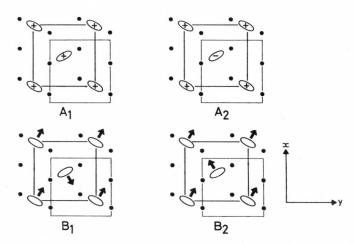

Figure 17. Schematic representations of some coupled vibrational modes of the two adsorbed CO molecules in the unit cell shown in Figure 16.

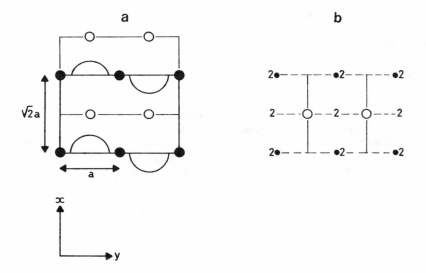

Figure 18. the proposed structure for CO/Ni(110) or CO/Pd(110)—(2 × 1) $p2mg$ showing adsorbate molecules (−) tilted with respect to the metal atom rows. The symmetry elements, twofold axis (2), glide planes (– – – –), and reflection planes (——) are also shown.

Table 8. The Symmetry Species, Number of Vibrations, and Spectroscopic Activities for Two Diatomic Molecules Adsorbed in the Bridging Mode so as to Give Rise to the Unit Cell Depicted in Figure 18, as for CO/Ni(110)—(2 × 1) p2mg

Site symmetry $S = C_s$	Factor group symmetry $F = C_{2v}$	Number of modes	Activity
A'	A_1	4	$z; xx, yy, zz$
	A_2	2	xy
A''	B_1	4	$x; xz$
	B_2	2	$y; yz$

(A' correlates to A_1 and B_1; A'' correlates to A_2 and B_2)

7.5. (2 × 1) Structure for CO Adsorbed on Ni(110) and Pd(110)

The structure proposed for the three adsorbate/adsorbent systems is indicated by the LEED analysis which shows the presence of a glide line through systematic absences.[60] For the structure, shown in Figure 18, the space group is $p2mg$ (C_{2v}^{III}), thought often mistakenly referred to as $p1g1$. The EEL spectra on both Ni(110)[61–63] and Pd(110)[64] surfaces indicate bridged CO molecules. The coverage is $\theta = 1$ and the glide line is interpreted as reflecting CO molecules tilted in alternate directions to relieve steric crowding.[60,65,66] The tilting causes the site symmetry to be C_s, but the factor group, appropriate to the unit cell vibrational analysis, is C_{2v}. The correlation is given in Table 8. The ν_{CO} stretch and ν_{CM_2} stretching modes which belong to the A_1 representation have been observed at 1985 and 420 cm^{-1}, respectively, on the Ni surface[61] and at 2008 and 380 cm^{-1} in the Pd surface.[64] In addition, for this array we expect some modes involving frustrated rotation displacements to belong to the A_1 representation (see Figure 19) and hence show dipole activity. A band at 315 cm^{-1} has been observed in specular EELS and assigned to such a mode.[61] For off-specular EELS along the $\langle 011 \rangle$ direction, A_1 and B_1 modes resort to A' symmetry and should be spectroscopically active, while along the $\langle 110 \rangle$ direction the A_1 and B_2 modes have the required A' symmetry. The B_2 mode has not yet been observed.

8. Adsorbent Surface Modes

In all the previous sections, we have concentrated our attention on the new degrees of freedom which are provided by the adsorbate

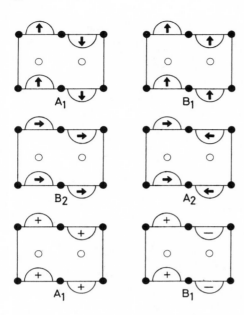

Figure 19. Schematic representations of some coupled vibrational modes for CO molecules in the structure of Figure 18.

molecules. In this section, we address ourselves to those modes that are predominantly adsorbent in character with displacements of the top layer adsorbent atoms. Apart from the intrinsic interest in investigating the vibrational properties of clean metal surfaces for comparison with the bulk, the behavior of the top layers of atoms following chemisorption is important.

Recently, it has become possible to measure the dispersion of the surface phonon modes of clean metals using either thermal He atoms[33-35] or electrons.[48] For simple bulk lattice terminations, the only dipole-allowed mode of the surface atom is that in which the surface atoms move together, in-phase against the second layer atoms. For off-specular EELS in a mirror plane, a second mode involving longitudinal parallel displacements becomes allowed. In this section, we shall discuss examples of adsorbate-covered and reconstructed surfaces. For both these situations, new surface phonon modes are to be expected. In the case of adsorbate-induced charges, we demonstrate that spectroscopic activity depends not only on the overlayer symmetry but also on the adsorption site.

Adsorption or reconstruction frequently, indeed usually, leads to a surface unit-cell larger than that of the clear surface. Consequently there are more adsorbent atoms associated with the unit cell and more vibrational degrees of freedom associated with the unit cell and each point in the SBZ. One can derive the form and symmetry of these phonon modes by considering the new unit cell. In all cases, one finds that one set of phonons corresponds to those of the clean surface while other sets arise by folding back phonons from the larger, clean SBZ into the new, smaller SBZ by the new G vectors.[67] However, the symmetry of the phonon modes, and hence their spectroscopic activity, depends critically on the adsorption site.[67,68]

We illustrate these effects by reference to the adsorption systems $c(2 \times 2)$ $CH_3O/Cu(100)$ and $c(2 \times 2)$ $CO/Cu(100)$ and the (1×2) reconstruction of Pt(110).

8.1. $c(2 \times 2)$ Overlayers on Cu(100)

Both the methoxy species, CH_3O, and carbon monoxide adsorb on a Cu(100) surface with the formation of a $c(2 \times 2)$, or primitive ($\sqrt{2} \times \sqrt{2}$) R45°, LEED pattern.[68,69] Figure 20 shows the relevant unit cells and SBZ for the (1×1) clean surface structure and the overlayer structure. The CO is believed to occupy on-top sites on the basis of LEED current/voltage analysis[70] while the CH_3O species is thought to occupy fourfold hollow sites.[68] In both cases, the space group is $p4mm$ (C_{4v}^I) and the site and factor groups are C_{4v}, if the influence of the CH_3 hydrogens is ignored. Note that, for the unit cell chosen to have an adsorbate species at the center, the disposition of metal atoms is different for the two adsorbates (Figure 20). This has important consequences for the correlation between factor group and metal atom site group.

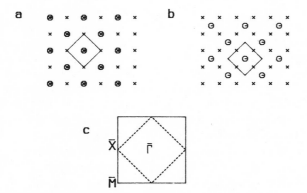

Figure 20. The primitive unit cells appropriate to the ($\sqrt{2} \times \sqrt{2}$) R45° structure of (a) CO/Cu(100) and (b) $CH_3O/Cu(100)$. ⊗ denotes CO molecules, ⊖ CH_3O species, and × metal atoms. (c) shows the SBZ appropriate to the clean (——) and adsorbate-covered ($\cdots$) surfaces.

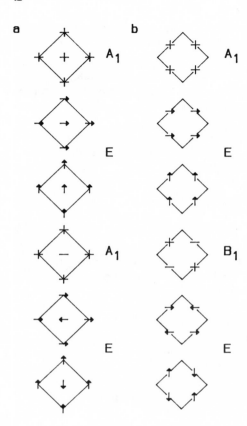

Figure 21. The six phonon modes of the copper lattice as expected at $k = 0$ ($\bar{\Gamma}$) for the fourfold hollow adsorption site (a) and "on-top" site (b). The symmetry of the modes within the relevant C_{4v} point group are also shown. (+) and (−) represent displacements perpendicular to the surface.

There are two Cu atoms per unit cell following adsorption, so six phonon modes based on the top layer Cu atoms are to be expected for each point in the SBZ. The modes at $k = 0$ are depicted in Figure 21 for both adsorption sites. The symmetry of the modes in the C_{4v} point group are also given. In each case the upper three modes come from the $\bar{\Gamma}$ point of the clean SBZ and the two Cu atoms of the unit cell move in-phase. The lower three modes are found at $\bar{\Gamma}$ in the new SBZ by virtue of being folded back from $\bar{M}$ by the new G vector. For these the two Cu atoms move out-of-phase. The significant difference lies in the new symmetry of the surface mode of perpendicular displacement, which derives from the $\bar{M}$ point of the clean SBZ. For on-top adsorption, the mode has A_1 symmetry and therefore is dipole active. For fourfold hollow site adsorption, the same mode has B_1 symmetry and is dipole inactive. The EEL spectrum of $c(2 \times 2)$ $CH_3O/Cu(100)$, recorded in the specular direction, shows no new loss features in the neighborhood of the elastic peak.[68] For CO/Cu(100), however, a new feature appears at 14.5 meV (116 cm^{-1}) and is closely associated with the $c(2 \times 2)$ LEED pattern.[68,69] The feature is assigned to the new A_1 perpendicular displacement mode in line with the above discussion. The EELS results in the loss region, therefore, support the earlier LEED experimental conclusions regarding the adsorption site.

8.2. (1×2) Reconstruction of Ir(110), Pt(110), and Au(110)

The clean surfaces of the third row transition elements Ir, Pt, and Au exhibit reconstructions on their clean (110) surfaces with longer-range periodicity than that of the corresponding bulk plane. The lattice spacing in the $\langle 001 \rangle$ direction is doubled.[71,73] The favored model for the reconstruction involves alternate missing close-packed rows of metal atoms as shown in Figure 22a.[71] An alternative, paired-row model is shown in Figure 22b. The surface unit cell contains two second layer metal atoms but either one (Figure 22a) or two (Figure 22b) top layer atoms. The space group is *pmm* (C_{2v}^I) and the factor group C_{2v}. In the paired row model, however, the surface atom site group is only C_s.

Figure 23a shows the phonon modes at $k = 0$ based on first and

a

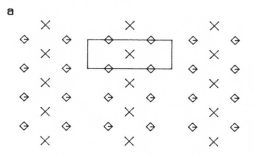

b

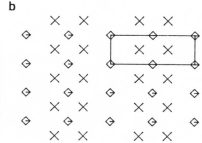

Figure 22. Possible models of the (1×2) reconstruction of Ir(110), Pt(110), and Au(110) surfaces. (a) Favored missing row model; (b) paired row model. The primitive unit cells are also shown. ×, first-layer atom; ◇, second-layer atom.

second layer atoms for the missing row model. The corresponding modes for the paired row model are given in Figure 23b. Looking more closely at the dipole active, A_1 modes, the concerted perpendicular motion of the top layer or second layer atoms belongs to this representation. For the missing row model, an A_1 mode can also arise from parallel displacement of the second layer atoms (Figure 23a). This mode will be strongly coupled with the perpendicular motion of the top layer atoms. For the paired-row model, an analogous A_1 mode arises in the top layer atoms (Figure 23b). While the missing row model has only three active modes, the paired new model generates a fourth A_1 mode involving perpendicular displacements of the second layer atoms probably coupled strongly to the A_1 mode with parallel displacements in the top layer discussed above. In principle information is available in the dipole activity of the surface phonon modes on the nature of surface reconstructions.

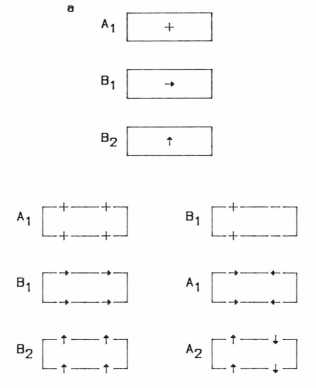

Figure 23. The phonon modes at $k = 0$ ($\bar{\Gamma}$) for the (1 × 2) structures of Figure 22, together with their symmetries. (a) Missing row model; (b) paired row model.

b

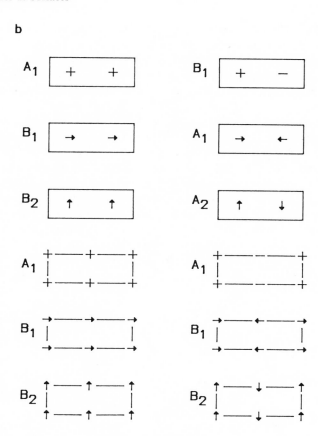

Figure 23. (*Continued*)

9. Conclusions

During the past decade or so, vibrational spectroscopy has begun to contribute greatly to our knowledge of surface adsorbed species. At a simple level, species can be identified using the background database provided by studies of gases, liquids, and solids. At a more sophisticated level, adsorption-induced bonding changes can be assessed, as can the strength and nature of adsorbate lateral interactions. All these developments are covered in much greater detail in other chapters of this book. In parallel with the development of experimental techniques for studying surface vibrations, there has been an improvement in our understanding of spectroscopic activity and vibrational selection rules, as they relate to surface species. The selection rules differ according to the technique and the nature of the adsorbent, but nevertheless they play a crucial role in spectral interpretation, and hence, in extending our understanding of the adsorption above the trivial level. Recently, a start has been made on the vibrational analysis of regular arrays of surface-adsorbed molecules. We are confident that much greater efforts will be successfully made in this area and also in obtaining vibrational frequencies for parallel modes which are currently difficult to observe. The details of band-splittings corresponding to coupled vibrations of different molecules within a unit cell are more likely to be observed for the modes perpendicular to the surface that can be observed with the higher resolution of the infrared and Raman optical spectroscopies. The modes parallel to the surface within the unit cell are at present first observable by off-specular EELS experiments.

In this chapter, we have attempted to outline the basic principles that govern the vibrations of adsorbed species and arrays of adsorbates. We have shown how these ideas can be applied to the detailed analysis of spectra already in the literature, placing particular emphasis on the analysis of regular arrays. It is hoped that this helps provide the foundation for an increasingly important and exciting area of surface science.

References

1. L. H. Little, *Infrared Spectra of Adsorbed Species,* Academic, New York (1966).
2. M. L. Hair, *Infrared Spectroscopy in Surface Chemistry,* Dekker, New York (1967).
3. A. V. Kiselev and V. I. Lygin, *Infrared Spectra of Surface Compounds,* Wiley, New York (1975).
4. *Chemical Applications of Thermal Neutron Scattering* (B. T. M. Willis, ed.), Oxford Univ. Press, Oxford (1973).
5. H. Seki, *J. Electron Spectrosc. Relat. Phenom.* **30,** 287 (1983).
6. *Vibrational Spectroscopy of Adsorbates* (R. F. Willis, ed.), Springer-Verlag, Berlin (1980).
7. *Vibrational Spectroscopies for Adsorbed Species* (A. T. Bell and M. L. Hair, eds.), American Chemical Society, Washington, D.C., (1980).
8. J. Pritchard, in *Chemical Physics of Solids and their Surfaces,* Vol. 7 (M. W. Roberts and J. M. Thomas, eds.), Specialist Periodical Reports, The Chemical Society, London (1978), p. 157.
9. H. Ibach and D. L. Mills, *Electron Energy Loss Spectroscopy and Surface Vibrations,* Academic, New York, (1982).
10. F. Hoffman, *Surf. Sci. Rep.* **3**(2,3) 107 (1983).
11. Chemical applications of LEED are covered in G. A. Somorjai, *Chemistry in Two-Dimensional Surfaces,* Cornell Univ. Press, Ithaca and London, (1981); a more

formal account of LEED is given in J. P. Pendry, *Low Energy Electron Diffraction*, Academic, London (1974).

12. N. Sheppard and J. Erkelens, *Appl. Spectros.* **38**, 471 (1984).
13. W. H. Smith and H. C. Eckstrom, *J. Chem. Phys.* **46**, 3657 (1967).
14. H. Nichols and R. M. Hexter, *J. Chem. Phys.* **75**, 3126 (1981).
15. H. Nichols and R. M. Hexter, *Surf. Sci.* **118**, 597 (1982).
16. N. V. Richardson and J. K. Sass, *Chem. Phys. Lett.* **62**, 267 (1979).
17. R. M. Hexter and M. G. Albrecht, *Spectrochim. Acta* **35A**, 233 (1979).
18. H. Nichols and R. M. Hexter, *J. Chem. Phys.* **73**, 965 (1980).
19. N. V. Richardson and A. M. Bradshaw, In *Electron Spectroscopy—Theory, Techniques and Applications*, Vol. 4 (C. R. Brundle and A. D. Baker, eds.), Academic, New York (1981).
20. G. Herzberg, *Infrared and Raman Spectra of Polyatomic Molecules*, Van Nostrand, New York (1945).
21. R. G. Greenler, *J. Chem. Phys.* **44**, 310 (1966).
22. H. A. Pearce and N. Sheppard, *Surf. Sci.* **59**, 205 (1976).
23. H. Ibach, *Surf. Sci.* **66**, 56 (1977).
24. J. E. Demuth and D. E. Eastman, *Phys. Rev. Lett.* **32**, 1123 (1974).
25. N. V. Richardson and P. Hofmann, *Vacuum* **33**, 793 (1983).
26. J. E. Demuth and H. Ibach, *Surf. Sci.* **78**, L238 (1978).
27. N. V. Richardson, *Vacuum* **33**, 787 (1983).
28. W. Ho, R. F. Willis, and E. W. Plummer, *Phys. Rev. Lett.* **40**, 1463 (1978).
29. C. H. Li, S. Y. Tong, and D. L. Mills, *Phys. Rev. B* **21**, 3057 (1980).
30. A. M. Baro, H. Ibach, and H. D. Bruchmann, *Surf. Sci.* **88**, 384 (1979).
31. G. Aers, T. B. Grimley, J. B. Pendry, and K. L. Sebastian, *J. Phys. C* **14**, 3995 (1981).
32. *Inelastic Electron Tunnelling Spectroscopy* (T. Wolfman, ed.), Springer-Verlag, Berlin (1978); see also Chap. 5 of Ref. 6 and Chap. 11 of Ref. 7.
33. G. Brusdeylins, R. B. Doak, and J. P. Toennies, *Phys. Rev. B* **27**, 3662 (1983).
34. M. Cates and D. R. Miller, *Phys. Rev. B* **28**, 3615 (1983).
35. R. B. Doak, V. Harten, and J. P. Toennies, *Phys. Rev. Lett.* **51**, 578 (1983).
36. A. P. Cracknell, *Thin Solid Films* **21**, 107 (1974).
37. *International Tables for X-Ray Crystallography*, Vol. 1 (N. F. M. Henry and K. Lonsdale, eds.), Kynoch Press, Birmingham, England (1965), 2nd Ed.
38. M. W. Howard, U. A. Jayasooriya, S. F. A. Kettle, D. B. Powell, and N. Sheppard, *J. Chem. Soc. Chem. Commun.* **18** (1979).
39. See the discussion of the related case of a hydrogen atom bonded to three metal atoms in J. A. Andrews, U. A. Jayasooriya, I. A. Oxton, D. B. Powell, N. Sheppard, P. F. Jackson, B. F. G. Johnson, and J. Lewis, *Inorg. Chem.* **19**, 3033 (1980).
40. N. V. Richardson and A. M. Bradshaw, *Surf. Sci.* **88**, 255 (1979).
41. D. C. Harris and M. D. Bertolucci, *Symmetry and Spectroscopy*, Oxford University Press, New York (1978).
42. W. Erley, A. M. Baro, and H. Ibach, *Surf. Sci.* **120**, 273 (1980).
43. J. P. Mathieu, *Spectres de Vibration et Symétrie des Molécules et des Cristaux*, Hermann, Paris (1945).
44. P. M. A. Sherwood, *Vibrational Spectroscopy of Solids*, Cambridge Univ. Press, London (1972).
45. W. G. Fateley, F. R. Dollish, N. T. Devitt, and F. F. Bentley, *Infrared and Raman Selection Rules for Molecular and Lattice Vibrations*, Wiley, Interscience, New York (1972).

46. J. C. Decius and R. M. Hexter, *Molecular Vibrations in Crystals,* McGraw-Hill, New York (1977).
47. H. Ibach and D. Bruchmann, *Phys. Rev. Lett.* **44,** 36 (1980).
48. S. Lehwald, J. M. Szeftel, H. Ibach, T. S. Rahman, and D. L. Mills, *Phys. Rev. Lett.* **50,** 518 (1983).
49. S. Andersson, in *Vibrations at Surfaces* (R. Caudano, J. M. Gilles, and A. A. Lucas, eds.), Plenum Press, New York (1982), p. 169.
50. J. M. Szeftel, *Surf. Sci.* **152/153,** 797 (1983).
51. S. Andersson, *Surf. Sci.* **79,** 385 (1979).
52. K. Griffiths, D. A. King, G. C. Aers and J. B. Pendry, *J. Phys. C* **15,** 4921 (1982).
53. N. V. Richardson and J. K. Sass, *Surf. Sci.* **103,** 496 (1981).
54. J. M. Szeftel, S. Lehwald, H. Ibach, T. S. Rahman, J. E. Black, and D. L. Mills, *Phys. Rev. Lett.* **51,** 268 (1983).
55. H. Froitzheim, H. Ibach, and S. Lehwald, *Phys. Rev. Lett.* **36,** 1549 (1976).
56. R. F. Willis, W. Ho, and E. W. Plummer, *Surf. Sci.* **80,** 593 (1979).
57. H. Froitzheim, H. Hopster, H. Ibach, and S. Lehwald, *Appl. Phys.* **13,** 47 (1977).
58. H. Steininger, S. Lehwald, and H. Ibach, *Surf. Sci.* **123,** 264 (1983).
59. W. Erley, H. Wagner, and H. Ibach, *Surf. Sci.* **80,** 612 (1979).
60. R. M. Lambert, *Surf. Sci.* **49,** 325 (1975).
61. B. J. Bandy, M. A. Chesters, P. Hollins, J. Pritchard, and N. Sheppard, *J. Mol. Struct.* **80,** 203 (1982).
62. J. C. Bertolini and B. Tandy, *Surf. Sci.* **102,** 131 (1981).
63. M. Nichijuna, S. Masuda, Y. Sakisaka, and M. Onchi, *Surf. Sci.* **107,** 31 (1981).
64. M. A. Chesters, G. S. McDougall, M. Pemble, and N. Sheppard, *Surf. Sci.* **164,** 425 (1985).
65. S. R. Bare, K. Griffiths, P. Hofmann, D. A. King, G. L. Nyberg, and N. V. Richardson, *Surf. Sci.* **120,** 367 (1982).
66. P. Hoffmann, S. R. Bare, N. V. Richardson, and D. A. King, *Solid State Commun.* **42,** 645 (1982).
67. N. V. Richardson, *Surf. Sci.* **126,** 337 (1983).
68. M. Persson and S. Andersson, *Surf. Sci.* **117,** 352 (1982).
69. S. Andersson and M. Persson, *Phys. Rev. B* **24,** 3659 (1981).
70. S. Andersson and J. B. Pendry, *Phys. Rev. Lett.* **43,** 363 (1979).
71. D. G. Fedak and N. A. Gjostein, *Surf. Sci.* **8,** 77 (1967).
72. H. Wolf, H. Jagodzinski, and W. Moritz, *Surf. Sci.* **77,** 265 (1978).
73. A. M. Lahee, W. Allison, R. F. Willis, and K. H. Rieder, *Surf. Sci.* **126,** 654 (1983).

2

Excitation Mechanisms in Vibrational Spectroscopy of Molecules on Surfaces

J. W. Gadzuk

1. Introduction

Today it is possible to answer questions concerning the details, on an atomic scale, of the bonding of atoms and molecules to solid surfaces, that were not even realistically being asked 15 years ago. The existence of this four-volume series on "Methods of Surface Characterization" stands as a testament to this incredible progress. Four areas of characterization which stand out nowadays are the following:

1. Long-range order, typically investigated with some diffraction method.
2. Local order or geometry, probed with many techniques such as EXAFS, angle-resolved photoemission, ESDIAD, and vibrational spectroscopy.
3. Elemental analysis, in which advantage is taken of some characteristic signature such as core electron binding energy (XPS, Auger).
4. Chemical state, inferred from core level binding shifts, valance level photoemission spectra, and molecular vibrational spectra.

Many of these methods have been discussed at great length in this series. Here we will focus on the vibrational spectroscopies with particular

J. W. Gadzuk • National Bureau of Standards, Gaithersburg, Maryland 20899.

emphasis on the various excitation mechanisms and physical processes which occur in a "vibrational spectroscopic event," guided by a strong belief that an enlightened user of a technique will be in a better position to utilize it in a creative manner. In an attempt to facilitate this goal, the present chapter will be concerned more with transparent ways to illustrate the necessary points than with excessive mathematical rigor or with exhaustive review of previously published work on a particular topic. Although sufficient references will be given so that the stimulated reader can gain access to the remaining vast literature, the particular choices here are highly subjective. Since the chapter by Richardson on normal modes at surfaces provide an in-depth perspective on the properties of the systems observed with vibrational spectroscopy, our emphasis is on how these properties make themselves known in the experiment and on how one makes connections between what is observed and the system properties.

The plan of this chapter is the following. Under the heading of introductory remarks, the why's of vibrational spectroscopy are considered, followed by a qualitative overview of the various techniques covered in this volume. A number of possible theoretical strategies for understanding the spectroscopies are then introduced. Section 2 deals with simple models of the excitation process. Similarities and differences between direct electron and photon vibrational excitation are presented. Excitation due to electron resonances, and neutron collisions are considered. The information content in a vibrational lineshape is taken up in Section 3. In particular the origins of frequency shifts and linewidths is discussed. Many of the points raised in Sections 1–3 are illustrated via specific examples throughout the presentation. The chosen examples, namely, (1) dipole/off-specular electron scattering; (2) shape resonance induced vibrational excitation; (3) localized/delocalized adsorbed hydrogen excitation with neutrons; and (4) vibrational linewidths, were selected not only because of their intrinsic interest but also because they are good examples of synergism between experimental and theoretical inquiry. More comprehensive presentations of existing, technique-specific studies appear in the other chapters of this volume. The hope here is that a unity among these techniques will be apparent.

1.1. Why Vibrational Spectroscopy?

Vibrational spectroscopy is based on the experimental fact that the constituent atoms of a molecule, solid, or combination of the two execute multidimensional, quasiperiodic motion over potential energy surfaces that are determined by the electronic/chemical state of the system, at least when the energy content of the motion is substantially less than typical electronic excitation or chemical dissociation energies of the bound system.[1-4] When the requirements of quantum mechanics are imposed upon the quasiperiodic motion, it is found that the low-lying vibrationally excited eigenstates of the molecular system occur with characteristic energies in the $\sim$50–250 meV (400–2000 cm^{-1}) range for vibrations involving the light atoms such as H, C, N, and O which are of obvious chemical importance. Transitions between these states give rise to a discrete emission or absorption spectrum which is also a characteristic signature of both the chemical species being interrogated and its local environment. Although the corresponding quasiperiodic motion of the (usually heavy) atoms comprising the substrate can be represented in terms of a bounded phonon continuum,[5] the maximum single-phonon energy or frequency is usually much less than the intramolecular frequencies. Because of this frequency mismatch, there is little mixing between the modes. Consequently the vibrational modes of the molecule retain much of their free space character, which fact permits species identification. The small deviations from free space behavior, such as in transition energies, linewidths, and split degeneracies, are the measured quantities that are used to ascertain both the local environment such as bond site or molecular orientation and the characteristics of the adsorbate–substrate interaction.[6] Without going into details yet, consider as an illustrative example the results of Andersson,[7] in which electron energy loss spectra (EELS) were obtained for various coverages of CO adsorbed on Ni(100), as shown in Figure 1. The points to be made here are that there are two types of losses, one a low-energy entity $\sim$50–60 meV, basically associated with the molecule–substrate bond, and the other(s) in the vicinity of 250 meV derived from the intramolecular vibration at 260 meV of the free-space CO. The presence of this loss permits identification of nondissociatively adsorbed CO and the small shifts and splittings allow for bond site determination of the "carbon-down" CO. It is in this spirit that much of surface vibrational spectroscopy is carried out. For a detailed view of the field, in addition to the chapters in this volume, a number of excellent review articles, collections, and books are highly recommended.[8-16] Furthermore, the popularity of vibrational spectroscopy has spawned what has become a bi/triennial international meeting entitled "Vibrations at Surfaces," and the proceedings of these meetings are also excellent representations of the state-of-the-art of vibrational spectroscopy at surfaces.[17-20]

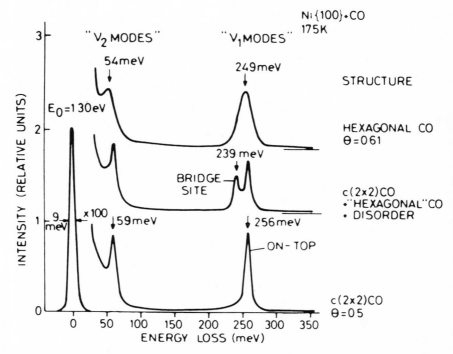

Figure 1. Electron energy loss spectra of CO chemisorbed on a Ni{100} surface showing the frequency variations of the C–O stretch (v_1) and the Ni–CO stretch (v_2) vibrational modes for different surface structures.[7]

1.2. Experimental Methods of Surface Vibrational Spectroscopy

An almost universal theorem of modern surface science is that no single experimental technique provides the answer to all questions. As is apparent from the seven chapters on experimental methods in this volume, a corollary to this theorem is that no single vibrational spectroscopic technique is a universal panacea. Different methodologies and probes have varying dynamic ranges, realistically possible resolutions, selection rule limitations (or strengths), sensitivities, and environmental restrictions. An overview of the situation is shown in Table 1,[21] where the most widespread techniques and selected characteristics are displayed. In this section, brief introductory commentary on each of these will be given, with emphasis placed on the relative roles of each spectroscopy.

Table 1. Techniques in Surface Vibrational Spectroscopy

Characteristic	Technique					
	IRAS	SERS	EELS	IETS	NIS	AIS
Resolution (FWHM) cm⁻¹	1–5	1–5	25–100	1–5	50–75	1–5
Spectral range (approx) cm⁻¹	1500–4000	1–4000	240–5000+	240–8000	16–1600+	80–500
Sensitivity (% monolayer)	0.1	0.1	0.1	0.1	0.1	0.1
Sample area (mm²)	10	1	1	1	Very large	1
Substrate	Crystals, metals, and insulators	Roughened Ag, Cu, Au, other good reflectors	Crystals conducting	Oxide–metal	Finely divided metals and insulators	Crystals, metals, and insulators
Adsorbate	Few, mainly CO	CN⁻, pyridine, other polarizable molecules	Many	Many	Mainly H	Xe
Ambient pressure	≤1 atm	Sample preparation ≤1 atm	<10⁻⁶ Torr	Sample preparation ≤1 atm	≤Several atm	≤10⁻⁶ Torr
UHV compatible	√	√	√	√	√	√
Theoretical situation	Classical	Evolving	Developed (in principle)	Developed (in principle)	Developed	Evolving

1.2.1. Infrared-Absorption Spectroscopy

The granddaddy of all surface vibrational spectroscopies involves direct excitation of the vibrational modes by absorption of infrared photons provided either from an incoherent continuum source or a coherent laser. Since the dominant interaction between long-wavelength radiation and molecules is of dipole character, only transitions involving vibrational states coupled by a dipole operator are observable. Because of the relatively weak coupling (and thus small cross sections) and the small number of surface molecules, some sort of gimmick is required in order to enhance the absorption loss signal. The two most popular schemes involve operation in the transmission mode using highly dispersed samples to increase the number of surface molecules, as discussed in the chapter by Bell, or in the multireflection mode in which the incident photon beam is forced to make several passes at the possibly perfect planar surface, as detailed by Hayden. The transmission mode provides a technique for *in situ* study of messy catalytic systems but cannot readily be adapted to the characterizable idealized surfaces of the surface science world. Other schemes for enhanced surface sensitivity involve manipulations of angle-of-incidence, polarization, and dielectric properties, for which the reader is referred to the landmark review article on surface electrodynamics[22] by Feibelman for further information.

In my opinion, the most desirable feature of IAS on surfaces is the $\sim$1–5 cm^{-1} resolution that is routinely possible with this technique.[23-24] Under these circumstances, meaningful line shape/width measurements are possible, and it is in this type of data that extremely useful information on dynamical processes exists.[6]

Finally the potential role of pump–probe time domain experiments should be mentioned.[25,26] Here the molecule is vibrationally excited by an ir laser tuned to a chosen vibrational transition and the population of the depleted ground state is probed as a function of time after the initial excitation, thus providing a direct measure of the vibrational relaxation or decay time. Undoubtedly this type of ir experiment will provide a unique and valuable type of data required for any meaningful understanding of the dynamics of molecular processes at surfaces[27-29] or of laser surface chemistry.[14,30,31]

1.2.2. Surface-Enhanced Raman Spectroscopy

An almost unprecedented focusing of research effort in surface science followed the discovery ten years ago of a giant enhancement ($\sim10^6$) in the Raman signal associated with vibrational excitation of the molecule pyridine adsorbed on silver electrodes in an electrochemical cell. The results of this activity are documented in many review articles[12,32–35] and in the chapter by Campion.

The original excitement generated by this discovery had at least a twofold origin. First, because of the very small intrinsic Raman cross sections, ordinary Raman spectroscopy did not look very promising as a potential surface characterization tool. With the prospects of 10^6 enhancements over gas phase cross sections, this obstacle might be surmountable. Second, since the enhancement was totally unexpected, there was no existing theory that accounted for the result, and this void opened up a very fertile area for theoretical modeling and speculation. The down side of the initial wave was that the existence of the effect appeared to be fairly system-specific, showing up mainly for a few molecules (CN, pyridine, other highly polarizable organics) adsorbed on roughened Ag surfaces—hardly the precursor of a generalized analytic tool.

The theoretical evolution was characterized by either/or debates on THE mechanism. As usually seems to be the case, when the question is finally resolved, the answer contains a bit of all reasonable propositions with the degree of admixture varying from system to system. It is probably fair to say though that everybody now agrees that the following effects are more or less important in generating the enhanced Raman signal.

1. Roughness. Some form of spatial inhomogeneity in the plane of the surface is necessary as a momentum source for **K**-conservation in the optical transitions. The "roughness" could appear as a natural product of electrochemical etching, as a controlled grating scratched onto the surface, small particle shapes, or whiskers (antenna effect).

2. Surface Electrodynamics. The presence of the substrate modifies the relevant electrodynamics in a number of crucial ways. First is the trivial mirror effect, which could enhance the field at the site of the adsorbate, simply by reflection. Second, the irregular geometry of the surface provides regions of high field concentration at points of high curvature. Third, the modulation of the surface by both the laser and the vibrating molecule gives rise to an enhanced polarization field at the adsorbate. In most theories dealing with these effects, some sort of surface response function that depends upon $\varepsilon(\mathbf{r}, \mathbf{r}'; \omega)$, the nonlocal frequency-dependent dielectric function (or optical constants) of the substrate, is required. It is here that the dominant effects of substrate specificity enter.

3. Chemical Effects. From chemisorption theory,[36] it is well known that the interaction between an adsorbed atom or molecule and a surface both broadens and shifts the discrete electronic states of the isolated species. For most molecules, this effect will give rise to some partially unoccupied levels that straddle the substrate Fermi level. It is now believed that the existence of these broadened unoccupied levels plays a significant role in the Raman enhancement for at least two reasons. First, it makes resonance Raman processes possible that are forbidden in the gas phase, in effect increasing the polarizability of the molecule, and second, it facilitates charge transfer processes between molecule and substrate that ultimately manifest themselves in the Raman spectrum.

Although there are still many unresolved issues in SERS, the area continues to develop and evolve. Presently the effect has been confirmed for substrates of the highly reflective metals Ag, Cu, Au, Li, K, and Na and a very large class of polarizable molecules. Measurement techniques have improved to the point where there are now papers appearing on "SURS" (U = unenhanced). These issues are discussed in detail in the chapter by Campion.

Finally it should be mentioned that perhaps the greatest promise for SERS lies in the study of surfaces such as those in electrochemical environments,[37] which are inaccessible via particle probes.

1.2.3. Electron Energy Loss Spectroscopy

By far, the current most prevalent technique used in surface vibrational spectroscopy is EELS. A monochromatic electron beam (primary energy ~ 1–10 eV) is directed onto a surface and the energy distribution of the scattered beam is measured, perhaps varying the primary energy, angle of incidence, or scattering angle. Loss features appear in the spectrum at energies corresponding to the ro-vibrational excited states of the composite surface. Among the reasons for the popularity of EELS is high sensitivity, making possible the investigation of partial monolayers on perfect single-crystal faces, large dynamic range, and availability of commercial apparati. From the physics point of view, a number of different excitation modes are possible that provide different information. For instance, as will be demonstrated in Section 2, the dominant interaction between the electron beam and the oscillating molecule is via a dipole coupling. This produces a scattered electron beam in a near specular direction, which should have the same loss features as in IRS. On the other hand, higher multipole short-range scattering (often referred to as impact scattering) shows loss features not constrained by dipole selection rules which should contain information about the molecular charge distribution. Since the angular distribution is fairly isotropic, this is a hard measurement to make. Lastly, resonance scattering gives rise to significant overtone losses which permit study of anharmonic effects, combination bands, and intramolecular dynamics over intermediate state potential surfaces.

The principal negative aspects of EELS are the ultrahigh vacuum requirements and the relatively poor resolution. At present, the best resolution that has been obtained is ~2–3 meV while routine surface analysis is carried out with this degraded to ~10 meV. While this is fine for species identification and fingerprinting, it is inadequate for lineshape studies in which dynamics is studied. Avery, in his chapter, presents many archival examples of this extremely powerful technique in action.

1.2.4. Inelastic Tunneling Spectroscopy

Somewhat after the dramatic confirmation of the BCS theory of superconductivity by Giaver, who observed the predicted gap in the excitation spectrum by measuring the current–voltage characteristics through a metal–insulator–metal tunnel junction, above and below the transition temperature, another important observation was made. Lambe and Jaklevic[38] detected loss features in current–voltage (or derivative) curves for M–I–M junctions with deposited molecular impurities within the oxide layers, that corresponded to the ir absorption spectrum of the molecule, and IETS was born.[13] In effect, the tunneling electron excited the dipole active modes of the molecule in a way quite analogous to dipole scattering in EELS. The resolution, however, was potentially better than EELS. A major limitation of the technique, though, is that one does not study the vibrational properties of a molecule adsorbed on a perfect crystal face. Instead, one is restricted by the necessary geometry of the method to molecules embedded within oxide matrices upon deposited metal films. The relationship between these systems and those of the UHV surface world is problematic. This is treated in depth by Hansma in this volume.

1.2.5. Neutron Inelastic Scattering

Neutron inelastic scattering has provided the largest body of data on bulk phonon dispersion relations due both to the compatibility of neutron energies and momenta with those of the phonon excitations of solids, and to the similarity between thermal neutron wavelengths and crystal lattice spacings. The good news concerning applications of NIS to vibrational spectroscopy of molecules on surfaces is that owing to specific features of the nuclear forces between neutrons and protons, the inelastic cross sections for incoherent neutron scattering by the large-amplitude vibrational and/or hindered rotational modes of hydrogen-containing molecules are much greater than for any other vibrational excitations. The bad news is that the absolute magnitude of even these cross sections is very small. Consequently NIS has found its most significant role as a tool for studying hydrogen modes of "adsorbates" on the surfaces of high-area porous materials such as Raney nickel. While NIS provides a unique capability for such studies, the extent to which this provides insights into the properties of the corresponding perfect single-crystal systems is still a controversial issue, as discussed in the chapter by Cavanagh, Rush, and Kelley.

Another potentially exploitable and unique feature of NIS derives from the fact that exceedingly small ($\sim\mu$eV) energy losses can be observed. This makes possible a spectroscopy of the tunnel-split hindered rotational modes of hydrogen groups (e.g., the hydrogen umbrella in NH_3), which in turn can be used as a diagnostic tool for ascertaining local geometry.[39] It is probably a fair assessment to say that NIS will remain a technique for very special applications.

1.2.6. Atomic Inelastic Scattering

As with neutrons, the compatibility between the energy, momentum, and wavelength of light, inert gas atomic beams, and substrate phonon energies, momenta, and lattice plane spacings have conspired to make AIS an optimal technique for the study of surface phonon excitation.[40,41] Fortunately, unlike neutrons, the atomic beam is highly sensitive to surface atoms. To date the only work published in which AIS has been used to study vibrations associated with adsorbates pertains to He scattering from physisorbed Xe.[42] Although a promising technique, its current mode of application takes it outside the scope of this volume.

1.3. Theoretical Strategies

One of the most appealing aspects of vibrational spectroscopy is that in its most primitive form, at least some of the data can be understood and analyzed without recourse to any theory of the spectroscopic process. Since the position of a loss feature is purely a consequence of an energy conservation law ($\varepsilon_{initial} = \varepsilon_{final} + \varepsilon_{excitation}$), excitation energies, and presumably species identification, are readily determined. However, without a theoretical picture, it is not possible to conclude that the absence of a loss feature characteristic of a particular species implies that the species is not present. The role of the ubiquitous spectroscopic "matrix element effects" and selection rules must be considered, and thus a theory of the excitation process is required. In other words, if one is to access the information contained in both the absolute and relative intensities of loss features, then theory must enter. Likewise the inelastic differential cross section can only be understood and exploited through a theoretical model.

In confronting the issue of how to present the theoretical/conceptual basis of surface vibrational spectroscopies, I have opted for an expository approach in which simple and (hopefully) transparent models are developed rather than a brute-force, formal scattering theory construction. While this approach may not give a cookbook recipe for doing large-scale computational work, it should provide accessible insights. For instance, the general strategy adopted here is based on the following logic. For the most part, the loss features observed in present-day surface vibrational spectroscopy involve transitions between the ground and low-lying vibrational states. In this regime, the quasiperiodic motion of the vibrational modes is usually well approximated by harmonic oscillations, and this simplification will be exploited. More importantly, since the dynamics of classical and quantum mechanical harmonic oscillators are identical,[43,44] classical mechanics will be used as much as possible. Connections with quantum mechanics will be made through the correspondence principle. In my opinion, there exists considerable obfuscation within the literature of the inherent simplicity of vibrational excitation processes. To balance this tendency, the remainder of this chapter attempts to present some of the important issues in a physically suggestive manner, as simply as possible and in a unified way.

2. Excitation Mechanisms

In the construction of vibrational excitation models, the position is adopted here that most *direct* excitation processes are little more than the creation of low-lying excited staes of a harmonic oscillator. Since the quantum and classical dynamics of the forced harmonic oscillator are easily related via Ehrenfest's theorem and the correspondence principle, the initial presentation will remain classical. The reader is referred to the representative sample of oscillator theory articles which relate to many areas of physics and chemistry[43-51] for "sophisticated" but equivalent treatments and more importantly, to the excellent classical mechanics book by Marion.[52]

In this section, once the required properties of the driven oscillator are demonstrated, the microscopic basis of dipole excitation of vibrational models is then discussed. Implications concerning so-called surface-selection rules will be indicated. Next the response to a driving force supplied by an external electromagnetic field (photons) will be investigated. Similarities and differences between photon and electron beam excitation will be pointed out and major features of EELS highlighted. Considerable attention will be directed towards the role of electronically excited states and resonances in the production of enhanced vibrational excitation. Some thought will also be given to the excitation process in NIS.

2.1. Harmonic Oscillator Mechanics

2.1.1. Free Oscillator

Newton's equation of motion for a mechanical system characterized by a displacement coordinate q and mass m, a restoring force proportional to the displacement $F_r = -kq$ (derived from a quadratic potential energy), and a nonconservative damping proportional to the velocity $F_d = -b\dot{q}$, is

$$\ddot{q} + 2\eta\dot{q} + \omega_0^2 q = 0 \tag{1}$$

where $\eta \equiv b/2m$, b is the viscosity, and $\omega_0 = (k/m)^{1/2}$ is a characteristic frequency of the motion if $\eta = 0$. The solution of equation (1), when $\omega_0^2 > \eta^2$, is termed underdamped motion (the only case of importance in vibrational spectroscopy) and is given by

$$q(t) = q_{am}e^{-\eta t}\cos(\omega_1 t - \delta) \tag{2}$$

where $\omega_1 \equiv (\omega_0^2 - \eta^2)^{1/2}$ and q_{am} and δ are "situationally dependent" initial conditions which must be supplied. If $\eta = 0$, the motion is periodic with the total energy

$$\varepsilon_{\text{tot}} = \frac{p^2(t)}{2m} + \tfrac{1}{2}kq^2(t) = \tfrac{1}{2}m\omega_0^2 q_{am}^2 \tag{3}$$

a constant of the motion, where $p(t) = m\dot{q}(t)$.

2.1.2. Forced Oscillator

Vibrational spectroscopy requires that the molecular vibrational modes/harmonic oscillators are excited or decay. Two important classes of excitation are considered.

2.1.2a. Time-Dependent Forcing Function. In many cases, the oscillator is exposed to some time-dependent interaction potential which is both separable and linear in oscillator coordinates, in the fashion

$$V_f(q, t) = q\lambda(t) \tag{4}$$

where $\lambda(t)$ is an arbitrary function of time. In analogy with equation (1), the corresponding equation of motion is

$$\ddot{q} + 2\eta\dot{q} + \omega_0^2 q = \frac{1}{m}\lambda(t) \tag{5}$$

The solution of equation (5) for an oscillator originally at rest, is given by[52]

$$q(t) = \int_{-\infty}^{+t} G(t, t')\lambda(t')\, dt' \tag{6}$$

where

$$G(t, t') = \frac{1}{m\omega_1} e^{-\eta(t-t')} \sin \omega_1(t - t'), \qquad t \geq t'$$

$$= 0, \qquad\qquad\qquad\qquad\qquad t < t' \tag{7}$$

is the oscillator Green's function or response function to an impulsive force.

For undamped motion ($\eta = 0$, $\omega_1 = \omega_0$) and a temporally localized forcing function, in the large-t limit where $\lambda(t) \to 0$, equations (6) and (7) can be written as

$$q(t) \simeq \mathrm{Im}\left\{ \frac{e^{i\omega_0 t}}{m\omega_0} \int_{-\infty}^{\infty} \lambda(t')e^{-i\omega_0 t'}\, dt' \right\}$$

$$= \sin \omega_0 t \frac{\lambda(\omega_0)}{m\omega_0} \tag{8}$$

where $\lambda(\omega_0) = \lambda(-\omega_0)$ is the Fourier transform of $\lambda(t)$. The total excitation energy deposited in the oscillator by the time-dependent force which follows from equation (4) is

$$\Delta\varepsilon_{\mathrm{classical}} = \tfrac{1}{2} k \frac{|\lambda(\omega_0)|^2}{(m\omega_0)^2} = \frac{|\lambda(\omega_0)|^2}{2m} \tag{9}$$

a very useful result.

In the special case of the harmonic oscillator, the correspondence principle provides an elegant and exact connection between the energy gain of the forced classical harmonic oscillator, as specified by equation (9), and the vibrational excitation probability distribution of the equivalent quantum mechanical harmonic oscillator subjected to the same forcing function. In terms of the parameter

$$\beta = \Delta\varepsilon_{\text{class}}/\hbar\omega_0 = |\lambda(\omega_0)|^2 / 2m\hbar\omega_0 \tag{10}$$

it has been demonstrated many times that the probability for a $n \rightarrow m$ vibrational transition induced by $V_f(q, t) = q\lambda(t)$, is given by[43,44,53]

$$P_{m \rightarrow n} = m!n!e^{-\beta}\beta^{(m+n)}(S_{m, n})^2 \tag{11}$$

where

$$S_{m,n} = \sum_{k=0}^{\min(m, n)} \frac{(-1)^k \beta^{-k}}{(m - k)!k!(n - k)!}$$

For events in which the oscillator is initially in its ground state, equation (11) reduces to

$$P_{0 \rightarrow n} = e^{-\beta}(\beta^n/n!) \tag{12}$$

a Poisson distribution. The dynamics of the particular process enter into equation (12) solely through equation (10), which expresses the functional dependence of β on the Fourier transform of $\lambda(t)$.

2.1.2b. Trajectorized Particle Forcing Function. Yet another means of vibrational excitation is through the influence of a moving particle. Suppose that this particle with mass M and coordinate $\mathbf{R}$ moves in some static potential field $V_p = V_p(\mathbf{R})$ and is coupled to the oscillator via some $V_{\text{int}} = V_{\text{int}}(q, \mathbf{R})$. The classical dynamics of the particle–oscillator system is then specified by the set of coupled equations of motion

$$\ddot{\mathbf{R}} = -\frac{1}{M}[\nabla_{\mathbf{R}}V_p(\mathbf{R}) + \nabla_R V_{\text{int}}(q, \mathbf{R})] \tag{13}$$

and

$$\ddot{q} + 2\eta\dot{q} + \omega_0^2 q = -\frac{1}{m}\frac{\partial}{\partial q}V_{\text{int}}(q, \mathbf{R}) \tag{14}$$

If in equation (13), the variable q is set equal to q_0, its mean value, then equations (13) and (14) decouple and equation (13) can in principle be solved for any prescribed V_p and V_{int} to yield $\mathbf{R} = \mathbf{R}(t)$. This is known as the trajectory approximation in which the time dependence of the oscillator motion does not influence the motion of the particle.

With regard to the oscillator, if a small displacement expansion of V_{int} is made,

$$V_{\text{int}}(q, \mathbf{R}) \simeq V_{\text{int}}(q_0, \mathbf{R}) + q\frac{\partial V_{\text{int}}}{\partial q}(q, \mathbf{R})\bigg|_{q=q_0} + \cdots \tag{15}$$

and the trajectory obtained from equation (13) inserted, then with $\partial V_{int}/\partial q \,|_{q=q_0} \equiv V'_{int}$, equations (14) and (15) become

$$\ddot{q} + 2\eta\dot{q} + \omega_0^2 q = \frac{-1}{m} V'_{int}[R(t)] \tag{16}$$

where V'_{int} is a functional of the trajectory. Equation (16) is exactly of the form of equation (5). Thus with

$$\lambda_{int}(\omega_0) \equiv \int_{-\infty}^{\infty} V'_{int}[R(t)]e^{-i\omega_0 t} \, dt \tag{17}$$

the excitation probabilities given by equations (10)–(12) are equally applicable to the incident particle excitation. It should be cautioned that the trajectory theory is inherently non-energy-conserving, since although energy was delivered to the oscillator, the incident particle was not allowed to slow down. Consequently, this approximation is best when $\Delta\varepsilon_{class}/\varepsilon_{trans} \ll 1$. This question has been discussed at great length elsewhere and the reader is pointed to Refs. 53–57 for further discussion.

2.2. Dynamic Dipoles

The total charge distribution associated with a given molecule adsorbed upon a surface can be expressed as a multipole expansion. By total charge distribution, we mean the sum of that of the molecule plus all induced or polarization charges, i.e., "image" charge distributions within the substrate.

Consider first the monopole contribution. In general there is some overall charge transfer $\equiv fe$ between substrate and adsorbate when a chemisorption bond is formed, where f is the fraction of an electron charge that is transferred.[36] The net charge distribution that results from the combination of adsorbate plus its induced image within a metallic substrate is shown in Figure 2a. The total dipole moment due to the molecular monopole is

$$\mu = 2fe(s_0(\omega) + q(t)) \tag{18}$$

where s_0 is the equilibrium location of the static charge centroid from the effective image plane[22,58,59] and q is a possibly oscillatory small displacement about this equilibrium point. Note that the image plane location has been written as a function of frequency. Because of the finite time lag of the substrate screening response to a dynamic perturbation, if one wishes to structure the substrate response in terms of image charge distributions, then the location of the effective image plane depends upon the frequency of the perturbing force.[22,59]

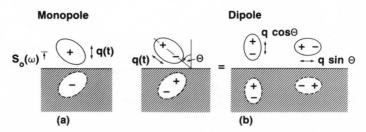

Figure 2. Molecular plus induced image charge distributions for (a) adsorbed molecule with permanent monopole charge; (b) adsorbed molecule with dynamic dipole moment.

Since a dipole moment can be expressed as the product of an effective charge times an oscillatory separation, $\mu = e^* q(t)$, it is reasonable to introduce the concept of a dynamic dipole charge associated with the time-dependent component of q. Identifying the dynamic dipole charge as

$$e_m^* = \frac{\partial \mu}{\partial t} \bigg/ \frac{dq}{dt}$$

if follows from equation (18) that

$$e_m^* = 2fe$$

and the dynamic dipole associated with the time dependent motion about s_0 is

$$\mu_{\mathrm{dyn}} = e_m^* q(t) = 2feq(t) \tag{19}$$

As we will see shortly, this is the dipole that is responsible for the $\sim 50\,\mathrm{meV}$ Ni–CO excitations shown in Figure 1, due to the vibrations of the "rigid molecule" within the mostly one-dimensional well formed by the chemisorption bond.

Both the static and dynamic dipole moment associated with the intrinsic molecular charge distribution also induce image dipoles within the substrate as shown in Figure 2b. However, if the dipole is decomposed into components perpendicular and parallel to the surface, as also depicted in Figure 2b, it is noticed that the perpendicular component is doubled, as in equation (14), whereas the parallel component is canceled by its instantaneous image. More rigorously, the total dynamical dipoles should be written in terms of dynamic retarded response functions

$$\mu_\perp^{\mathrm{tot}}(t) = \mu_\perp(t) + \int_{-\infty}^{t} R_\perp(t, t') \mu_\perp(t') \, dt'$$

and

$$\mu_{\parallel}^{\text{tot}}(t) = \mu_{\parallel}(t) - \int_{-\infty}^{t} R_{\parallel}(t, t')\mu_{\parallel}(t') \, dt'$$

which for the usual cases where $\mu(t) \sim e^{i\omega_0 t}$ and $R(t, t') = R(t - t')$, easily reduce to

$$\mu_{\perp}^{\text{tot}}(t) = \mu_{\perp}(t)[1 + R_{\perp}(\omega_0)] \tag{20a}$$

and

$$\mu_{\parallel}^{\text{tot}}(t) = \mu_{\parallel}(t)[1 - R_{\parallel}(\omega_0)] \tag{20b}$$

Since typical vibrational frequencies $\omega_0 \ll \omega_p$, where $\omega_p \sim 3\text{--}15$ eV is the substrate plasmon frequency which sets the time scale of the substrate screening response,[21] the dynamic intramolecular dipole is screened almost as a static dipole, with $R_{\perp}(\omega_0) = R_{\parallel}(\omega_0) = 1$, and the limit depicted by the image pictures in Figure 2b prevails. It is this image doubling of perpendicular dipole components and cancellation of parallel ones which are responsible for the "surface selection rules" governing dipole excitation processes at surfaces.[10,60]

2.3. Radiative Excitation

The most common mode of vibrational spectroscopy using infrared excitations is the measurement of the spectral composition of the change of reflectance of a surface due to the presence of adsorbed molecules.[11,15,22,61-63] The wavelength dependence of the absorption/ reflection bands more or less directly yields the spectral distribution of the ir-active vibrational modes of the adsorbed molecules.

The dynamic interaction between a point dipole and time-dependent electric field associated with the radiation incident upon the dipole is

$$\begin{aligned} H_{\text{int}}(t) &= \boldsymbol{\mu}(t) \cdot \mathbf{E}(t) \\ &= e^* \mathbf{q}(t) \cdot \mathbf{E}(t) \\ &= e^* q(t) E(t) \cos \alpha \end{aligned} \tag{21}$$

with α the angle between the oscillator coordinate and the polarization vector of the radiation. Since H_{int} is separable in the manner of equation (4), the radiatively driven oscillator equation of motion is given by equation (5). The Fourier transform of $E(t)$, normalized according to the conventions in Jackson,[64] is

$$E_{\alpha}(\omega) = \frac{\cos \alpha}{\sqrt{2\pi}} \int E(t) e^{i\omega t} \, dt \tag{22}$$

If $E(\omega) \approx$ const over the width of the absorption band ($\Delta E_{\text{FWHM}} \simeq 2\hbar\eta$)
then $E_\alpha(\omega) \simeq E_\alpha(\omega_0)$, in which case equations (5), (9), (21), and (22)
yield the expression

$$\Delta\varepsilon = \pi \frac{e^{*2} |E_\alpha(\omega_0)|^2}{m} \qquad (23)$$

for the energy absorption by the oscillator. In terms of the flux
$f_\alpha(\omega_0) = (c/2\pi) E_\alpha(\omega_0)^2$, in units of incident energy per area per unit
frequency, the average energy absorption is

$$\Delta\varepsilon = \frac{2\pi^2 e^{*2}}{mc} f_\alpha(\omega_0) \qquad (24)$$

which is proportional to the energy content of the radiation pulse.
Equation (23) or (24), in conjunction with equations (10) and (11), yield
the excitation probability and spectral distribution of excitations.

The crucial issue that still must be addressed is the determination of
the field strength at the site of the oscillator/adsorbed molecule, for a
given incident field. As Greenler has noted,[61] "It would appear that the
surface of the metal would be an excellent place to locate a thin layer of
material which was to be hidden from detection by the infrared
spectroscopist." The reason behind this can be understood in terms of the
infrared reflection properties of a "typical" metal, as sketched in Figure
3. The polarization of an electromagnetic wave incident upon a surface at
an angle θ can be decomposed into a component parallel to the surface
($\hat{\varepsilon}_s$) and a component within the plane of incidence ($\hat{\varepsilon}_p$), as shown in
Figure 3a. Maxwell's equation demand that at "the boundary" between
the two optically different media, the parallel or s-polarized E field is
continuous. Furthermore, as a consequence of the fact that infrared fields
do not penetrate into metals (on a distance scale set by the wavelength),
the net incident plus reflected s-polarized wave must vanish at the
surface. This condition requires a phase change $\delta_s = \pi$ between the
incident and reflected wave for a perfect conductor, and near π for good
reflecting metals, as shown in Figure 3b. On the other hand, the
boundary conditions of continuous electric displacement normal to the
surface lead to no (or small for real metals) phase change between the
normal components of the incident and reflected p-polarized waves. As a
result, the net field at the oscillator is the constructively interfering sum
of incident plus reflected p-polarized waves, which is enhanced as the
propagation direction is oriented more towards grazing incidence, as
shown in Figure 3d.

The points just raised can be algebraically summarized as follows.
Owing to the electrodynamics of the surface, the total oscillating field felt

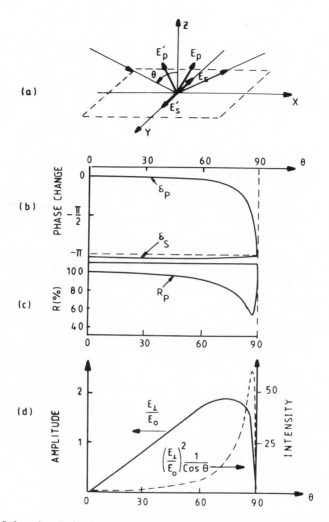

Figure 3. Infrared optical reflectivity of metal surface showing (a) the incident and reflected S- and P-Polarized electric field vectors; (b) the angular dependence of phase change; (c) the reflectance of the P-polarized component; and (d) the normalized amplitude (left side) and intensity (right side) variation of the component of the P-polarized electric field acting normal to the surface.[15]

at the site of the adsorbed molecule is, to a first approximation, given by

$$E_{tot}(\omega_0) = E_{in}(\omega_0)[1 + g_p(\theta, \omega)]\sin \theta$$

where $E_{in}(\omega_0)$ is the incident strength and $g(\theta, \omega)$ is the frequency-dependent reflection amplitude for p-polarized radiation, incident at angle θ (assuming azimuthal symmetry). Thus for an array of noninteracting oscillators distributed over the surface, the energy absorption out of an incident beam of fixed width is

$$\Delta\varepsilon = \frac{\pi e^{*2}\cos^2 \alpha}{m}|E_{in}(\omega_0)|^2 \times |1 + g_p(\theta, \omega)|^2 \frac{\sin^2 \theta}{\cos \theta} \qquad (25)$$

where the $\cos^{-1} \theta$ term arises from the increase in illuminated surface area with increasing θ. For present purposes we have made arguments presuming that the surface was a perfect conductor. The generalization to real metals characterized by finite optical constants follows from the Fresnel equations. Further generalization in terms of nonlocal surface electrodynamics has been carried out.[22] Aside from some special circumstances such as the silver specificity in SERS or radiation in which $\omega \simeq$ plasmon frequency of the substrate, the semiquantitative points raised here remain intact.

The basic message is that the presence of a metal surface requires certain electromagnetic boundary conditions to be satisfied, as illustrated in Figure 4. In this presentation, we have placed the boundary conditions on the electromagnetic field driving the oscillator (Figure 4a). An alternative approach (Section 2.2) would be to impose the boundary conditions on the oscillator (image dipoles) and then calculate the interaction energy of this array with the incident field (Figure 4b). Either way yields the following identical relevant conclusions:

1. The ir activity of a mode depends only upon the component normal to the surface,

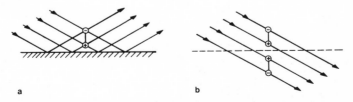

a   b

Figure 4. (a) A dipole near a surface is affected by the fields of both the incident and reflected radiation. (b) An alternate way to view the interaction of 4a: the radiation interacts with the dipole and its optical image. There is an abrupt phase shift as the waves cross the dashed boundary.[66]

2. Only the normal component of a p-polarized wave is important in ir excitation,
3. Energy absorption and vibrational excitation probabilities are enhanced as the angle of incidence is moved toward grazing.

2.4. Electronic Excitation

Several different modes of vibrational excitation by ($\sim$1–10 eV) electron beams are important in surface spectroscopy. These include dipole, "impact," and resonance scattering. When operative, each of these modes permits access to different information since different selection rules and potential energy surfaces control the outcome of the spectroscopic process.

2.4.1. Dipole Scattering

There have appeared in the literature many theoretical treatments of dipole scattering of electrons by adsorbed atoms which can be accessed through several excellent review articles.[9,10,15,16,67] Inevitably, the most significant conclusions of these studies are reproduced by both classical and quantum mechanical models. Therefore we will here continue the exposition in terms of the driven oscillator model presented in Section 2.1.

Consider the scattering geometry shown in Figure 5. A "classical electron" characterized by spatial coordinates (ρ, z), velocity components

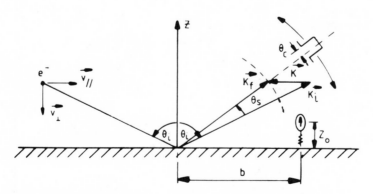

Figure 5. Inelastic electron scattering geometry for a molecular excitation located a distance b from the point of impact and a distance z_0 above the surface plane of reflection. $\vec{k}_i$, $\vec{k}_f$, and $\vec{k}$ refer to the incident, final scattered, and excitation wave vectors, respectively; θ_s is the scattered angle and θ_c the analyzer collection angle.[15]

$(v_{\parallel}, v_{\perp})$, with polar angle of incidence $\theta_i = \tan^{-1} (|v_{\parallel}| / v_{\perp})$, is elastically reflected from a surface at an impact point a distance b from an adsorbed molecule which possesses a dynamic dipole moment. In addition, the positive image of the incident electron is located at $(\rho, -z)$. The total electric field from the incident electron plus its image, at the site of the oscillator (with $z_0 \ll r = (\rho^2 + z^2)^{1/2}$) is

$$E_z = \frac{2ze}{(\rho^2 + z^2)^{3/2}}$$

$$E_{\parallel} = 0$$
(26)

which is yet another consequence of the perfect-screening boundary conditions of an ideal metal. For the moving electron, both ρ and z are functions of time, hence $E_z = E_z(t)$ within the trajectory approximation. The dynamic dipole interaction between the electron and the oscillator is given by equations (21) and (26) as

$$H_{int}(t) = \frac{2e^* e q_{\perp} z(t)}{[\rho^2(t) + z^2(t)]^{3/2}}$$
(27)

where $q_{\perp}$ is the component of the oscillator motion normal to the surface. Since equation (27) is of the separable form of equation (4), the oscillator equation of motion is equation (16), whose solution is given by equation (8). It is expedient to express the dipole potential $(\sim z/r^3)$ in equation (27) as a two-dimensional Fourier transform[10]

$$z/(\rho^2 + z^2)^{3/2} = \int \frac{d^2 Q_{\parallel}}{(2\pi)} e^{i Q_{\parallel} \cdot \rho} e^{-Q_{\parallel} z}$$
(28)

which when combined with equations (17) and (27) yields

$$\lambda_{int}(\omega_0) = \frac{e^* e}{\pi} \int d^2 Q_{\parallel} \, dt \, e^{i Q_{\parallel} \cdot \rho(t)} e^{-Q_{\parallel} z(t)} e^{+i\omega_0 t}$$
(29)

The excitation spectrum is obtained from equations (10), (12), and (29) and a prescribed trajectory. A reasonable and analytically tractable choice is the specularly reflecting straight line trajectory:

$$z(t) = v_{\perp} |t|$$

$$\rho(t) = b + v_{\parallel} t$$
(30)

with the electron hitting the surface at $t = 0$. Inserting equation (30) in (29) and performing the time integral gives

$$\lambda_{int}(\omega_0) = \frac{2e^* e}{\pi} \int d^2 Q_{\parallel} e^{i Q_{\parallel} \cdot b} \frac{Q_{\parallel} v_{\perp}}{(\Omega_Q^2 + Q_{\parallel}^2 v_{\perp}^2)}$$
(31)

with $\Omega_\mathbf{Q} \equiv \omega_0 + \mathbf{Q}_\| \cdot \mathbf{v}$. Consequently the Poisson parameter β which is equivalent to the excitation probability [equation (12)] is

$$\beta_{\text{dip}} = \left(\frac{2e^*e}{\pi}\right)^2 \frac{1}{2m_0\hbar\omega_0} \int d^2Q_\| \, d^2Q'_\| \, e^{i(\mathbf{Q}_\| - \mathbf{Q}'_\|) \cdot \mathbf{b}} F(\mathbf{Q}_\|)F(\mathbf{Q}'_\|) \quad (32)$$

where $F(\mathbf{Q}_\|)$ is the remaining integrand in equation (31). The quantity actually measured is obtained by integrating equation (32) over all impact parameters $\mathbf{b}$. This easily leads to an expression for the excitation cross section

$$\sigma_x = \int d^2\mathbf{b} \, \beta_{\text{dip}}(\mathbf{b}) = \gamma^2 \int d^2Q_\| \frac{4Q_\|^2 v_\perp^2}{(\Omega_\mathbf{Q}^2 + Q_\|^2 v_\perp^2)^2} \quad (33)$$

with $\gamma \equiv (2e^*e/\hbar)(\hbar/2m_0\omega_0)^{1/2}$. For a distribution of independent oscillators on a surface, the effective Poisson parameter is

$$\beta_{\text{dip}} = \sigma_x/A \quad (34)$$

where A is the average surface area per oscillator or molecule. In the limit in which $v_\| = 0$, the differential cross section given by equation (33) is quite peaked around momentum transfer $Q_\| \simeq \omega_0/v_\perp$. Since $v = 5.9 \times 10^{15} \times \varepsilon_{\text{in}}^{1/2}$ Å/sec (ε_{in}, the incident primary energy in eV) with $\hbar\omega_0 \simeq 0.25$ eV, and $\varepsilon_{\text{in}} \sim 2$ eV, it follows that $Q_\| \simeq 0.05$ Å^{-1} whereas $k_{\text{in}} \simeq 1$ Å^{-1}. Two implications of this follow: First, since only the long-wavelength Fourier components of the dipole potential are important, from equation (28) it is apparent that most of the inelastic dipole scattering occurs when $z \approx 20$ Å, thus justifying the point dipole and perfect flat surface approximations, both of which smooth over atomic scale details. Second, since $Q_\|/k_{\text{in}} \ll 1$, inelastic dipole scattering is basically slightly deflected forward scattering. Elastic scattering from the substrate provides the momentum to reflect and/or diffract[68] the electrons.

In order to make further contact with experiment, equation (33) must be expressed in terms of primary electron energy, angle of incidence, scattering angle, and detector acceptance angle. This is a tedious but straightforward exercise in kinematics, which has appeared frequently[15,69–71] and will not be repeated here, other than to note that the final answer is purely a consequence of equation (33), energy conservation via $\hbar^2(k_{\text{in}}^2 - k_f^2)/2m_e = \hbar\omega_0$, parallel momentum conservation $\mathbf{Q}_\| = \mathbf{k}_{\text{in}}^\| - \mathbf{K}_f^\|$, and the small scattering angle approximation. The resulting differential cross section which follows from these procedures is

$$d\sigma_x = \frac{4\gamma^2 m_e \cos\theta_i f(\theta_s, \psi_s)\theta_s \, d\theta_s \, d\psi_s}{\varepsilon_{\text{in}}(\theta_s^2 + \theta_0^2)^2} \quad (35)$$

with

$$f(\theta_s, \psi_s) \equiv (\theta_s \cos \psi_s - \theta_0 \tan \theta_i)^2 + \theta_s^2 \sin^2 \psi_s \sec^2 \theta_i$$

$$\theta_0^2 \equiv \hbar\omega_0/2\varepsilon_{\text{in}}$$

and θ_s, ψ_s the polar and azimuthal angle of the scattered electron with respect to the outgoing elastic beam. Integrating equation (35) over θ_c, the aperture angle of the detector (typically $\sim 3°$) yields the loss intensity

$$\beta_{\text{dip}} \simeq \frac{\pi m_e \gamma^2}{\varepsilon_{\text{in}} A \cos \theta_i} \left[(1 + \cos^2 \theta_i)\ln\left(1 + \frac{\theta_c^2}{\theta_0^2}\right) + (1 - 3\cos^2 \theta_i)\frac{\theta_c^2}{\theta_0^2 + \theta_c^2} \right]$$

(36)

The principal conclusions concerning dipole scattering that can be drawn from these considerations and equations (35) and (36) are the following:

1. Only ir active modes with a component of dynamic dipole moment perpendicular to the surface are observed in loss spectra.
2. Inelastic dipole scattering involves little momentum transfer. Hence forward scattering prevails.
3. The angular distribution of inelastically scattered electrons is in the form of a narrow lobe, slightly displaced in reflection angle from the direction of elastically scattered electrons.
4. The width of the lobe, characterized by θ_0, increases with decreasing primary energy.

The definitive and oft-cited confirmation of the dipole scattering theory from adsorbates was due to Andersson and co-workers, who reported on the energy and angular dependence of inelastic loss features from the vibrational modes of CO adsorbed on Cu(100).[72] Two loss features, one at 43 meV and the other at 260 meV, are associated with the molecule–substrate bond and the intramolecular stretch, respectively. The loss intensity as a function of primary energy, both observed and calculated from an expression equivalent to equation (36), are shown in Figure 6. The dynamic dipole charge e^* was used as a fitting parameter. The required value $e_{\text{CO}}^* = 0.67e$ for the 260-meV loss is in excellent accord with the free CO value of $0.64e$. The angular distributions for two primary energies observed and calculated from equation (35) are shown in Figure 7. The obviously excellent agreement between theory and experiment provides strong support for the mechanisms of dipole scattering, as put forward in this section.

2.4.2. Impact Scattering

Since the basic interaction between the incident electron and the vibrational system of the adsorbed atom is Coulombic, the potential can

Cu(100)c(2×2)CO

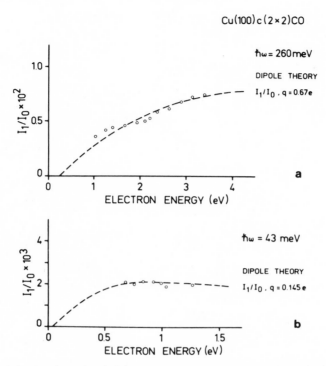

Figure 6. Relative loss intensity, (I_1/I_0), vs. primary electron energy of the fundamental (a) C–O and (b) Cu–CO stretching vibrational modes for $c(2 \times 2)$CO on Cu(100) (O) experimental data. The dashed curves represent the dipole theory predictions.[72]

be expressed in terms of a multipole expansion. Because of the long-range nature of the monopole and dipole potentials, the inelastic scattering is quite insensitive to the particular atomic-scale details, and as a result, the basic theory of dipole excitation could be presented in a rather general (and simple) form. In the case of "higher poles" and direct scattering off the ion cores of the molecule, the incident electron feels the full atomic details of the molecule, and hence a complete quantum scattering theory description is required. This has the unfortunate consequence of not easily lending itself to quantitative generalizations.

In order to establish a theoretical construction, let the quantum system of the incident electron plus vibrating molecule be characterized by the Hamiltonian

$$H_{tot} = H_{el}(\mathbf{r}) + H_v(\mathbf{q}) + V_c(\mathbf{r}, \mathbf{q}) \tag{37}$$

where $H_{el}(\mathbf{r})$, the electronic Hamiltonian, includes all static potentials

$$Cu(100)c(2\times2)CO$$

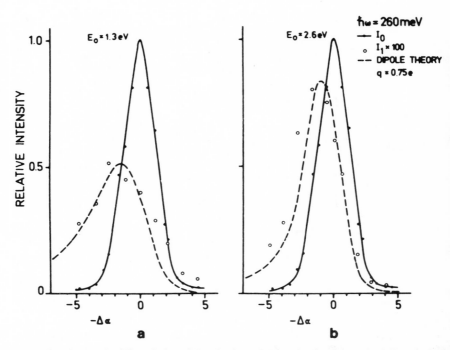

Figure 7. The angular dependence of the elastic peak intensity ($-\bullet-$) and the 260-meV loss peak intensity ($\bigcirc$) versus the change, ($-\Delta\alpha$), in the angle between the analyzer and the specimen surface normal for (a) 1.3 eV and (b) 2.6 eV incident electrons. The dashed curves are dipole theory calculations for an effective dynamic charge $q = 0.75e$.[72]

provided by the substrate, $H_v(\mathbf{q})$ is the molecular vibrational Hamiltonian, and $V_c(\mathbf{r}, \mathbf{q})$ the coupling between the two. Because of the different time scales for electronic and vibrational motion, the Born–Oppenheimer approximation is reasonable, which allows for the total wave function to be written in the product form

$$\psi_{tot}(\mathbf{r}, \mathbf{q}) = \psi_{el}(\mathbf{r}; \mathbf{q})\chi_v(\mathbf{q}) \tag{38}$$

where $\mathbf{q}$ is treated as a parameter, not a dynamic variable, within the electronic Hilbert space. The amplitude for inelastically scattering an electron from an incident plane wave state $e^{i\mathbf{k}\cdot\mathbf{r}}$ into an asymptotic state $\psi_{k'}(\mathbf{r}; \mathbf{q})$ accompanied by a $v \to v'$ vibrational transition is

$$f(\mathbf{k}v \to \mathbf{k}'v') = (-m_e/2\pi\hbar^2)\langle \psi_{k'}(\mathbf{r}; \mathbf{q})\chi_{v'}(\mathbf{q}) \mid V_c(\mathbf{r}, \mathbf{q}) \mid e^{i\mathbf{k}\cdot\mathbf{r}}\chi_v(\mathbf{q})\rangle$$

or selectively rearranged in terms of explicit integrals:

$$f(kv \to k'v') = (-m_e/2\pi\hbar^2) \int d^3q\, \chi_{v'}^*(\mathbf{q}) \left\{ \int d^3r\, \psi_{\mathbf{k}'}^*(\mathbf{r}; q) V_c(\mathbf{r}, q) e^{i\mathbf{k}\cdot\mathbf{r}} \right\} \chi_v(\mathbf{q})$$

(39)

If $\psi_{\mathbf{k}'}(\mathbf{r}; \mathbf{q}) \simeq \psi_{\mathbf{k}'}(\mathbf{r}; \mathbf{q}_{eq})$ with q_{eq} the vibrational ground state equilibrium separation, and $V_c(\mathbf{r}, \mathbf{q}) \simeq V_c(\mathbf{r}, \mathbf{q}_{eq}) + \mathbf{q} \cdot \nabla_\mathbf{q} V_c(\mathbf{r}, \mathbf{q})|_{q=q_{eq}} + \cdots$, then the lowest-order nonvanishing term leads to dipole excitation discussed earlier, since

$$f_d(kv \to k'v') \sim \int d^3q\, \chi_{v'}^*(\mathbf{q}) \mathbf{q} \chi_v(\mathbf{q}) \cdot \int d^3r\, \psi_{\mathbf{k}'}^*(\mathbf{r}) \nabla_\mathbf{q} V_c(\mathbf{r}, \mathbf{q})|_{q=q_{eq}} e^{i\mathbf{k}\cdot\mathbf{r}}$$

is a product of a vibrational dipole matrix element times the electronic factor. For a harmonic oscillator, the vibrational matrix element yields the selection rule $v' = v \pm 1$.

In the case of "impact scattering," not only does the electron scattering state feel the q dependence, but more importantly, the high-order terms in the expansion of $V_c(\mathbf{r}, \mathbf{q})$ are significant. Thus separability and useful classification in terms of multipoles is thwarted. However, in terms of the electron scattering amplitude

$$f_{\mathbf{k},\mathbf{k}'}(\mathbf{q}) \equiv (-m_e/2\pi\hbar^2) \int d^3\mathbf{r}\, \psi_{\mathbf{k}'}^*(\mathbf{r}; q) V_c(\mathbf{r}, q) e^{i\mathbf{k}\cdot\mathbf{r}}$$

(40)

expanded in partial waves (assuming $|\mathbf{k}| \simeq |\mathbf{k}'|$)

$$f_{\mathbf{k},\mathbf{k}'}(\mathbf{q}) = \sum_{\substack{l,m \\ l',m'}} Y_{l,m}(\hat{k}) f_{\substack{l,m \\ l',m'}}(q) Y_{l',m'}(\hat{k}')$$

(41)

the differential cross section that follows from equations (39)–(41) is

$$\frac{d\sigma}{d\Omega} = |f(kv \to k'v')|^2 = \sum_{\substack{l,m \\ l',m'}} |Y_{l,m}(\hat{k})\langle v'| f_{\substack{l,m \\ l',m'}}(q) |v\rangle Y_{l',m'}(\hat{k}')|^2$$

(42)

which has two major consequencies on observables in EELS. First, since $f(q) \neq q$, overtone excitation is expected to be enhanced in the impact regime. Second, owing to the fact that the intimate short-range details of the molecular potential are responsible for the scattering [equation (40)], the momentum transfer to the electron will be substantial. Thus a large range of scattering directions $\hat{k}'$ is to be expected [equation (42)], in contrast to the narrow, near specular lobes characteristic of dipole scattering. That the scattered wave is diffuse rather than directed is a

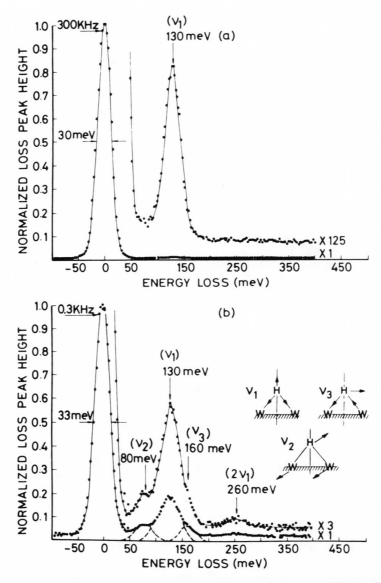

Figure 8. Normalized electron energy-loss spectra for saturation coverage (β_1 phase) of H chemisorbed on W(100) for $\theta_i = 23°$ incident angle and an impact energy $E_0 = 9.65$ eV: (a) specular beam direction: (b) +17° off the specular direction towards the surface. The elastic beam count rate (in kilohertz) and the energy resolution are indicated in the figure. The fundamental vibrational modes (inset Figure 8b) correspond to the bridge-site C_{2v} symmetry bonding. The incident beam is along the [100] crystal direction.[77]

hindrance in experimental detection. On the other hand the fact that it appears in directions where it is not overwhelmed by the elastic and dipole contributions should be a help. For further theoretical discussion, one should consult Refs. 73–76.

The definitive experiments that confirmed the basic ideas of dipole versus impact vibrational excitation of adsorbed atoms and molecules were carried out by Ho, Willis, and Plummer in their angle-resolved EELS studies of atomic H adsorbed on W(100).[77] Their major observations are shown in Figures 8 and 9. In Figure 8a, the specular loss

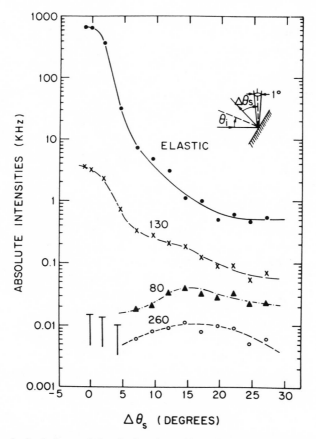

Figure 9. Angle dependence of the absolute intensities as a function of the collection angle $\Delta\theta_s$. The scattering geometry is shown in the upper inset. The impact energy $E_0 = 9.65$ eV and angle of incidence $\theta_i = 23°$. The partial error bars for the data near the specular direction ($\Delta\theta_s = 0°$) for the 260-meV peak indicate the upper limit in intensity. (See Figure 8a for the low signal-to-background ratio at 260 meV.)[77]

spectrum, for incident electron kinetic energy = 9.65 eV, and angle of incidence θ_i = 23°, shows a solitary loss at $\Delta\varepsilon$ = 130 mV, corresponding to a single quantum excitation of the dipole-allowed v_1 symmetric stretch mode normal to the surface. Figure 8b shows the loss spectrum for a collection angle = +17° off the specular direction, still in the scattering plane. Not only are the dipole-forbidden v_2 and v_3 modes (see inset) excited, but also a v_1 overtone is excited, in accord with expectations from impact scattering theory. The angular dependence of the intensity in the various loss features is shown in Figure 9. The dipole-excited single v_1 loss fairly well tracks the "specular" peak, again in accord with dipole theory, whereas the 80-mV v_2 loss and the v_1 overtone peak at angles substantially different from specular are also in agreement with expectations from impact scattering theory. With confirmation of the principles of impact scattering in hand, it is possible to use the special angular properties of the inelastically scattered electrons to further determine the geometry of chemisorption complexes. This will be further discussed by Avery in the chapter on EELS in this volume.

2.5. Resonance Excitation

So far we have been concerned with excitation mechanisms in which the incident probe couples directly to the vibrational degree of freedom. Another important class of vibrational excitation mechanisms involves the temporary formation of virtually excited electronic states of the molecule–surface complex. We have in mind processes such as photon excitation of neutral molecules basic to resonance Raman scattering,[78,79] or temporary negative ion creation due to electron capture in a shape resonance during an EELS event.[80–83] Within the context of vibrational spectroscopy, resonance excitation is important for the following reason. The Born–Oppenheimer separation of electronic and vibrational degrees of freedom [equation (38)] implies that the electronic state of the molecular system determines the potential energy surface over which the intramolecular motion occurs.[84] From the point of view of the vibrational degrees of freedom, an electronic transition looks like a sudden change from the initial ground state PES to that one determined by the excited electronic state. Chances are that at the time of the PES switch, the nuclei find themselves in a nonequilibrium position with respect to the new PES and thus they respond to the nonvanishing forces by moving. However, after an excited state lifetime $\equiv \tau_R$, the electronic system is returned to its ground state, and thus the original PES is restored, but now with the vibrational coordinates displaced from the ground state equilibrium point. This leaves the molecule in a distribution of vibrationally excited states that is strongly dependent upon the magnitude of τ_R relative to the characteristic vibrational time associated with the excited electronic state. The substantial enhancement of overtone loss features is very useful for a number of reasons. First, the intensity distribution serves as a probe of the intramolecular dynamics over the excited state PES.[4,78,79] Secondly, departures from equal level spacing observed in the overtone spectrum provide input needed to determine anharmonic corrections to the potential, via a Birge–Spooner presentation.[85]

A semiquantitative theory of the excited vibrational state distribution can be given a parallel development in terms of Franck–Condon (Figure 10a)[86] or wavepacket dynamics (Figure 10b)[4,57] arguments. For specificity, Figure 10 illustrates the intramolecular dynamics that occur during the life of a temporary negative diatomic molecular ion. At time $t = 0$, the molecule in $|0\rangle$, the vibrational ground state of some Hamiltonian H_{A_2}, suddenly finds itself time-evolving according to $H_{A_2^-}$ until $t = \tau_R$ when H_{A_2} is returned. Solving the time-dependent Schrödinger equation for this piecewise continuous time-dependent Hamiltonian,[86] the vibrational wave function for $t > \tau_R$ is

$$\chi_v(t > \tau_R) = \exp[-iH_{A_2}(t - \tau_R)/\hbar]e - iH_{A_2R}/\hbar\,|0\rangle$$

The probability that the molecule ends up in the nth excited vibrational state is

$$P_n = |\langle n \mid \chi_v(t > \tau_R)\rangle|^2 = |\langle n| e - iH_{A_2^-R}/\hbar\,|0\rangle|^2 \qquad (43)$$

or in terms of $\varepsilon_{\tilde{m}}$ and $|\tilde{m}\rangle$, the molecular ion vibrational eigenvalues and eigenstates,

$$P_n = \left| \sum_{\tilde{m}} \langle n \mid \tilde{m}\rangle e^{-i\varepsilon_{\tilde{m}}\tau_R/\hbar} \langle \tilde{m} \mid 0\rangle \right|^2 \qquad (44)$$

where $\langle n \mid \tilde{m}\rangle$ and $\langle \tilde{m} \mid 0\rangle$ are overlap integrals of the neutral and molecular ion vibrational states. From the Franck–Condon point of view illustrated in Figure 10a and quantified by equation (44), the lifetime-dependent excitation distribution arises from the interfering paths connecting $|0\rangle$ with $|n\rangle$, introduced here through the coherent $\tilde{m}$ sum in equation (44). Although equation (44) is quite general, it is especially convenient when both potentials are harmonic oscillators (perhaps with different frequencies), since analytic recursive relations exist for the overlap integrals.[87]

An illuminating wavepacket description is rigorously possible when both potentials are harmonic and either $\omega_0 = \bar{\omega}$ (the molecular ion frequency) or $\bar{\omega}_0\tau_R \ll 1$ (short time dynamics).[88] In general the

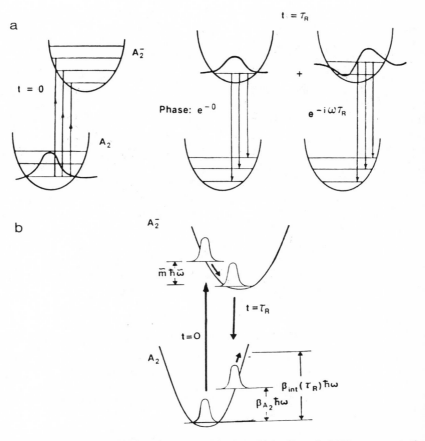

Figure 10. (a) Displaced oscillator potentials in which G and RES represent the intramolecular potentials of A_2 and A_2^-, respectively. Representative vibrational wavefunctions and phase factors acquired at time of switch off are shown. (b) Propagating wavepacket realization of the sequence shown in 10a.[57]

excitation can be thought of as propagating an initially prepared wavepacket (determined by $|0\rangle$) over the A_2^- potential and then returning it to A_2 at $t = \tau_R$, as shown in Figure 10b, where it oscillates. In the case of the displaced oscillator potentials for which $V_{A_2} = \frac{1}{2}kq^2$ and $V_{A_2^-} = \frac{1}{2}k(q - q_0)^2$ with q_0 the separation between well minima, a time-dependent interaction potential is identified as

$$V_f(q, t) = V_{A_2^-} - V_{A_2} = -qkq_0 + \frac{1}{2}kq_0^2 \qquad (0 < t < \tau_R) \qquad (45)$$

which is of the form of equation (4), aside from the irrelevant constant term. Thus the results of Section 2.1.2a are immediately applicable. From equations (8)–(10) and (45),

$$\lambda(\omega_0) = -kq_0 \int_0^{\tau_R} e^{-i\omega_0 t'} \, dt' = -\frac{2kq_0 e^{-i\omega_0 \tau_R/2}}{\omega_0} \sin(\omega_0 \tau_R/2),$$

$$\Delta\varepsilon_{\text{classical}} = 2\left(\frac{kq_0^2}{2}\right)(1 - \cos \omega\tau_R)$$

and

$$\beta(\tau_R) = 2(1 - \cos \omega\tau_R)\beta_{A_2} \qquad (46)$$

where $\beta_{A_2} \equiv kq_0^2/2\hbar\omega_0$ is the usual Poisson parameter associated with a permanent switch of potential surfaces. Equation (46) in conjunction with equation (12) yields the semiclassical wavepacket dynamics vibrational distribution, which is equivalent to the Franck–Condon result given by equation (44). Although the arguments given here relied on certain technical assumptions ($\bar{\omega} = \omega_0$, no wavepacket spreading, etc.), a more general procedure has been established by Heller[88] and Stechel and Schwartz.[89] As is readily apparent from equation (46), if the resonance lifetime is such that $\omega\tau_R = N\pi$ (N an odd integer), then $\beta(\tau_R) = 4\beta_{A_2}$ and maximal resonance excitation follows. However if $N =$ even integer, the molecule is returned to its ground state. Real systems will show distributions intermediate between these extremes. Furthermore, environmental changes could alter τ_R and hence the intensity distributions of resonance excited vibrational overtones. Finally we note that although it may be begging the issue to argue Franck–Condon versus wavepacket dynamics for the simple one-dimensional diatomic molecules, in polyatomic molecules in which the orientation of multidimensional normal coordinates is different for different electronic states,[90] the wavepacket approach quickly becomes computationally far more efficient than the Franck–Condon theory.

A very illuminating example of significant overtone excitation due to the formation of temporary negative molecular ions is the EELS study of N_2 physisorbed and condensed on Ag surfaces by Demuth, Schmeisser, and Avouris.[16,91] The N_2 loss spectrum for a primary electron energy = 1.5 eV is shown in Figure 11a, where substantial overtone excitation of the 0.29 eV, $0 \rightarrow 1$ fundamental is to be noted. The existence of an N_2^- shape resonance in gas phase electron scattering, for incident electron kinetic energy $\sim$2–4 eV, has been long known and understood.[81–83] The gas phase $0 \rightarrow 1$ cross section, as a function of primary energy, is shown in Figure 11b. It is argued that the loss spectrum shown in Figure 11a is due to the formation of a temporary negative ion, as in the gas phase, but

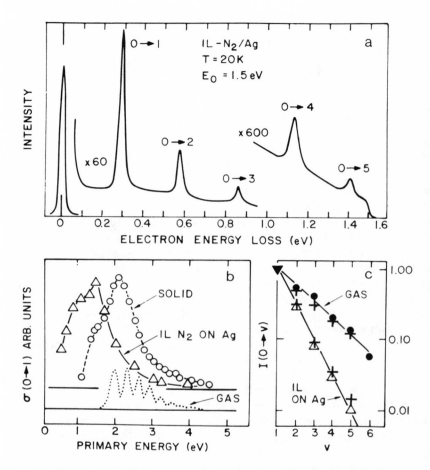

Figure 11. (a) The vibrational spectrum (fundamental and several overtones) of N_2 on a silver surface at 20 K obtained via resonance electron scattering. (b) Resonance profiles for the fundamental vibrational excitation of N_2 in the gaseous, solid, and adsorbed (on Ag) phases. (c) Normalized intensity of the vibrational overtones of a monolayer of N_2 on Ag ($\triangle$) compared to the corresponding intensities for gaseous N_2 ($\bigcirc$), and theoretical values ($+$) from equation (44). Adapted from Refs. 16 and 86.

with the threshold moved lower in energy due to image potential reductions in the electronic bound state energies. This is supported by the observed cross sections for both the physisorbed and condensed N_2, also shown in Figure 11b. The physisorbed cross section appears as a smoothed gas phase version shifted down in primary energy ~1.5 eV, which would be a reasonable image shift. In the case of condensed multilayers that are not in intimate contact with the Ag, the resulting cross sections are just those of the gas phase, appropriately broadened.

Finally, the observed intensity distribution versus overtone excitation number are shown in Figure 11c for both gas phase and physisorbed N_2. Also shown are theoretical values[86] obtained from equation (44) evaluated with appropriate N_2 parameters taken from Hartree–Fock calculations.[92] It was originally suggested that the presence of the surface should decrease negative ion resonance lifetime,[91] and this could be responsible for the reduced overtone excitation of the physisorbed N_2 compared to that of the free molecule. By using the calculated free N_2/N_2^- parameters and a physisorbed resonance lifetime $= 0.4\tau_R$(free), equation (44) provides an excellent representation of both the free and physisorbed intensity distributions. Hopefully the success in this study of resonance excitation will stimulate further work along these lines as there is a large wealth of important information in overtone spectra which cannot be accessed through straightforward studies of fundamental excitations alone.

2.6. Neutron Excitation

As already mentioned, the fact that neutrons scatter very weakly and are quite selective in what they scatter from provides both the pluses and minuses of the technique. For present purposes, weak scattering means that Born-approximation scattering theory is a totally adequate conceptual framework.[93,94]

This nonmagnetic interaction between a neutron and the atoms of the target is due to nuclear forces whose origins are beyond the scope of discussion here. We merely note that for an array of atoms, the interaction is modeled by the Fermi pseudopotential

$$V(r) = \frac{2\pi\hbar^2}{m_n} \sum_i b_i(T)\delta(\mathbf{r} - \mathbf{R}_i(t)) \tag{47}$$

where $b_i(T)$ is the scattering length of the ith nucleus located at the fluctuating position $\mathbf{R}_i(t) = \mathbf{R}_{i0} + \mathbf{q}_i(t)$, defined here so as to include Debye–Waller factors.

The amplitude for inelastic $\mathbf{k} \to \mathbf{k}'$ neutrons scattering, accompanied by $v \to v'$ vibrational excitation (deexcitation?) is given by equations (39) and (47) as

$$f(\mathbf{k}v \to \mathbf{k}'v') = \sum_i b_i \int d^3r\, d^3q_i\, e^{-i\mathbf{k}\cdot\mathbf{r}} \chi_{v'}^*(\mathbf{q}_i)\delta(\mathbf{r} - \mathbf{R}_{i0} - \mathbf{q}_i(t))e^{i\mathbf{k}\cdot\mathbf{r}}\chi_v(\mathbf{q}_i)$$

$$= \sum_i b_i e^{i\Delta\mathbf{k}\cdot\mathbf{R}_{i0}} F_{v,\,v'}^i(\Delta\mathbf{k}, t) \tag{48}$$

where $\Delta\mathbf{k} \equiv \mathbf{k} - \mathbf{k}'$ is the momentum transfer and

$$F^i_{v,\,v'}(\Delta\mathbf{k}; t) \equiv \int d^3q_i\, \chi^*_{v'}(\mathbf{q}_i) e^{i\Delta\mathbf{k}\,\cdot\,\mathbf{q}_i(t)} \chi_v(\mathbf{q}_i) \tag{49}$$

is the "form factor" associated with the $v \to v'$ vibrational transition. The inelastic differential cross section per unit energy range is

$$\frac{d^2\sigma}{d\Omega\,d\varepsilon(\mathbf{k}')} = \frac{|\mathbf{k}'|}{|\mathbf{k}|}\, |f(\mathbf{k}v \to \mathbf{k}'v')|^2 \delta(\varepsilon(\mathbf{k}') + \varepsilon_{v'} - \varepsilon(\mathbf{k}) - \varepsilon_v)$$

$$= \frac{|\mathbf{k}'|}{|\mathbf{k}|} \sum_{i,j} b_i b_j^* e^{i\Delta\mathbf{k}\,\cdot\,(\mathbf{R}_{i0}-\mathbf{R}_{j0})} F^i_{vv'}(\Delta\mathbf{k}, t) F^{*j}_{vv'}(\Delta\mathbf{k}, t)$$

$$\times\, \delta(\varepsilon(\mathbf{k}') + \varepsilon_{v'} - \varepsilon(\mathbf{k}) - \varepsilon_v) \tag{40}$$

Equation (50) can be written as a sum of coherent plus incoherent contributions. The coherent cross section provides information on relative positions of atoms and on collective excitations such as phonons whereas the incoherent contribution contains information about single-particle dynamics. To make this explicit, first let $F^i = F^j$ if the elemental species (but not necessarily the isotope) are the same at sites i and j. Then if the isotope identity at the various sites is uncorrelated, the ensemble average over sites is

$$\langle b_i b_j^* \rangle = |\langle b \rangle|^2 + \delta_{i,j}(\langle |b|^2 \rangle - |\langle b \rangle|^2) \tag{51}$$

where $\langle b \rangle$ is the ensemble averaged scattering length and the delta function term in equation (51) is a measure of the fluctuations about the mean. The differential cross section can thus be written as

$$\frac{d^2\sigma}{d\Omega\,d\varepsilon(\mathbf{k}')} = \frac{|\mathbf{k}'|}{|\mathbf{k}|}\left\{ |\langle b \rangle|^2\, |F_{v,\,v'}(\Delta\mathbf{k}, t)|^2 \left| \sum_i e^{i\Delta\mathbf{k}\,\cdot\,\mathbf{R}_{i0}} \right|^2 \right.$$

$$+ N(\langle |b|^2 \rangle - |\langle b \rangle|^2)|F_{v,\,v'}(\Delta\mathbf{k}, t)|^2 \bigg\}\delta(\varepsilon(\mathbf{k}')$$

$$+\, \varepsilon_{v'} - \varepsilon(\mathbf{k}) - \varepsilon_v) \tag{52}$$

the sum of a coherent plus incoherent term. As seen in the table of neutron cross sections in the chapter by Cavanagh et al., the whims of nature dictate that the total incoherent cross section per H atom ($\sigma_{\text{inc}} = 4\pi b^2_{\text{inc}} = 79.7 \times 10^{-8}\,\text{Å}^2$ with $b^2_{\text{inc}} \equiv (\langle |b|^2 \rangle - |\langle b \rangle|^2)$ is far greater than for any other likely element, hence the utility of NIS for the study of hydrogen-containing systems. Furthermore, for highly dispersed, nonperiodic systems, coherent scattering is expected to be small, independent of scattering length arguments.

The usual expression for the incoherent inelastic differential cross section follows from a few additional algebraic steps, namely, letting $\exp(i\Delta\mathbf{k} \cdot \mathbf{q}) \simeq 1 + i(\Delta\mathbf{k} \cdot \hat{\varepsilon}_q)q$ in equation (49) ($\hat{\varepsilon}_q$ is a unit vector along the vibrational coordinate) and expressing the energy conserving delta function in its Fourier-time representation. It is then a standard exercise[95] to manipulate equations (49) and (52) into the form

$$\frac{d^2\sigma}{d\Omega\,d\omega} = N\frac{|\mathbf{k}'|}{|\mathbf{k}|}\frac{\sigma_{inc}}{4\pi}e^{-2W(\Delta\mathbf{k})}(\Delta\mathbf{k} \cdot \hat{\varepsilon}_q)^2$$

$$\times \int \frac{dt}{2\pi}e^{-i\omega t}\langle v| q(t) \cdot q(0) |v\rangle \qquad (53)$$

$[\omega = \varepsilon(\mathbf{k}) - \varepsilon(\mathbf{k}')]$ which depends upon the temporal correlations of the "dipole matrix element." Since $q = (\hbar/2m\omega_0)^{1/2}(b^+ + b)$, where b^+, b are Boson operators, further standard reductions of the correlation function yield an expression $(\hbar/2m\omega_0)G(T)$, where $G(T)$ are the temperature-dependent terms given by Cavanagh et al. in their equation (1) in this volume.

An interesting possibility arises when doing vibrational spectroscopy on adsorbed H, the mainstay of NIS. As has been known for a long time in bulk hydride studies,[94] owing to the quantum nature of H (large zero point energy, tunneling possibilities, etc.) even if the ground state is localized at one site, the first vibrationally excited state could be delocalized, in which case the discrete eigenvalue $\varepsilon_{v'}$ would be replaced by a band of energies. Recent calculations have suggested that this might be the case for H adsorbed on Ni(100).[96] As demonstrated by Casella,[94] there is no problem in treating discrete to continuum vibrational transitions within the context of the model outlined here. All that is necessary is to express the excited state continuum wave function as a coherent sum of localized functions

$$\chi_{v'}(\mathbf{q}; \boldsymbol{\kappa}) = \frac{1}{\sqrt{N}}\sum_i e^{i\boldsymbol{\kappa} \cdot \mathbf{R}_{i0}}\chi_{v'}(\mathbf{q} - \mathbf{R}_{i0})$$

calculate multicenter form factors, and add a summation on excited Hydrogen wave vectors $\boldsymbol{\kappa}$. If adsorption measurements are done in which individual $\boldsymbol{\kappa}$'s are not resolved, the effects of this delocalization are to broaden the particular absorption line due to the dispersion. Puska et al. in their stimulating paper have suggested that there may be some experimental EELS evidence for this effect,[96] although on the basis of NIS data, this suggestion has not received universal acceptance.[97] The ensuing discussion on the relationship between perfect single-crystal EELS experiments and those of a highly dispersed substrate via NIS illustrate the current state of controversy concerning the relevance of NIS in surface science.

3. Line-shape Information Content

In the introductory section of this chapter, it was emphasized that the principal motivations for carrying out vibrational spectroscopy of adsorbed molecules were twofold. First, within the limitations set by selection rules, as discussed throughout Section 2, the presence of a loss feature with approximately the same energy as in a free molecule spectrum permitted identification of the molecules existence on the surface. Second, the small shifts in this energy carry information about the local chemistry of the adsorption bond such as coordination, degree of back bonding from the substrate, mechanical hybridization with substrate modes, and "image potential shifts." These possibilities have been nicely enumerated by Willis, Lucas, and Mahan.[15] Furthermore if the observed absorption linewidth can be decomposed into an in-homogeneous part due to different nonequivalent bonding sites plus a homogeneous portion, then the latter contains valuable information pertinent to the dynamics of elastic and inelastic interaction between the localized oscillator and the continua of substrate excitations. By reason of causality, any inelastic interaction giving rise to a level width also produces a related level shift. Thus in order to *properly interpret* even the level shifts, one must be able to deal with the dynamics.

Historically it can be seen that surface vibrational spectroscopy analysis has proceeded, based on the (oversimplistic) presumption that the observed linewidth of an absorption band is a direct measure of some energy decay mechanism between an excited oscillator and a continuum heat bath [the $2\eta\dot{q}$ term in equation (1)].[98-101] Rather inexplicably, it has only recently been noticed in the surface world[6] that elastic collisions between the localized oscillator and the thermally fluctuating substrate/heat bath give rise to dephasing linewidths, which are well known in condensed phase vibrational spectroscopy to frequently dominate the linewidth.[102-106] As will be discussed shortly, when both elastic and inelastic collisions are possible, it is customary to partition the dynamic interactions into two types: those that cause level decay due to energy dissipation and those that cause decay of the temporal correlations of the oscillations (dephasing). In the language of condensed phase spectros-copy, the linewidth associated with a transition between vibrational levels a and b is given as

$$\Delta\omega_{ab} = \frac{2}{T_2} = \left(\frac{1}{T_{1a}} + \frac{1}{T_{1b}}\right) + \frac{2}{T_2'} \tag{54}$$

where T_{1a} and T_{1b} are the lifetimes of levels a and b, while T_2' accounts for the additional broadening due to pure dephasing processes.

To build upon these opening remarks, the remainder of this section will be structured as follows. In Section 3.1, line-shape generalities are discussed in terms of temporal correlation functions and some simple models are presented, in conjunction with 2.1, which illustrate the respective roles of level decay and dephasing in producing nontrivial lineshapes. Section 3.2 is devoted to the general problem of a discrete vibrational state in interaction with a continuum of boson excitations (electron–hole pairs, phonons, or photons) characteristic of T_1 decay processes. As will be seen, qualitatively different shifts, for given widths, due to coupling with the three mentioned continua are expected, and this fact could be useful in data interpretation. Finally, in Section 3.3, two recent experimental studies that have addressed these issues will be discussed.

3.1. Line-Shape Generalities/Dephasing

There is a vast literature on spectroscopic line shapes.[102–107] Invariably, the starting point of any physical discussion is recognition of the fact that the line shape is the Fourier transform of the correlation function of a physical variable that is active in the particle spectroscopic transition. From the previous section it is apparent that the oscillator coordinate $q(t)$ is the relevant variable.

Thus the line-shape expression is written as

$$I(\omega) = \int_{-\infty}^{\infty} dt\, e^{i\omega t} \langle q(0) q^*(t) \rangle / \langle q^2 \rangle \tag{55}$$

For an ergodic system, the ensemble average correlation function can be replaced by a single-particle time average:

$$f(t) \equiv \langle q(0) q^*(t) \rangle = \lim_{T \to \infty} \frac{1}{T} \int_{-T/2}^{T/2} dt'\, q(t') q^*(t' + t) \tag{56}$$

where $q(t)$ could be one of the expressions obtained in Section 2 augmented by the influence of the vibrational coupling to the fluctuating background provided by the substrate. Although it is not obvious that the assumption of ergodicity is appropriate for the low-lying excited states of nearly harmonic systems, the desired inquiries into these subtleties are outside the scope of this particular chapter.[2–4] To proceed from here, one must consider specific ways to describe the interaction of the molecule with the thermally fluctuating substrate background. A standard procedure in line-shape theory[107] is to make the substitution

$$\omega_0 \rightarrow \omega_0' + \Delta\omega(t) \tag{57}$$

in which the bare oscillator frequency is shifted to ω_0', and the effects of the background are accounted for by a random modulation of the frequency. With $\dot{q} = i\omega_0 q$ for a bare oscillator, the substitution given by equation (57) and integration yields

$$q(t) = q_0 e^{i\omega_0' t} \exp\left[i\int_0^t \Delta\omega(t')\, dt'\right]$$

which with equations (55) and (56) leads to the stochastic line-shape expression

$$I(\omega) = \int_{-\infty}^{\infty} dt \exp[i(\omega - \omega_0')t]\left\langle \exp\left[i\int_0^t \Delta\omega(t')\, dt'\right]\right\rangle \tag{58}$$

in terms of the correlation function of the frequency modulation caused by fluctuations in the background.

In order to relate physical "reality" with the consequent line shape, consider the three qualitatively different oscillator trajectories shown in Figure 12. An undamped monochromatic oscillator is pictured in Figure 12a. In this case, $\Delta\omega(t') = 0$, ω_0' is pure real, and equation (58) yields a delta function. The oscillator experiencing amplitude decay is shown in Figure 12b. The simplest description is provided by adding an imaginary part to ω_0' in equation (58) [still with $\Delta\omega(t') = 0$], in which case the line shape broadens out into a Lorentzian. Some microscopic details of this effect will be discussed in Section 3.2. The subtlest effect is pure dephasing shown in Figure 12c. Here the oscillator propagates in an undamped manner for a time duration $= 2\tau_c$, at which point it undergoes an impulsive elastic collision which induces a phase shift $= \delta$. The position coordinate outside the $2\tau_c$ time window is uncorrelated with the coordinate within owing to the acquisition of the "random" phase factor. If the correlation function in equation (58) decays exponentially as $\exp(-t/T_2')$ with $T_2' \equiv (\langle\Delta\omega^2\rangle)^{-1/2}$ and $\langle\Delta\omega^2\rangle \equiv 1/(2\tau_c)^2$, then the line shape is also Lorentzian. Although the time interval $2\tau_c$ was defined as that time duration over which positional correlation is displayed before complete randomization by a single impulsive collision, the same qualitative end result could occur from a sequence of weaker inelastic events.[108] Thus another working specification of τ_c could follow from $\dot{n}\delta_0 2\tau_c = \phi_c$, where $\dot{n}$ is the actual collision frequency, δ_0 the phase shift per collision, and ϕ_c some critical phase accumulation needed for the demise of correlations, say $\phi_c \simeq \pi/2$. With this choice, the FWHM takes

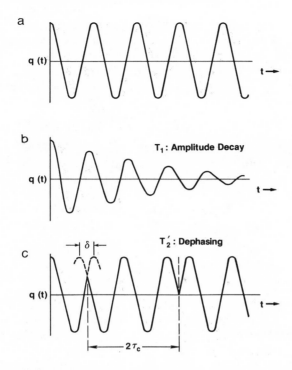

Figure 12. Oscillator trajectories illustrating various time dependences. (a) Undamped, collision-free motion resulting in delta-function line shapes. (b) Damped motion showing amplitude decay resulting in T_1 broadening. (c) Motion that includes impulsive elastic collisions leading to T_2', pure dephasing broadening.[6]

the appealing form

$$1/2\tau_c = 2\dot{n}\delta_0/\pi = 1/T_2' \tag{59}$$

which goes to zero when either (or both) the collision strength (δ_0) or frequency ($\dot{n}$) does, in agreement with intuition, which suggests that a zero strength, infrequent dephasing collision should have no observable consequences.

A particular microscopic model for dephasing that has proven useful in describing the line shapes of high-frequency modes in molecular crystals and solutions is the so-called exchange coupling model due to Harris, Shelby, and Cornelius.[105] In this model, anharmonic coupling between the ($\sim 2000 \text{ cm}^{-1}$) high-frequency intramolecular stretch modes and the large-amplitude, low-frequency ($\equiv \omega_l$) librational modes associated with the molecule–surface bond could give rise to apparent

frequency modulations in the upper mode due to thermal fluctuations in
the population/amplitude of the ω_l mode which is near-degenerate with
substrate phonons. This arises in the following manner. Suppose that a
set of coordinates $\{\delta R\}$ are identified with the librational mode
displacements. The true anharmonic potential for the coupled system can
be written as

$$V(q, \delta R) = V_{quad}(q, \delta R) + q^2 F(\delta R) + q^3 G(\delta R) + \cdots$$

where V_{quad} includes all combinations of q and δR up to quadratic.
These can be dealt with by arguments presented in Section 2. If
$\delta R = \delta R(t)$ is now regarded as a stochastic variable with $\langle \delta R \rangle = 0$, and
$\langle \delta R^2 \rangle = (\hbar/2M\omega_l)\ [1 + 2n(T)] = (\hbar/2M\omega_l)\coth(\hbar\omega_l/2kT)$, upon power series expansion of $F(\delta R)$ on δR, the lowest-order term that can be
responsible for a pure dephasing effect is the fourth-order one

$$V_4(q, \delta R) = q^2\delta R^2\, d^4V/d^2q\, d^2\delta R\big|_{\substack{q=q_0 \\ \delta R=\delta R_0}}$$

$$\equiv \frac{q^2 n_l(t, T)C_{anh}}{2} \tag{60}$$

where n_l is the temperature-dependent, fluctuating population of ω_l and
C_{anh} is the anharmonic coupling constant. The resulting (undamped)
oscillator equation of motion is

$$\ddot{q} + \omega_0^2 q = -\frac{C_{anh}n_l(t, T)q}{m}$$

which can be rearranged as

$$\ddot{q} + [\omega_0 + \Delta\omega(t)]^2 q \approx 0 \tag{61}$$

with

$$\Delta\omega(t) \simeq \frac{C_{anh}n_l(t, T)}{m\omega_0}$$

when $\langle \Delta\omega^2(t) \rangle \ll \omega^2$.

The resulting picture is one in which the stochastically fluctuating
population of the librational or hindered translation mode which is
anharmonically coupled to the intramolecular stretch causes a time-
dependent modulation of the oscillator frequency, without any energy
transfer, and hence a linewidth due to pure dephasing. Comparing
equations (59) and (61), it is apparent that $n_l(t, T) \sim \exp(-\varepsilon_l/kT)$ plays
the role of collision frequency and $C_{anh}/m\omega_0$ that of collision strength
and thus phase shift. An example of this important effect will be given in
Section 3.3.

3.2. Localized Oscillator–Continuum Systems

The abstract problem of a discrete state interacting with a continuum has been addressed in many different physical contexts, as enumerated elsewhere.[6,100] For present purposes, the discrete state is considered to be the localized oscillator discussed throughout this chapter, and the continuum either the electron–hole pair excitations of the sub-strate,[98–101] the phonon bands,[110–111] or the radiation field.[112] Quite generally, the spectral composition of a localized harmonic oscillator *linearly coupled* to any of these continua can be obtained in an exact, closed form by performing a normal mode analysis of the coupled system[100] in the manner due to Fano.[113] The resulting spectral function for the localized oscillator is

$$\rho_{\text{osc}}(\omega_l) = \frac{1}{\pi} \frac{2\omega_0^2 \Delta(\omega_l)}{[\omega_l^2 - \omega_0^2 - \omega_0 \Lambda(\omega_l)]^2 + [\omega_0 \Delta(\omega_l)]^2} \tag{62}$$

with the frequency shift and width functions given by

$$\Lambda(\omega_l) \equiv P\left(\sum_{\mathbf{k}} \frac{2\omega_{\mathbf{k}} |\lambda_{\mathbf{k}}|^2}{\omega_l^2 - \omega_{\mathbf{k}}^2} \right) \tag{63}$$

and

$$\Delta(\omega_l) \equiv 2\pi \sum_{\mathbf{k}} |\lambda_{\mathbf{k}}|^2 2\omega_{\mathbf{k}} \delta(\omega_l^2 - \omega_{\mathbf{k}}^2) \tag{64}$$

where P denotes a principal part sum, $\lambda_{\mathbf{k}}$ the oscillator-field coupling constant, and $\mathbf{k}$ the field quantum number or momentum state. In effect, the coupling has renormalized ω_0, the real part of the oscillator frequency and added a nonvanishing imaginary part to account for the fact that the energy associated with an initial externally imposed displacement of the localized oscillator (the spectroscopic excitation) will disperse throughout the entire system on a time scale set by $\sim \hbar / \Delta(\omega_l)$. In short, this is the microscopic basis of the phenomenological replacements suggested by equation (57).

For present illustrative purposes, we can take advantage of the considerable simplifications that occur in the "density-of-states limit"; that is, the $\mathbf{k}$ sums in equations (63) and (64) are replaced by $\int \rho(\omega)\, d\omega$ and $\lambda_{\mathbf{k}}$ by $\lambda(\omega)$, where $\rho(\omega)$ is the field density of states. Furthermore, a frequently invoked assumption is to take $\lambda(\omega) = \bar{\lambda} = \text{const}$ for $\omega < \omega_c$, $= 0$ for $\omega > \omega_c$ where ω_c is a characteristic cutoff frequency whose magnitude depends upon the specific continuum. Within these limits, equations (63) and (64) reduce to

$$\Lambda(\omega_l) \simeq |\bar{\lambda}|^2 \int_0^{\omega_c} \frac{2\omega\rho(\omega)\,d\omega}{\omega_l^2 - \omega^2} \tag{65}$$

and

$$\Delta(\omega_l) \simeq \pi\rho(\omega_l)|\bar{\lambda}|^2, \qquad \omega_l \le \omega_c$$
$$= 0, \qquad \omega_l > \omega_c \tag{66}$$

To proceed, the details of the continua must be supplied.

3.2.1. Electron–Hole Pairs

First consider the localized oscillator coupled to the electron–hole pair excitations (density fluctuations) associated with the filled Fermi sphere of conduction band electrons of a metallic substrate. It is well established[114] that for excitation frequencies less than $\approx \varepsilon_F/\hbar$, the Fermi frequency, the pair density of states can be approximated by

$$\rho(\omega) \simeq \hbar^2 \rho_{\varepsilon_F}^2 \omega \tag{67}$$

where ρ_{ε_F} is the Fermi level electron density of states. A common approximation, based on detailed studies, is to take $\hbar\omega_c \approx$ the Fermi energy, thus permitting equal accessibility to all states within the bounded continuum. This being done, equations (65) and (66) immediately reduce to

$$\Lambda(\omega_l) = \frac{2}{\pi} C\omega_c \left(-1 + \tfrac{1}{2}\frac{\omega_l}{\omega_c} \ln \left| \frac{\omega_c + \omega_l}{\omega_c - \omega_l} \right| \right)$$

and

$$\Delta(\omega_l) = \begin{cases} C\omega_l & \text{for} \quad \omega_l \le \omega_c \\ 0 & \text{for} \quad \omega_l < \omega_c \end{cases}$$

with

$$C \equiv \pi\hbar^2 \rho_{\varepsilon_F}^2 |\bar{\lambda}_{e\text{-}h}|^2$$

Typically, $\hbar\omega_c \approx 1\text{--}10\,\text{eV}$, whereas $\hbar\omega_l = \hbar\omega_0[1 + \Lambda(\omega_l)/\omega_0]^{1/2} \le 0.25\,\text{eV}$ so the level shift $\Lambda(\omega_l) \simeq -2C\omega_c/\pi$ is always negative (red shifted) in the frequency range appropriate to vibrational spectroscopy. Interestingly, the dominant term in the ratio

$$\Lambda(\omega_l)/\Delta(\omega_l) \simeq -\tfrac{1}{2}\pi\omega_c/\omega_l$$

is not only independent of $\bar{\lambda}$, but also shows that the pair-induced frequency red shift is a factor of ~10–100 times larger than the causality-related widths. In fact, what has just been presented is the microscopic basis for what is commonly called an image dipole shift. Red shifts by a factor of several tens of linewidths are routinely observed.

Finally, the essential features of the model are shown in the top panel of Figure 13. On the left is shown the discrete state at ω_0 plus the linear density of states pair continuum. Since for parameters appropriate to vibrational spectroscopy, $\omega_0/\omega_c \ll 1$, the discrete state is correctly placed near the bottom of the continuum, where it experiences a significant redshift due to the overwhelmingly predominant repulsion from the continuum states with $\omega > \omega_0$. In spite of a substantial shift, the level width could be very small as the width function is proportional to ω_l. The resulting oscillator spectral function is shown on the top right part of Figure 13.

3.2.2. Phonons

The simplest example of oscillator–phonon coupling involves a Debye solid in which the one-phonon-continuum density of states is given by

$$\rho_{1p}(\omega) = \begin{cases} 3\omega^2/\omega_d^3 & \text{for} \quad \omega < \omega_d \\ 0 & \text{for} \quad \omega > \omega_d \end{cases} \tag{68}$$

where ω_d, the Debye frequency, is usually ≤ 0.05 eV, considerably smaller than the intramolecular vibrational energies of interest. Again taking a uniform coupling constant, equations (65), (66), and (68) yield

$$\Lambda_{1p}(\omega_l) = \frac{3 |\bar{\lambda}_{1p}|^2}{\omega_d} \left[-1 - \frac{\omega_l^2}{\omega_d^2} \ln(1 - \omega_d^2/\omega_l^2) \right] \simeq \frac{3 |\bar{\lambda}_{1p}|^2 \omega_d}{2\omega_l^2} \tag{69}$$

and

$$\Delta_{1p}(\omega_l > \omega_d) = 0$$

Coupling to the single-phonon continuum bounded by $\omega_d < \omega_l$ could produce a significant oscillator shift to the blue, but with no broadening since the discrete state and continuum are nondegenerate. Coupling with a multiphonon continuum is necessary for a phonon-produced width. For instance, a two-phonon continuum density of states could be taken as the convolution

$$\rho_{2p}(\omega) = \int_0^{\omega_d} \rho_{1p}(\omega_1) \, d\omega_1 \int_0^{\omega_d} \rho_{1p}(\omega_2) \delta(\omega_1 + \omega_2 - \omega) \, d\omega_2$$

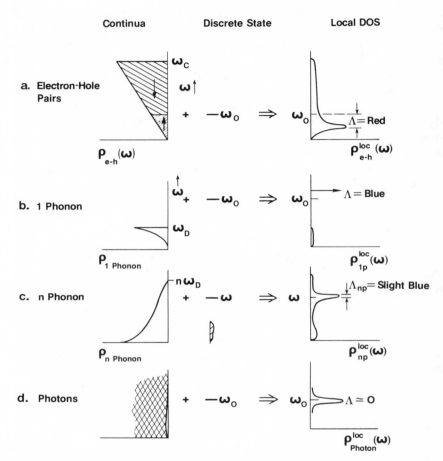

Figure 13. Illustration of local spectral function (right) resulting from various continua (left) couplings to discrete oscillator states (middle). (a) Electron–hole pair continuum. Characterized by the linearly increasing density of states produces a substantial oscillator red shift together with broadening. (b) One-phonon continuum, within quadratic Debye model, produces blue shift when $\omega_l > \omega_d$ with no broadening, together with some spectral weight in the region $0 < \omega_l < \omega_d$. (c) n-Phonon continuum coupled to one-phonon-renormalized localized oscillator, produces slight broadening but with blue shift dominated by the one-phonon renormalization. (d) Photon continuum produces little change from free oscillator spectral function since "bare oscillator" already contains electromagnetic field renormalization.[6]

together with a two-phonon coupling constant $\bar{\lambda}_{2p} \simeq \bar{\lambda}_{1p}^2$. Generalization to n-phonon continua is straightforward. Since the coupling constant is presumed small (otherwise the zero-order discrete oscillator would be a bad starting point), the dominant blue shift arises from the single-phonon continuum, given by equation (69), whereas the dominant phonon width is due to the n-phonon continuum, where n satisfies $(n - 1)\omega_d < \omega_l < n\omega_d$ and $\bar{\lambda}_{np} \simeq \lambda_{1p}^n$. The essential features of the phonon coupling are displayed in the second panel of Figure 13 where on the left are shown the discrete state at ω_0 plus the blue-shifting one-phonon continuum. As with the pair continuum, in spite of a substantial shift (blue rather than red), the level width could be very small since the width function is proportional to $|\bar{\lambda}_{1p}|^{2n}$. The effect of the one-phonon localized spectral function coupled to an n-phonon continuum are shown on the third panel of Figure 13. A slight additional blue shift plus a broadening of the sharp feature result.

3.2.3. Photons

The final oscillator-continuum coupling we consider is to the electromagnetic field. As with the phonon field, the photon density of states is quadratic with frequency. The shifts and widths can still be thought of in terms of equations (65) and (66), but with one important difference. From quantum electrodynamics we know that physical properties are already renormalized with respect to the electromagentic field. In the present context, this means that the observable free oscillator frequency ω_0 already contains the frequency shift $\Lambda(\omega_l)$. Consequently, in contrast with the pair and phonon continua coupling, the dominant effect on vibrational line shapes due to the phonon continuum is the production of a width, but with no shift, as displayed in the bottom panel of Figure 13. Note, however, that higher-order frequency shifts, small relative to level widths, are possible.[64]

3.2.4. Summary and Generalizations

Within a single formalism, we have considered the modes of a localized oscillator linearly coupled to three different boson fields which could produce an apparent linewidth in vibrational spectroscopy of the oscillator due to T_1 processes. Causality requires that for every width, there is a shift. If a given width Δ is due to coupling (or "decay") into (i) electron–hole pair, (ii) phonon, or (iii) photon continua, level shifts (i) ~ 10–100 Δ to the red, (ii) far to the blue, or (iii) much less than Δ must also be experienced, as shown schematically in Figure 13. These necessary connections between shifts and widths should be of use in the interpretation of actual vibrational line shapes of molecules adsorbed or embedded in a solid state environment.

3.3. Experimental Realization

Two important studies have recently been published that confront the line-shape issue in a direct way.

Trenary, Uram, Bozso, and Yates have studied the absorption spectrum of CO adsorbed on Ni(111) using IRAS.[115] In particular they were concerned with the temperature dependence of the line shape in the range from 8 to 300 K. Within this range, CO adsorbs both in twofold bridge sites and on top sites, producing two intramolecular stretch absorption lines centered at ~1840 and 2050 cm^{-1}, respectively. Working with an instrumental resolution ~14 cm^{-1} FWHM, the two lines displayed qualitatively different temperature-dependent widths as shown in Figure 14. The ontop line shows a mild increase in width with temperature, whereas the more tightly bound bridge site CO displays a dramatic increase with temperature, from the low-temperature value of ~14 cm^{-1} set by resolution and other temperature-independent mechanisms up to ~50 cm^{-1} at 300 K.

Upon consideration of the various possible broadening mechanisms elucidated in Sections 3.1 and 3.2, it appears that the exchange-coupling dephasing mechanism is the only one that could produce such a temperature dependence. Electron–hole pair damping is fairly temperature independent. Since $\omega_0 \sim 2000$ cm^{-1}, whereas the substrate Debye frequency ~300 cm^{-1}, T_1 phonon damping would require creation of ~7 phonons, which not only is very unlikely, but would also show a different temperature dependence. The observed temperature variation is exactly that which is implied by equation (60), the anharmonic coupling between the 2000 cm^{-1} mode and a lower-frequency (~700 cm^{-1}) libration in contact with the substrate. Furthermore, Trenary and co-workers argue that owing to increased involvement of the CO $2\pi^*$ orbital in the bridge site bond, the anharmonic coupling and hence dephasing width should be greater than for the ontop geometry, as observed. Thus it is quite reasonable to believe that the observed temperature-dependent width in this 80–300 K range is due to T_2' dephasing processes.

A cautionary point, which has been emphasized by Ryberg,[116] should be made. The total observed linewidth can be written as

$$\Delta_{tot}(T) = [\Delta_{res}^2 + \Delta_{other}^2 + \Delta_{dep}^2(T)]^{1/2}$$

where Δ_{res} is instrumental resolution, Δ_{other} is temperature-independent

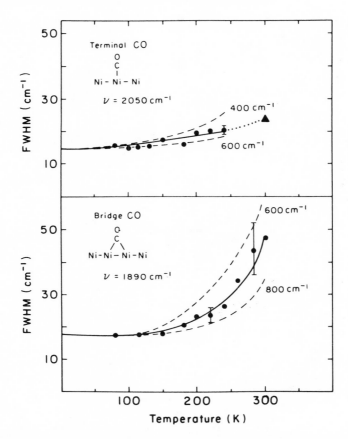

Figure 14. Plots of the line widths (FWHM) as a function of temperature. The dashed lines represent Arrhenius functions with the indicated activation energies in cm^{-1}. Deconvolution of the instrumental line shape would reduce the measured low-temperature linewidths by less than $2\,cm^{-1}$. For terminal CO, the triangular point has been obtained at $0.1\,Torr$ CO pressure.[115]

other mechanisms, and $\Delta_{dep}(T)$ is the dephasing contribution. With data such as those shown in Figure 14, nothing can be said that sheds light on the mechanisms responsible for the $14\text{-}cm^{-1}$ background width. While the temperature-dependent part does seem due to dephasing, other "interesting" mechanisms could also be operative on the level $\lesssim 10\,cm^{-1}$.

The final work here focuses on the exciting new wave of time-domain pump-probe experiments which independently measure the T_1 energy relaxation time, as mentioned in Section 1.2.1. Heilweil, Cassasa, Cavanagh, and Stephenson have measured transmission ir spectra, as a

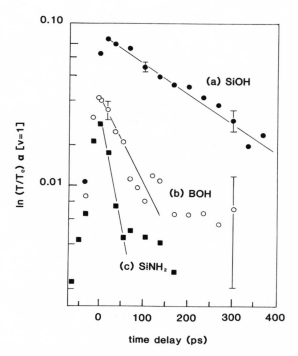

Figure 15. Experimental vibrational population ($v = 1$) decays at room temperature for adsorbate modes on colloidal silica. (a) OH-stretch at $3745\,\mathrm{cm}^{-1}$ of SiOH on dried pressed disk in vacuum, $T_1 = 194 \pm 12\,\mathrm{ps}$, (b) OH-stretch at $3705\,\mathrm{cm}^{-1}$ of BOH on dried, BCl_3 modified silica disk, $T_1 \simeq 70\,\mathrm{ps}$, and (c) NH-stretches at 3420 and $3500\,\mathrm{cm}^{-1}$ of NH_3 modified silica in CCl_4 exhibiting measurable saturation, but with pulse width limited decays, $T_1 \simeq 10\,\mathrm{ps}$.

function of probe time delay after $0 \rightarrow 1$ vibrational excitation, for a number of hydroxyl-containing molecules "adsorbed" on silica surfaces. Some of their results are shown in Figure 15. The slope of these rather straight lines (plotted semilogarithmically) directly yields T_1, the energy relaxation time. Rather surprisingly, almost all systems displayed $T_1 \sim 10^{-10}\,\mathrm{s}$, which implies that the energy deposited in the local mode remains localized for thousands of oscillator vibrations—very good news for the future of laser surface chemistry.[14,30,31] From the point of view of vibrational spectroscopy, these results suggest negligible T_1 contributions to observed absorption linewidths, at least for the systems studied by Heilweil and co-workers. An important factor must be kept in mind though. For the insulating systems studied, electron–hole pair damping is not important, whereas for the metallic substrates of more widespread interest, pair excitation could produce considerably more damping. In any event, time domain laser experiments of this sort certainly are the direction for the next generation of new intellectual excitement in the area of vibrational spectroscopy and reaction dynamics at solid surfaces.

References

1. W. H. Flygare, *Molecular Structure and Dynamics*, Prentice-Hall, Englewood Cliffs, New Jersey (1978).
2. D. W. Noid, M. L. Koszykowski, and R. A. Marcus, Quasiperiodic and stochastic behavior in molecules, *Ann. Rev. Phys. Chem.* **32**, 267–309 (1981).
3. D. M. Wardlaw, D. W. Noid, and R. A. Marcus, Semiclassical and quantum vibrational intensities, *J. Phys. Chem.* **88**, 536–547 (1984).
4. E. J. Heller, The semiclassical way to molecular spectroscopy, *Acc. Chem. Res.* **14**, 368–375 (1981).
5. J. M. Ziman, *Electrons and Phonons*, Clarendon Press, Oxford (1960).
6. J. W. Gadzuk and A. C. Luntz, On vibrational lineshapes of adsorbed molecules, *Surf. Sci.* **144**, 429–450 (1984).
7. S. Andersson, Vibrational excitations and structure of CO adsorbed on Ni(100), *Solid State Commun.* **21**, 75–81 (1977).
8. A. T. Bell and M. L. Hair, ed., *Vibrational Spectroscopies for Adsorbed Species*, American Chemical Society Symposium Series 137, (1980).
9. R. F. Willis, ed., *Vibrational Spectroscopy of Adsorbates*, Springer-Verlag, Berlin (1980).
10. H. Ibach and D. L. Mills, *Electron Energy Loss Spectroscopy and Surface Vibrations*, Academic, New York (1982).
11. A. M. Bradshaw, Vibrational spectroscopy of adsorbed atoms and molecules, *Appl. Surf. Sci.* **11/12**, 712–729 (1982).
12. R. K. Chang and T. E. Furtak, eds., *Surface Enhanced Raman Scattering*, Plenum Press, New York (1982).
13. P. K. Hansma, ed., *Tunneling Spectroscopy*, Plenum Press, New York (1982).
14. F. R. Aussenegg, A. Leitner, and M. E. Lippitsch, eds., *Surface Studies with Lasers*, Springer-Verlag, Berlin (1983).
15. R. F. Willis, A. A. Lucas, and G. D. Mahan, in: *The Chemical Physics of Solid Surfaces and Heterogeneous Catalysis*, Vol. 2, *Adsorption at Solid Surface* (D. A. King and D. P. Woodruff, eds.) pp. 59–163, Elsevier, Amsterdam (1983).
16. P. Avouris and J. Demuth, Electron energy loss spectroscopy in the study of surfaces, *Ann. Rev. Phys. Chem.* **35**, 49–73 (1984).
17. H. Ibach and S. Lehwald, eds., *Vibrations in Adsorbed Layers*, Berichte der Kernforschungsanlage Jülich-Jül-Conf-26 (1978).
18. R. Caudano, J. M. Gilles, and A. A. Lucas, eds., *Vibrations at Surfaces*, Plenum Press, New York (1982).
19. C. R. Brundle and H. Morowitz, eds., *Vibrations at Surfaces*, Elsevier, Amsterdam (1983).
20. D. A. King, N. V. Richardson, and S. Holloway, eds., *Vibrations at Surfaces*, Elsevier, Amsterdam (1986).
21. R. F. Willis, in Ref. 9.

22. P. J. Feibelman, Surface electromagnetic fields, *Prog. Surf. Sci.* **12**, 287–408 (1982).
23. R. Ryberg, Carbon monoxide adsorbed on Cu(100) studied by infrared spectroscopy, *Surf. Sci.* **114**, 627–641 (1982).
24. Y. J. Chabal and A. J. Sievers, Infrared study of hydrogen chemisorbed on W(100) by surface-electromagnetic-wave spectroscopy, *Phys. Rev. B* **24**, 2921–2934 (1981).
25. E. J. Heilweil, M. P. Casassa, R. R. Cavanagh, and J. C. Stephenson, Picosecond vibrational energy relaxation of surface hydroxyl groups on colloidal silica, *J. Chem. Phys.* **81**, 2856–2858 (1984).
26. N. Bloembergen and A. H. Zewail, Energy redistribution in isolated molecules and the question of mode-selective laser chemistry revisited, *J. Phys. Chem.* **88**, 5459–5465 (1984).
27. J. W. Gadzuk and H. Metiu, in Ref. 18.
28. K. Schönhammer and O. Gunnarsson, in Ref. 19.
29. D. Langreth and H. Suhl, eds., *Many-Body Phenomena at Surfaces,* Academic, Orlando (1984).
30. V. I. Goldanskii, V. A. Namiot, and R. V. Khokhlov, On the possibility of controlling surface phenomena by means of laser radiation, *Zh. Eksp. Teor Fiz.* **70**, 2349–2359 (1976) [English translation: *Sov. Phys. JETP* **43**, 1226–1232 (1976)].
31. T. J. Chuang, Laser-induced gas–surface interactions, *Surf. Sci. Rep.* **3**, 1–106 (1983).
32. T. E. Furtak and J. Reyes, A critical analysis of theoretical models for the giant Raman effect from adsorbed molecules, *Surf. Sci.* **93**, 351–382 (1980).
33. H. Metiu and P. Das, The electromagnetic theory of surface enhanced spectroscopy, *Ann. Rev. Phys. Chem.* **35**, 507–536 (1984).
34. A. C. Campion, Surface enhanced Raman scattering, *Comments Solid State Phys.* **3**, 107–123 (1984).
35. I. Pockrand, *Surface Enhanced Raman Vibrational Studies at Solid/Gas Interfaces,* Springer-Verlag, Berlin (1984).
36. M. Scheffler and A. M. Bradshaw, in Ref. 15.
37. J. Thietke, J. Billman, and A. Otto, in: *Dynamics on Surfaces* (B. Pullman, ed.) pp. 345–364, D. Reidel, Dordrecht (1984).
38. J. Lambe and R. C. Jaklevic, Molecular vibration spectra by inelastic electron tunneling, *Phys. Rev.* **165**, 821–832 (1968).
39. U. Landman, G. G. Kleiman, C. L. Cleveland, E. Kuster, R. N. Barnett, and J. W. Gadzuk, Hindered and modulated rotations of adsorbed diatomic molecules: States and spectra, *Phys. Rev. B* **29**, 4313–4326 (1984).
40. G. Benedek and U. Valbusa, eds., *Dynamics of Gas–Surface Interaction,* Springer-Verlag, Berlin (1982).
41. J. P. Toennies, Phonon inelastic scattering of He atoms from single crystal surfaces, *J. Vac. Sci. Technol.* **A2**(2), 1055–1065 (1984).
42. B. F. Mason and B. R. Williams, Inelastic atom scattering from a Cu(001) surface and an ordered adsorbed layer of Xe atoms at 16°K, *Phys. Rev. Lett.* **46**, 1138–1142 (1981).
43. E. H. Kerner, Note on the forced and damped oscillator in quantum mechanics, *Can. J. Phys.* **36**, 371–377 (1958).
44. R. P. Feynman and A. R. Hibbs, *Quantum Mechanics and Path Integrals,* McGraw-Hill, New York (1965).
45. P. W. Langhoff, S. T. Epstein, and M. Karplus, Aspects of time-dependent perturbation theory, *Rev. Mod. Phys.* **44**, 602–644 (1972).
46. A. P. Clark and I. C. Percival, Vibrational excitation and the Feynman correspondence identity, *J. Phys. B.* **8**, 1939–1952 (1975).

47. W. R. Gentry, in: *Atom-Molecule Collision Theory, A Guide for the Experimentalist* (R. B. Bernstein, ed.), pp. 391–425, Plenum Press, New York (1979).
48. V. P. Gutschick and M. M. Nieto, Coherent states for general potentials. V. Time evolution, *Phys. Rev. D.* **22**, 403–418 (1980).
49. H. D. Meyer, On the forced harmonic oscillator with time-dependent frequency, *Chem. Phys.* **61**, 365–383 (1981).
50. D. J. Tannor and E. J. Heller, Polyatomic Raman scattering for general harmonic potentials, *J. Chem. Phys.* **77**, 202–218 (1982).
51. H. Grabert, U. Weiss, and P. Talkner, Quantum theory of the damped harmonic oscillator, *Z. Phys. B* **55**, 87–94 (1984).
52. J. B. Marion, *Classical Dynamics of Particles and Systems,* pp. 128–164, Academic, New York (1965).
53. E. E. Nikitin, *Theory of Elementary Atomic and Molecular Processes in Gases,* pp. 52–58, Clarendon Press, Oxford (1974).
54. A. W. Kleyn, J. Los, and E. A. Gislason, Vibronic coupling at intersections of covalent and ionic states, *Phys. Rept.* **90**, 1–71 (1982).
55. A. A. Lucas and M. Sunjić, Fast-electron spectroscopy of collective excitations in solids, *Prog. Surf. Sci.* **2**, Part 2, 75–137 (1972).
56. J. W. Duff and D. G. Truhlar, Tests of semiclassical treatments of vibrational-translation energy transfer in collinear collisions of helium with hydrogen molecules, *Chem. Phys.* **9**, 243–273 (1975).
57. J. W. Gadzuk and S. Holloway, Charge transfer and vibrational excitation in molecule-surface collisions: Trajectorized quantum theory, *Physica Scripta* **32**, 413–422 (1985).
58. N. D. Lang, in: *Theory of the Inhomogeneous Electron Gas* (S. Lundqvist and N. H. March, eds.) pp. 309–389, Plenum Press, New York (1983).
59. G. Korzeniewski, T. Maniv, and H. Metiu, The interaction between an oscillating dipole and a metal surface described by a jellium model and the random phase approximation, *Chem. Phys. Lett.* **73**, 212–217 (1980).
60. K. L. Sebastion, The selection rule in electron energy loss spectroscopy of adsorbed molecules, *J. Phys. C.* **13**, L115–117 (1980).
61. R. G. Greenler, Infrared study of adsorbed molecules on metal surfaces by reflection techniques, *J. Chem. Phys.* **44**, 310–315 (1966).
62. J. D. E. McIntyre and D. E. Aspnes, Differential reflection spectroscopy of very thin surface films, *Surf. Sci.* **24**, 417–434 (1971).
63. J. Anderson, G. W. Rubloff, M. A. Passler, and P. J. Stiles, Surface reflectance spectroscopy studies of chemisorption on W(100), *Phys. Rev. B* **10**, 2410–2415 (1974).
64. J. D. Jackson, *Classical Electrodynamics,* Wiley, New York (1962).
65. B. N. J. Persson, Absorption of photons by molecules adsorbed on metal surfaces, *Solid State Commun.* **30**, 163–166 (1979).
66. R. G. Greenler, in Ref. 18.
67. M. Sunjić, in Ref. 40.
68. D. L. Mills, The scattering of low energy electrons by electric field fluctuations near crystal surfaces, *Surf. Sci.* **48**, 59–79 (1975).
69. B. N. J. Persson, Theory of inelastic scattering of slow electrons by molecules adsorbed on metal surfaces, *Solid State Commun.* **24**, 573–575 (1977).
70. D. M. Newns, in Ref. 9.
71. D. Sokcević, Z. Lenac, R. Brako, and M. Sunjić, Excitation of adsorbed molecule vibrations in low energy electron scattering, *Z. Phys. B* **28**, 273–281 (1977).

72. S. Andersson, B. N. J. Persson, T. Gustafsson, and E. W. Plummer, Vibrational excitation cross-section for adsorbed CO, *Solid State Commun.* **34,** 473–476 (1980).
73. J. W. Davenport, W. Ho, and J. R. Schrieffer, Theory of vibrationally inelastic electron scattering from oriented molecules, *Phys. Rev. B* **17,** 3115–3127 (1978).
74. C. H. Li, S. Y. Tong, and D. L. Mills, Large-angle inelastic electron scattering from adsorbate vibrations: Basic theory, *Phys. Rev. B* **21,** 3057–3073 (1980).
75. G. C. Aers, T. B. Grimley, J. B. Pendry, and K. L. Sebastion, Electron energy loss spectroscopy. Calculation of the impact scattering from W(100)$p(1 \times 1)$H, *J. Phys. C* **14,** 3995–4007 (1981).
76. R. F. Willis, in Ref. 9.
77. W. Ho, R. F. Willis, and E. W. Plummer, Observation of nondipole electron impact vibrational excitations: H on W(100), *Phys. Rev. Lett.* **40,** 1463–1466 (1978).
78. E. J. Heller, R. L. Sundberg, and D. Tannor, Simple aspects of Raman scattering, *J. Phys. Chem.* **86,** 1822–1833 (1982).
79. R. L. Sundberg and E. J. Heller, Preparation and dynamics of vibrational hot spots in polyatomics via Raman scattering, *Chem. Phys. Lett.* **93,** 586–591 (1982).
80. D. T. Birtwistle and A. Herzenberg, Vibrational excitation of N_2 by resonance scattering of electrons, *J. Phys. B* **4,** 53–70 (1971).
81. G. J. Schulz, Resonances in electron impact on diatomic molecules, *Rev. Mod. Phys.* **45,** 423–486 (1973).
82. W. Domcke and L. S. Cederbaum, Theory of the vibrational structure of resonances in electron molecule scattering, *Phys. Rev. A* **16,** 1465–1482 (1977).
83. C. W. McCurdy and J. L. Turner, Wave packet formulation of the boomerang model for resonant electron–molecule scattering, *J. Chem. Phys.* **78,** 6773–6779 (1983).
84. *Potential Energy Surfaces and Dynamics Calculations* (D. G. Truhlar, ed.), Plenum Press, New York (1981).
85. D. Schmeisser, J. E. Demuth, and Ph. Avouris, Electron-energy-loss studies of physisorbed O_2 and N_2 on Ag and Cu surfaces, *Phys. Rev. B* **26,** 4857–4863 (1982).
86. J. W. Gadzuk, Shape resonances, overtones, and electron energy loss spectroscopy of gas phase and physisorbed diatomic molecules, *J. Chem. Phys.* **79,** 3982–3987 (1983).
87. C. Manneback, Computation of the intensities of vibrational spectra of electronic bands in diatomic molecules, *Physica* **17,** 1001–1010 (1951).
88. E. J. Heller, Time-dependent approach to semiclassical dynamics, *J. Chem. Phys.* **62,** 1544–1555 (1975).
89. E. B. Stechel and R. N. Schwartz, Spreading and recurrence in anharmonic quantum-mechanical systems, *Chem. Phys. Lett.* **83,** 350–356 (1981).
90. R. T. Pack, Simple theory of diffuse vibrational structure in continuous uv spectrum of polyatomic molecules. I Collinear photodissociation of symmetric triatomics, *J. Chem. Phys.* **65,** 4765–4770 (1976).
91. J. E. Demuth, D. Schmeisser, and Ph. Avouris, Resonance scattering of electrons from N_2, CO, O_2 and H_2 adsorbed on a silver surface, *Phys. Rev. Lett.* **47,** 1166–1169 (1981).
92. M. Krauss and F. H. Mies, Molecular-orbital calculation of the shape resonance in N_2^-, *Phys. Rev. A* **1,** 1592–1598 (1970).
93. W. Marshall and S. W. Lovesey, *Theory of Thermal Neutron Scattering,* Clarendon Press, Oxfrod (1971).
94. J. P. McTague, M. Nielson, and L. Passell, in: *Chemistry and Physics of Solid Surfaces, Volume II,* (R. Vanselow, ed.,) CRC Press, Boca Raton, Florida (1979).
95. R. C. Casella, Generalized theory of neutron scattering from hydrogen in metals, *Phys. Rev. B* **28,** 2927–2936 (1983).

96. M. J. Puska, R. M. Nieminen, M. Manninen, B. Chakraborty, S. Holloway, and J. K. Nørskov, Quantum motion of chemisorbed hydrogen on Ni surfaces, *Phys. Rev. Lett.* **51**, 1081–1084 (1983).
97. R. R. Cavanagh, J. J. Rush, and R. D. Kelley, Comment on "Quantum motion of chemisorbed hydrogen on Ni surfaces", *Phys. Rev. Lett.* **52**, 2100 (1984).
98. J. W. Gadzuk, Localized vibrational modes in Fermi liquids. General theory, *Phys. Rev. B* **24**, 1651–1663 (1981).
99. V. P., Zhdanov and K. I. Zamaraev, Vibrational relaxation of adsorbed molecules. Mechanisms and manifestations in chemical reactions on solid surfaces, *Catal. Rev. Sci. Eng.* **24**(3), 373–413 (1982).
100. B. Hellsing and M. Persson, Electronic damping of atomic and molecular vibrations at metal surfaces, *Phys. Scr.* **29**, 360–371 (1984).
101. P. Avouris and B. N. J. Persson, Excited states at metal surfaces and their nonradiative relaxation, *J. Phys. Chem.* **88**, 837–848 (1984).
102. A. Laubereau and W. Kaiser, Vibrational dynamics of liquids and solids investigated by picosecond light pulses, *Rev. Mod. Phys.* **50**, 607–665 (1978).
103. D. A. Wiersma, Coherent optical transient studies of dephasing and relaxation in electronic transitions of large molecules in the condensed phase, *Adv. Chem. Phys.* **47**, 421–485 (1981).
104. D. W. Oxtoby, Vibrational population relaxation in liquids, *Adv. Chem. Phys.* **47**, 487–519 (1981).
105. R. M. Shelby, C. B. Harris, and P. A. Cornelius, The origin of vibrational dephasing of polyatomic molecules in condensed phases, *J. Chem. Phys.* **70**, 34–41 (1979).
106. A. H. Zewail, Optical molecular dephasing: Principles of and probings by coherent laser spectroscopy, *Acc. Chem. Res.* **13**, 360–368 (1980).
107. R. Kubo, A stochastic theory of line shape, *Adv. Chem. Phys.* **15**, 101–127 (1969).
108. P. W. Anderson, A mathematical model for the narrowing of spectral lines by exchange or motion, *J. Phys. Soc. Jpn.* **9**, 316–339 (1954).
109. G. P. Brivio and T. B. Grimley, Lifetimes of electronically adiabatic vibrational states of a chemisorbed atom, *J. Phys. C* **10**, 2351–2363 (1977).
110. T. S. Rahman, D. L. Mills, and J. E. Black, Low frequency surface resonance modes in electron energy loss spectroscopy, *J. Electron Spectrosc. Relat. Phenom.* **29**, 199–212 (1983).
111. J. C. Ariyasu, D. L. Mills, K. G. Lloyd, and J. C. Hemminger, Anharmonic damping of vibrational modes, *Phys. Rev. B* **28**, 6123–6126 (1983).
112. R. R. Chance, A Prock, and R. Silbey, Molecular fluorescence and energy transfer near interfaces, *Adv. Chem. Phys.* **37**, 1–65 (1978).
113. U. Fano, Effects of configuration interaction on intensities and phase shifts, *Phys. Rev.* **124**, 1866–1878 (1961).
114. E. Müller-Hartman, T. V. Ramakrishnan, and G. Toulouse, Localized dynamic perturbations in metals, *Phys. Rev. B* **3**, 1102–1119 (1971).
115. M. Trenary, K. J. Uram, F. Bozso, and J. T. Yates, Jr., Temperature dependence of the vibrational lineshape of CO chemisorbed on the Ni(111) surface, *Surf. Sci.* **146**, 269–280 (1984).
116. R. Ryberg, Vibrational line shape of chemisorbed CO, *Phys. Rev.* **B32**, 2671–2673 (1985).

3

Infrared Spectroscopy of High-Area Catalytic Surfaces

Alexis T. Bell

1. Introduction

Infrared spectroscopy has proven to be an enormously useful technique for obtaining information about the structure of catalysts and species adsorbed on catalyst surfaces. The widespread application of the method is a consequence of the ease with which samples can be prepared and analyzed, and the absence of a need for complex workup of the spectral data prior to interpretation. A further attraction is that spectra can readily be taken at high temperatures and pressures, and under conditions at which catalysts are used in industrial practice.

The application of infrared spectroscopy to the characterization of catalysts and adsorbed species was first carried out by Terenin and other Russian scientists in the late 1940s. These studies focused primarily on silicas and silica-aluminas. In the 1950s Eischens and co-workers pioneered the application of the technique to study supported metal catalysts. Since that time, the range of catalysts examined by infrared spectroscopy has widened significantly, and important advances have been made in the techniques for acquiring spectra. The reader interested in a review of the very large body of literature pertaining to infrared spectroscopy of catalysts and adsorbed species should refer to Refs. 1–11.

Alexis T. Bell • Materials and Molecular Research Division, Lawrence Berkeley Laboratory, and Department of Chemical Engineering, University of California, Berkeley, California 94720.

The purpose of this chapter is to review the experimental techniques used to obtain infrared spectra from catalysts in the form of high-surface-area powders. A discussion is presented of the principal methods of acquiring spectra and the problems associated with making quantitative measurements. The relationship of sample composition and method of preparation to the quality of spectra obtained is examined. The design of infrared cells is reviewed with special emphasis given to designs permitting *in situ* observations. Finally, the performance characteristics of dispersive and Fourier-transform spectrometers are compared and contrasted.

2. Techniques for the Acquisition of Spectra

There are three primary techniques for acquiring infrared spectra of high-surface-area powdered samples: transmission spectroscopy, diffuse-reflectance spectroscopy, and photoacoustic spectroscopy. Attenuated total reflection and emission spectroscopies have also been used, but these techniques suffer from a number of disadvantages that limit their applicability.[12] In this section we will outline the theory underlying the three principal methods of spectral acquisition and discuss the problems of obtaining quantitative information.

2.1. Transmission Spectroscopy

Transmission spectroscopy has been the most widely used technique to study the structure of adsorbed species and the structure of compound catalysts (e.g., metal oxides, metal sulfides, zeolites, etc.). The sample is usually prepared by pressing a fine powder of the catalyst into a self-supporting disk. The disk is then placed perpendicular to a beam of infrared radiation, and the spectrum is recorded by observing the transmitted beam.

Figure 1 illustrates the positions of the incident and transmitted beams in a typical transmission experiment. The transmittance of the sample at a given wave number, $T(\bar{\nu})$, is

$$T(\bar{\nu}) = I(\bar{\nu})/I_0(\bar{\nu}) \tag{1}$$

where $I_0(\bar{\nu})$ and $I(\bar{\nu})$ are the intensities of the incident and transmitted beams, respectively. As shown in Figure 1, $T(\bar{\nu})$ varies across the spectrum and is a minimum at the frequency of an absorption band. For a thin slab with uniform optical properties, the transmittance at the peak

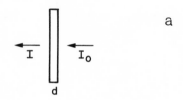

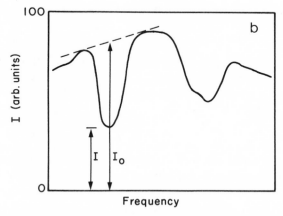

Figure 1. (a) A schematic illustration of a transmission experiment. (b) An illustration of the method used to determine the incident and transmitted radiation from a transmission spectrum.

minimum, $T(\bar{v}_p)$, can be related to the concentration of the absorber, c, and the sample thickness, d, by the Beer–Lambert law.

$$T(\bar{v}_p) = \exp(-\varepsilon cd) \qquad (2)$$

The quantity ε in equation (2) is known as the extinction coefficient and has units of cm^2/mol. Since $T(\bar{v}_p)$ is an exponential function of c, it is more convenient to work in terms of the absorbance for a peak, $A(\bar{v}_p)$. The absorbance is related to the transmittance by

$$A(\bar{v}_p) = -\ln[T(\bar{v}_p)]$$
$$= \varepsilon cd$$

For materials such as catalysts in which an absorbing functional group may be present in a variety of bonding environments, it is often preferable to work in terms of the integrated absorbance, $\bar{A}$, rather than the absorbance at the band maximum. The quantity $\bar{A}$ is defined by

$$\bar{A} = \int_{\bar{v}_1}^{\bar{v}_2} A(\bar{v}) \, d\bar{v}$$

$$= \int_{\bar{v}_1}^{\bar{v}_2} \ln[I_0(\bar{v})/I(\bar{v})] \, d\bar{v} \tag{4}$$

The integrated absorbance is related to the concentration of absorbers by

$$\bar{A} = \bar{\varepsilon} c d \tag{5}$$

where $\bar{\varepsilon}$ is known as the integrated absorption coefficient and has units of cm/mol. Equation (5) can be simplified by noting that the product cd is the concentration of absorbers per square centimeter of sample cross-sectional area, $\bar{c}$. Hence

$$\bar{A} = \bar{\varepsilon} \bar{c} \tag{6}$$

If one is interested solely in the spectrum of adsorbed species, then it is necessary to recognize that the observed absorbance is the sum of the absorbances for the catalyst and the adsorbate:

$$\bar{A} = \bar{A}_c + \bar{A}_a \tag{7}$$

To obtain $\bar{A}_a$, equation (7) is rearranged as follows:

$$\bar{A}_a = \bar{A} - \bar{A}_c$$

$$= \int_{\bar{v}_1}^{\bar{v}_2} \ln[I_c(\bar{v})/I(\bar{v})] \, d\bar{v}$$

$$= \bar{\varepsilon}_a \bar{c}_a \tag{8}$$

In equation (8), $I_c(\bar{v})$ is the intensity of the transmitted beam for the catalyst in the absence of the adsorbate, and $\bar{\varepsilon}_a$ and $\bar{c}_a$ are the integrated extinction coefficient and concentration of the adsorbate, respectively.

The extent to which $\bar{\varepsilon}$ is a constant depends on whether one considers the catalyst or species adsorbed on the catalyst surface. For vibrations of the catalyst lattice or functional groups present at the surface of the catalyst (e.g., OH, SH, etc.), $\bar{\varepsilon}$ is a constant, independent of the sample thickness. If one observes the spectrum of species adsorbed on a catalyst, then $\bar{\varepsilon}$ may not be constant with changing adsorbate concentration. An example of what can occur in the case CO adsorption on Ru/SiO_2 is illustrated in Figure 2.[13] At low CO coverages, $\bar{\varepsilon}_{CO}$ is constant, but then decreases as the coverage increases to unity. The

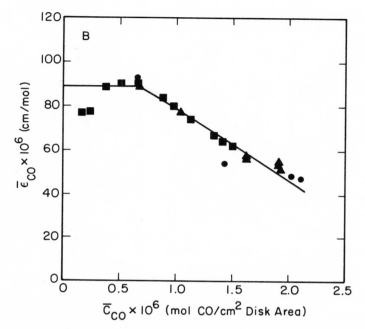

Figure 2. The influence of surface coverage on the integrated absorbance for CO adsorbed on Ru/SiO$_2$ (reproduced with permission from Ref. 13).

change in $\bar{\varepsilon}_{CO}$ with coverage is due to lateral interactions between the adsorbed CO molecules.

Scattering of the incident radiation can be significant for samples prepared from powders. The effect of scattering is to reduce the intensity of the transmitted beam, thereby making it harder to obtain a good spectrum. For Rayleigh scattering, the amount of scattering is given by

$$S \propto d_p^3(\bar{v})^4 \qquad (9)$$

where d_p is the particle diameter. Thus, scattering is most severe for large particles and high wave numbers. The effect of particle size on band intensity is clearly shown in Figure 3, for calcium carbonate particles with diameters between 55 and 5 μm.[14] It is evident from this figure that for samples of a fixed mass the apparent extinction coefficient will be less than the true extinction coefficient for particles larger than about 5 μm. To avoid such artifacts, it is desirable to work with particles less than 1 μm in diameter, whenever possible.

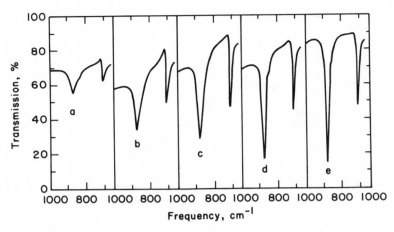

Figure 3. Effect of particle size on the infrared spectrum of calcite in KBr disks: (a) 55 μm; (b) 40 μm; (c) 23 μm; (d) 14.7 μm; (e) 5 μm (reproduced with permission from Ref. 14).

2.2. Diffuse-Reflectance Spectroscopy

Diffuse-reflectance spectroscopy has been used much less commonly than transmission spectroscopy in catalysis.[12,15,16] For the most part this has been due to the more complex optical arrangement required and the absence of simple, easy to use cell designs. As will be discussed below, both of these constraints have recently been eliminated. Since diffuse-reflectance spectroscopy does not require the compression of a powder into a self-supporting disk, materials difficult to compress (e.g., TiO_2, certain zeolites, etc.) can readily be examined.

To take the diffuse reflectance spectrum of a powder, the sample is placed in a shallow cup and exposed to a beam of infrared radiation. The incident radiation passes into the bulk of the sample, and undergoes reflection, refraction, and absorption before reemerging at the sample surface. The diffusely reflected radiation from the sample is collected by a spherical or elliptical mirror and focused onto the detector of the spectrometer.

The interpretation of diffuse-reflectance spectra is based on the phenomenological theory developed by Kubelka and Munk.[17] This approach assumes that all the radiation within the sample can be resolved into two fluxes, as shown in Figure 4. The flux I is taken to propagate in the negative x direction, and the flux J in the positive x direction. Fluxes in all other directions of the infinitely extended sample are assumed to be returned to the $\pm x$ direction by scattering without absorption. Under

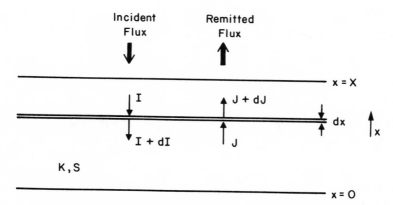

Figure 4. A diagrammatic illustration of the Kubelka–Munk approximations. All of the radiation is divided into two fluxes of opposite sign. The incident flux ($-x$ direction) is taken to be diffuse. The absorption and scattering of radiation is described by the two parameters K and S, respectively.

these conditions, the change in I over the differential distance dx is

$$\frac{dI}{dx} = -(K + S)I + SJ \tag{10}$$

The constants K and S account, respectively, for losses due to absorption and scattering of the incident radiation flux. The change in J, the upward traveling flux, in traversing the layer dx is

$$-\frac{dJ}{dx} = SI - (K + S)J \tag{11}$$

The boundary conditions on equations (10) and (11) are

$$I = I_0 \qquad \text{at } x = L$$
$$J = -R_g I \qquad \text{at } x = 0 \tag{12}$$

where R_g is the specular reflectivity of the cup surface.

Equations (10) and (11) can be solved to obtain the following expression for R, the reflectance at $x = L$:

$$R = \frac{J}{I} = \frac{1 - R_g[a - b\,\coth(bSL)]}{a - R_g + b\,\coth(bSL)} \tag{13}$$

where

$$a = 1 + (K/S)$$
$$b = (a^2 - 1)^{1/2} \tag{14}$$

For a very thick sample, the background reflectance is zero and since $\coth(bSL) \to 1$ as $L \to \infty$

$$R_\infty = 1 + (K/S) - [(K/S)^2 + 2K/S]^{1/2} \tag{15}$$

which, upon solving for K/S, gives

$$K/S = F(R_\infty) = (1 - R_\infty)^2/(2R_\infty) \tag{16}$$

Equation (16) defines the well-known Kubelka–Munk remission function relating the experimentally determined diffuse reflectance of an "infinitely" thick sample to the parameters K and S. This expression is useful for spectroscopy studies as long as S is independent of $\bar{\nu}$, so that the only dependence on $\bar{\nu}$ is through K. Since K can now be expressed as

$$K = \bar{\varepsilon}c \tag{17}$$

it follows that

$$F(R_\infty) = \bar{\varepsilon}c/S \tag{18}$$

2.3. Photoacoustic Spectroscopy

The application of photoacoustic spectroscopy to obtain infrared spectra of powdered samples is a relatively recent development.[15,18–20] The primary advantage of the technique over transmission and diffuse-reflectance spectroscopies is that it is virtually independent of catalyst particle size, thereby eliminating the concern about preparation of the sample in the proper physical form.

The origins of the photoacoustic effect can be understood with the aid of Figure 5. The sample is placed in a sealed cell filled with a nonadsorbing inert gas. A chopped beam of radiation incident upon the sample causes periodic heating of the sample through absorption of radiation. The sample in turn periodically heats the filler gas in the boundary layer $2\pi/a_g$, where $1/a_g$ is the thermal diffusion length of the gas. The periodic pressure change in the filler gas is detected by a sensitive microphone. Because radiation is preferentially absorbed at the absorption maxima of the sample, the acoustic output depends on the wavelength of the incident radiation and a photoacoustic spectrum is obtained.

Rosencwaig and Gersho[21] have shown that the time-dependent change in the gas pressure within the cell, $\Delta P(t)$, is related to the heat released to the gas, Q, by the following expression:

$$\Delta P(t) = Q \exp[i(\omega t - \pi/4)] \tag{19}$$

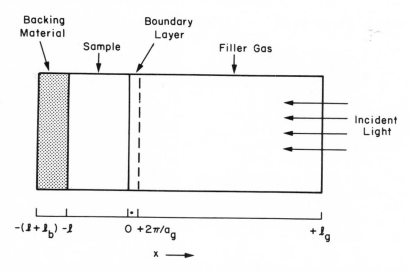

Figure 5. The photoacoustic cell geometry. Chopped light incident upon a sample in the closed cell causes periodic heating of the sample through absorption of radiation. The sample in turn periodically heats the filler gas in the boundary layer $2\pi/a_g$, where $1/a_g$ is the thermal diffusion length of the gas. The periodic pressure change in the filler gas is detected by a microphone (not shown). Because radiation is preferentially absorbed at the absorption maxima of the sample, the acoustic output depends on the wavelength of the incident radiation and a photoacoustic spectrum is obtained (reproduced with permission from Ref. 21).

where ω is the modulation frequency of the incident radiation. The quantity Q is a function of the optical and thermal properties of the sample, and of the thermal properties of the gas over the sample. For the case of thermally and optically thick solids,[20,21]

$$Q = -\frac{iP_0I_0\gamma}{4\sqrt{2}l_g}\,T_0\!\left(\frac{\mu_s^2\mu_g}{k_s}\right)\varepsilon c \tag{20}$$

where I_0 is the intensity of the incident radiation, P_0 is the ambient gas pressure, T_0 the ambient gas temperature, γ the ratio of specific heats of the gas at constant pressure and volume, l_g the height of the ambient gas layer, k_s the specific heat of the solid, and μ_s and μ_g are the thermal diffusion lengths of the solid and gas, respectively. Equations (19) and (20) indicate that when ε is independent of c the amplitude of $\Delta P(t)$ will be proportional to c. Further details concerning the extraction of quantitative data from photoacoustic spectra are discussed by Teng and Royce.[22]

3. Sample Preparation

The manner in which catalyst samples are prepared for infrared studies can influence the quality of the spectra obtained. As was already discussed in the preceding section, scattering of the incident radiation can be a serious problem when spectra are acquired by either transmission or diffuse-reflectance spectroscopy. The extent of scattering is a function of both particle size and the difference in index of refraction between the particles and the surrounding medium. For Rayleigh scattering the amount of scattering is proportional to the third power of the particle diameter [see equation (9)], and hence the extent of scattering decreases rapidly with decreasing particle size. In practice, the best results are achieved when the particle size is below the wavelength of the incident radiation ($\leq$1–5 μm). The extent of scattering can also be decreased by filling the voids between particles with a liquid or solid whose refractive index is close to that of the particles. Both paraffin oil and KBr have been used for this purpose.[6] The principal drawback of this approach is that the void-filling material may contain substances that react with the catalyst, thereby altering its characteristics. Catalyst pretreatment (i.e., reduction, oxidation, etc.) is also made more difficult, if not impossible, with samples prepared in this fashion.

The composition of catalysts that can be investigated by infrared spectroscopy is limited only by the degree to which the sample absorbs infrared radiation. For very strongly absorbing materials, the amount of material required to obtain useful spectra may be so small as to make it difficult to work with. The vast majority of the studies reported in the literature have involved silica, alumina, silica-alumina, zeolites, magnesia, zinc oxide, and titania.

Infrared spectra of species adsorbed on supported metals have been investigated very extensively. For such studies, the metal is usually dispersed on a finely divided oxide support (e.g., SiO_2, Al_2O_3, TiO_2, MgO, ZnO, Cr_2O_3, etc.). The metal is introduced onto the support either by incipient wetness impregnation with a solution of a metal salt (e.g., $PdCl_2$, $RhCl_3$, $FeNO_3$, etc.) or by ion exchange with either an anionic or cationic metal complex (e.g., $PtCl_6^{2-}$, $[Pt(NH_3)_4]^{2+}$, etc.).[23] The slurry is then evaporated to dryness. The precursor is converted to the metal either by direct reduction of the precursor in H_2 at elevated temperature, or by calcination in air followed by H_2 reduction. To avoid excessive absorption and scattering of the incident infrared radiation, the metal loading is usually kept below 10% by weight of the reduced metal on the support. It is also desirable to achieve a high dispersion of the metal in order to obtain a high concentration of adsorption centers per mass of finished catalyst. This objective is frequently achieved by lowering the metal loading to a few percent and using ion exchange, rather than impregnation with a salt solution, as the means for introducing the metal precursor.

If spectra are to be acquired by diffuse-reflectance or photoacoustic spectroscopy, no additional sample preparation is required. This, in fact, represents a significant advantage of these techniques. For transmission spectroscopy the catalyst is usually pressed into a self-supporting disk. The disk typically has a diameter of 10–25 mm and contains 10–50 mg/cm^2 of catalyst. The selection of disk thickness (~100 μm) is dictated by the need to have sufficient material in the beam of radiation to obtain a spectrum, but not so much that transmission is strongly limited by scattering and/or support absorption. Disks are produced using a stainless steel die, consisting of a cylinder and two removable pistons. A standard KBr die is also suitable for this purpose. The pressure required depends on the catalyst composition. Thus, for example, pressures of 100 to 12,000 psi are used for silica, but 15,000 to 80,000 psi are needed for alumina.[6] With some materials the production of a uniform disk that does not crack upon removal from the die can be a problem. Experience shows that these difficulties can often be overcome by humidification of the powder prior to pressing, polishing the die faces, and heating the die to ~65°C just prior to use.[11]

An alternative to producing a disk is to spread the catalyst powder out as a thin layer on an infrared transparent window. This can be achieved by spraying a suspension of the catalyst in a volatile solvent such as methanol onto a heated infrared-transparent window. Very thin catalyst layers can be obtained this way, but there is always the concern about the integrity of the film and its bond to the window.[24]

An important aspect of sample preparation not often given adequate attention is the effect of sample thickness on the dynamics of gas adsorption and reaction. The penetration of an absorbing or reacting gas into a catalyst disk or layer is usually by diffusion. If flow through the sample is achieved, then convection governs the rate of delivery of new material. For strongly adsorbing gases (i.e., where the sticking coefficient is close to unity) or very rapid reactions, the gas concentration within the disk or layer may be highly nonuniform. Since the incident infrared beam samples material from all depths, the spectrum in such cases will be an average of the profile in coverages. An extreme case of this problem is frontal, or chromatographic, adsorption which results in the progression of a sharp wave of adsorption from the outside of the sample to its center. If it is desired to obtain spectra truly representative of a fixed fractional coverage, then frontal adsorption must be avoided. This can be done by annealing the sample, following adsorption, at a temperature below which substantial desorption occurs. For reacting systems, the absence of diffusional transport effects can be assessed by application of the Weisz–Prater criterion[25]

$$\frac{d^2 r_m \rho_c}{c_r D_e} < 0.1 \tag{21}$$

where d is the disk thickness, r_m is the rate of reaction per weight of catalyst, ρ_c the catalyst (i.e., disk) density, c_r is the gas-phase concentration of reactant, and D_e is the effective diffusivity.

4. Cell Designs

4.1. Transmission Spectroscopy

A large number of cell designs have been described in the literature, for obtaining infrared spectra from high-surface-area powders (see, for example, Refs. 1–11). While the details of individual designs vary widely, there are a number of common features. The cell is an enclosure of small volume surrounding the sample, which is usually in the form of a self-supporting disk. Infrared transparent windows are attached to openings in opposing walls of the cell, to permit passage of an infrared beam through the cell. The cell is also provided with passages for the introduction and withdrawal of gas. If the sample is to be heated while in the infrared beam, then a small heater is built into the cell or the cell is heated externally. On the other hand, if the sample need not be heated while in the infrared beam, then heating is accomplished outside the portion of the cell used for acquiring spectra. In such instances, the sample must be transferred mechanically from one portion of the cell to another.

The body of transmission infrared cells is usually made of glass (i.e., Pyrex or quartz) or stainless steel. Glass is inert, easy to work with, and convenient for the construction of cells to be used at ambient and subambient pressures. For work at pressures above 1 atm, cells are usually built of stainless steel. If there is concern about corrosion or the catalytic activity of steel, then the interior surfaces of the cell can be gold plated or coated with an aluminum film, which is then oxidized to form an alumina barrier.

As indicated in Table 1, a wide range of materials can be used for the infrared-transparent windows. The selection of a particular material

Table 1. Optical Materials Useful in Infrared Studies

Material	Useful range (cm^{-1})	Comments
NaCl (rock salt)	>500	Widely used, cheap, and easily worked; must be kept dry, however.
KBr	>310	Soft and easily scratched. Similar to NaCl, but has greater range. Used as powder for pressed disk technique.
CsBr	>240	Easily worked, but harder than the other halides. Expensive. Not hygroscopic.
Quartz (SiO_2)	>2500	Insoluble and easy to work in fused form. Useful for high-temperature work and in the overtone region.
Sapphire (Al_2O_3)	>2600	Has high mechanical strength but is expensive.
MgO (Irtran 5[a])	>1200	Hard and costly. Can be sealed to a high-expansion glass.
LiF	>1200	Very useful in the near infrared because of good dispersion. Scratches easily.
CaF_2 (fluorite)	>900	Inert to most chemicals, but tends to be costly. Good from 73 to 373 K. Has low solubility in all except NH_4^+ salts.
AgCl	>400	Soft material. Has low melting point, is corrosive and photosensitive.
MgF_2 (Irtran 1[a])	1333	Strong. Will withstand temperature of 1173 K. Chemically durable.
ZnS (Irtran 2[a])	714	Chemically durable, good up to 1073 K, strong, very useful.
ZnSe (Irtran 4[a])	500	Useful up to 573 K. Soluble in acids.

[a] Eastman Kodak Co., Rochester, New York.

depends upon the spectral range to be observed, the strength of the window if studies are to be carried out under vacuum or at high pressure, sensitivity to fogging caused by moisture, and cost. The most commonly used materials are CaF_2 and KBr. The former is strong and inert, but limited to spectral observations above ~900 cm^{-1}. The latter material extends the spectral range down to ~310 cm^{-1}, but is soft and readily pitted by moisture.

Various methods have been used to attach the infrared windows to the cell body. If the windows need not be heated, or can be cooled, it is possible to use Viton O-rings to make a vacuum-tight seal. Where a permanent attachment is tolerable, it is also possible to use a low residual vapor pressure epoxy. Achieving a vacuum-tight seal at higher temperatures is more difficult. Kalrez O-rings can be used up to 573 K with all types of windows. Calcium fluoride windows can be sealed with silver chloride to silver tubes or flanges, which can then be attached to the cell body, providing a seal bakable in principle to 700 K.[26] Magnesium oxide windows can be sealed to soda glass, silicon windows to Pyrex, and various combinations employed to make cells that can be heated to 650 K.[27]

The distance between the interior surfaces of the two windows defines the optical path length through the cell. This distance should, ideally, be kept small to minimize absorption of infrared radiation by gas present in the cell. To illustrate the extent to which gas-phase adsorption might be a problem, let us consider the following example. We will assume that we have a 50% dispersion of 1 wt% of Pt on a high-surface-area support. At saturation coverage of the Pt crystallites by an adsorbate, the product of the gas pressure and gas path length at which infrared absorption in the gas phase and in the sample become comparable is given by

$$pd_g = 5.3 \times 10^2 \, \text{cm Pa} \tag{22}$$

where p is the gas pressure and d_g is the gas path length. In equation (22), the extinction coefficient for the adsorbate is taken to be the same in the gas phase and in the adsorbed phase. Thus, if the gas path length is 1 cm, the absorbance due to the gas phase will be comparable to that due to the adsorbed phase at 5.3×10^2 Pa.

In instances where the effects of gas absorption cannot be eliminated by cell design, the gas-phase spectrum can often be subtracted from that of the gas plus adsorbed phase. To do so, either the sample must be removed from the path of the infrared beam or the gas-phase spectrum recorded using a separate cell.

Figure 6 illustrates an infrared cell designed to permit movement of the sample to a separate furnace.[11] With such a cell, the gas-phase spectrum can be recorded with the sample in the furnace. When the sample is lowered into the path of the infrared beam, a spectrum of the sample and gas phase can be recorded. It should be noted that since the cell windows are not heated they can be bonded directly to the cell using a low-vapor-pressure adhesive.

Another very simple cell design is shown in Figure 7.[28] In this case, the sample can be heated *in situ*. To avoid overheating the window seals these are water cooled. This is disadvantageous, though, in that it creates a strong thermal gradient along the reactor making it difficult to heat the sample uniformly. An alternative design that provides more uniform sample heating is shown in Figure 8.[29] The heater in this case is enclosed within the cell itself and is in poor thermal contact with the cell body.

A UHV cell designed for simultaneous infrared and temperature-programmed desorption studies is shown in Figure 9.[30] The cell is assembled from commercially available CaF_2 windows which sandwich a $2\frac{3}{4}$ in. flange into which the sample holder is mounted. The holder is a

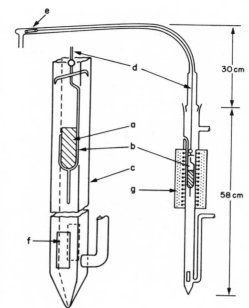

Figure 6. Infrared cell permitting sample movement. a, Sample; b, quartz sample holder; c, cell body (quartz, 2.5 cm square cross section, attached to 34/45 Vycor joint); d, fiber or wire of quartz, stainless steel, or gold; e, Pyrex-enclosed magnet; f, window openings; g, furnace. (Reproduced with permission from Ref. 27.)

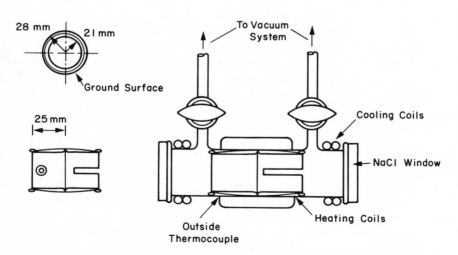

Figure 7. Infrared cell and sample holder (reproduced with permission from Ref. 28).

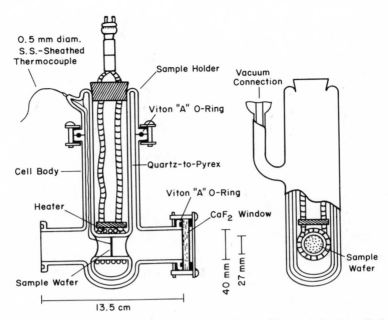

Figure 8. Infrared cell for catalyst studies. (Reproduced with permission from Ref. 29.)

copper ring coated with nickel and gold. A thin stainless-steel tube is soldered around the circumference of the ring to cool the sample using liquid N_2, and a 50-W cartridge heater is inserted into the ring mount to heat the sample.

The cells shown in Figures 7, 8, and 9 can be used for *in situ* studies of chemical reactions but are limited to operation at pressures of about 1 atm or less. Cells designed for operation at pressures up to 30 MPa have been described in the literature.[31–34] One example is shown in Figure 10.[32] The cell body consists of two stainless steel flanges. Knife edges are cut into each flange to permit the formation of a gas-tight seal when the cell is assembled. Polycrystalline CaF_2 windows are placed into wells cut into each flange. The seal between each window and its corresponding flange is formed by Kalrez O-rings (DuPont). The dead volume in the assembled reactor is $0.4 \, cm^3$ and the optical path length through the central cavity is 2.4 mm, including the thickness of the disk (~0.1 mm). The cell is heated externally by two etched foil heating elements (Thermal Circuits, Inc.). Operation of the cell at pressures up to 3 MPa and temperatures up to 573 K has been accomplished successfully.

Figure 11 illustrates a cell designed by Edwards and Schraeder.[33]

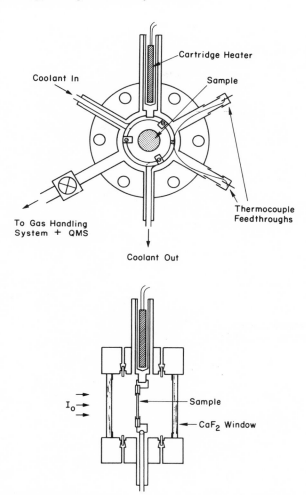

Figure 9. UHV cell for simultaneous infrared and thermal desorption spectroscopy. (Reproduced with permission from Ref. 30.)

The cell body is constructed of stainless steel and the windows are CaF_2. Heating is by four cartridge heaters inserted into holes drilled into the cell body. With Kalrez O-rings, the authors report that this cell can be used at pressures ranging from 10^{-4} Pa to 24 MPa and temperatures from ambient to 573 K.

A cell permitting operation at even higher pressure and temperatures than those shown in Figures 10 and 11 has been described by Penninger[34] and is illustrated in Figure 12. The unique feature of this

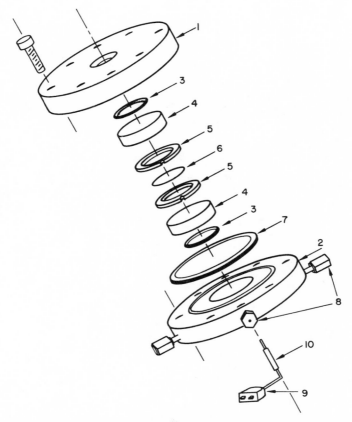

Figure 10. High-pressure infrared cell. 1, Top Flange; 2, bottom flange; 3, Kalrez O-ring; 4, CaF$_2$ window; 5, sample holder; 6, catalyst disk; 7, copper gasket; 8, Swagelok fitting; 9, sheathed thermocouple; 10, sleeve attached to thermocouple sleeve. (Reproduced with permission from Ref. 32.)

cell is the use of compressed Rulon-25 (Dixon Corporation, Bristol, R.I.) rings to seal the windows to the cell body. The maximum usable temperature and pressure for this cell are reported to be 723 K and 30 MPa, respectively.

4.2. Diffuse-Reflectance and Photoacoustic Spectroscopies

A number of different cells for carrying out diffuse-reflectance spectroscopy have been described in the literature.[12] The arrangement

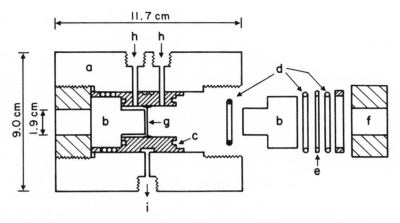

Figure 11. High-pressure infrared cell. a, Cell body; b, CaF_2 stepped windows; c, spacer; d, Viton or Kalrez O-rings; e, aluminum gasket; f, screw plug; g, wafer sample; h, inlet ports; i, outlet ports. (Reproduced with permission from Ref. 33.)

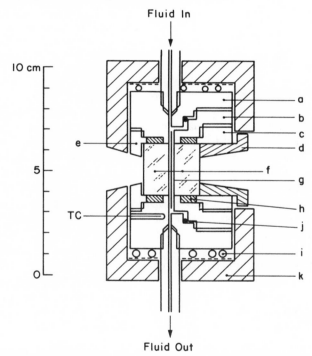

Figure 12. High-pressure infrared cell. a, Reactor body; b, removeable side of the reaction space; c, window seal compression plug; d, adjustable window support; e, window support/seal compression plug; f, optical windows; g, catalyst disk in reaction space; h, window seal; i, heating element; j, O-ring; k, thermal insulation box. (Reproduced with permission from Ref. 34.)

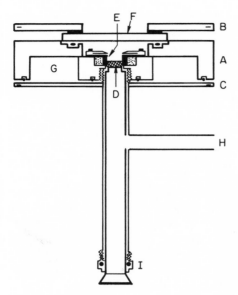

Figure 13. Diffuse reflectance cell. A, Main body; B, coverplate for sample compartment; C, coverplate for the cooling water cavity; D, frit to support the sample; E, sample heater; F, KCl window; G, cavity for cooling water; H, gas outlet line; I, nut to vacuum-tighten the sample holder to the housing tube. (Reproduced with permission from Ref. 35.)

shown in Figure 13, recently reported by Hamadeh *et al.*,[35] is particularly well suited for *in situ* studies of catalysts. Gases can be passed directly through the powdered sample permitting intimate contact, and the sample can be heated to temperatures as high as 873 K. The range of pressures over which the cell can be used is 10^{-4} Pa to 101 kPa. A commercially available cell, also designed for catalytic studies, has very recently been introduced by Spectra-Tech., Inc. (Stamford, Connecticut).

While photoacoustic spectroscopy cells are available commercially, these devices are not designed to allow catalyst pretreatment. These and similar designs have been adapted, though, to study adsorption on catalytic materials.[12,15] Very recently McGovern *et al.*[36] have described a design in which the microphone can be isolated from the cell itself so that catalyst pretreatment and reactions can be carried out in the cell. Acquisition of spectra is limited, though, to postadsorption, or postreaction, conditions. An illustration of the cell and the manner in which it is connected to the microphone is shown in Figure 14.

5. Ancillary Equipment

In most infrared studies of catalysts it is of interest to determine the effects of gas composition and temperature on the structure of the

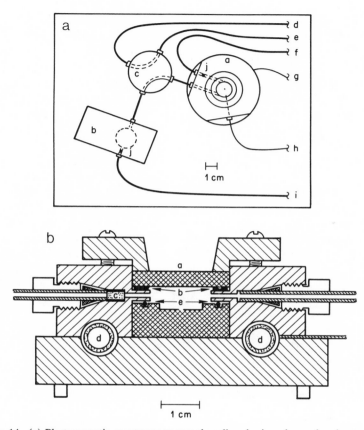

Figure 14. (a) Photoacoustic spectroscopy sample cell and microphone chamber. a, Sample cell; b, microphone chamber; c, isolation valve; d, microphone chamber outlet; e, sample cell inlet; f, sample cell outlet; g, sample heater power line; h, thermocouple; i, microphone chamber inlet; j, 5-m stainless-steel frit. (b) High-temperature photoacoustic spectroscopy; cell. a, KBr window; b, 0.8-mm graphite gasket; c, 5-m stainless-steel frit; d, spiral electric heater; e, hollow metal O-ring. (Reproduced with permission from Ref. 36.)

catalyst or the structure of adsorbates present in the catalyst surface. For such investigations it is necessary to connect the infrared cell to a gas manifold. If the cell is to be evacuated, then it must also be connected to a vacuum pump. The equipment used for such purposes is of a standard design and construction.

The gases contacting the catalyst sample should be pretreated to remove contaminants such as O_2, H_2O, and metal carbonyls. Hydrogen can be purified of O_2 and H_2O by passage through a palladium thimble or by passage through a Deoxo unit (Engelhard) which converts O_2 to H_2O and then through a liquid–nitrogen cooled trap filled with molecular sieve, to remove H_2O. Inert carriers, such as He, Ar, N_2, can also be purified of O_2 and H_2O using the latter technique. Metal carbonyls [e.g., $Fe(CO)_5$, $Ni(CO)_4$, etc.] are often present in high-pressure cylinders used for CO storage. Such substances if not removed will decompose on the catalyst sample, thereby contaminating it. An effective method for removing metal carbonyls is to decompose them in a trap heated to 600 K, filled with glass beads.

6. Spectrometers

Infrared spectra of high-surface-area catalysts can be recorded with either a dispersive or a Fourier-transform spectrometer. Most of the published work done prior to 1970 was carried out using dispersive spectrometers. However, the introduction of reliable commercial Fourier-transform spectrometers in the late 1960s has led to an increasing utilization of such spectrometers for infrared studies. Each type of instrument has its advantages and disadvantages. Since the optical principles governing the operation of dispersive and Fourier-transform spectrometers are different, we will review these prior to making a direct comparison of the two types of instrument.

6.1. Dispersive Spectrometers

The physics governing dispersive spectrometers will be illustrated in terms of the operation of a grating spectrometer. The optical layout for such an instrument is shown in Figure 15. Radiation emitted by the source is divided into two beams. One beam passes through the sample while the other serves as a reference. A rotating sectored mirror combines the two beams to form a single beam consisting of alternate pulses of radiation from the sample and reference beams. The combined beam enters the monochromator through a slit, and is dispersed into its spectral elements by a grating. The radiation appearing at the exit slit of the monochromator is determined by the diffraction relationship

$$n/\bar{v} = 2g \cos \phi \sin \theta \tag{23}$$

where $\bar{v}$ is the wave number for a given spectral element, $n = 0, 1, 2 \ldots$ is the order of the diffraction, g is the spacing between rulings on the grating, ϕ is the angle between the incident and diffracted beam, and θ is the angle of rotation of the grating from the zero order ($n = 0$). For any

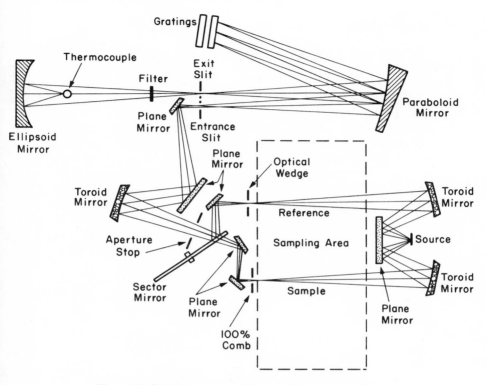

Figure 15. Optical arrangement for a grating spectrometer.

value of θ, equation (23) is satisfied by a number of values of $\bar{\nu}$ corresponding to different values of n. Therefore, radiation of different wave numbers corresponding to different orders of diffraction from the grating emerge from the exit slit. An optical filter is used to reject radiation from all but the desired order of diffraction.

After filtering, radiation from the monochromator is focused onto a thermocouple detector. The alternating signal at the detector is amplified and fed to a servo-motor, which moves a reference beam attenuator to equalize the intensity of the sample and reference beams. The alternating signal is thereby reduced, producing a state of equilibrium. The absorbance of a sample placed in the sample beam is determined by the extent of movement of the reference beam attenuator. It is significant to note that modern dispersion spectrometers eliminate the "comb" servo system and employ instead a digitized ratioing method giving much better reproducibility and performance.

6.2. Fourier-Transform Spectrometers

The principles governing the operation of a Fourier-transform spectrometer can be understood best by following a beam of monochromatic radiation through the optical layout shown in Figure 16. The beam enters the source side of a Michelson interferometer where part of the beam is reflected to the fixed mirror by the beam splitter and part is transmitted to the movable mirror. Following reflection, both beams recombine at the beam splitter. Since the path length of the beam reflected from the movable mirror is in general slightly different from that of the beam reflected from the fixed mirror, the two beams interfere either constructively or destructively. Displacement of the movable mirror at a fixed velocity, v, modulates the beam exiting from the interferometer at a frequency $2v$. For a mirror velocity of 0.2 cm/s, modulation frequencies between 1.6 kHz and 160 Hz are produced over the spectral range of 4000–400 cm^{-1}.

The modulated beam is directed through either the sample or reference side of the sample compartment and is finally focused on the detector. For most midinfrared work, a triglycine sulfate (TGS) pyro-

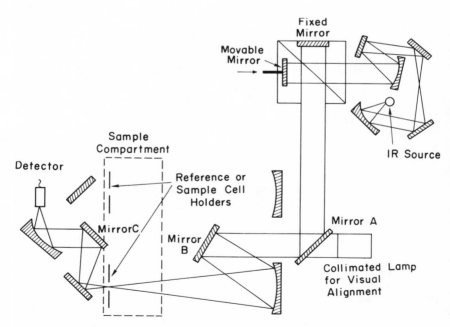

Figure 16. Optical arrangement for a Fourier-transform spectrometer.

electric bolometer is used as the detector because of its very high frequency response (>1 MHz).

The signal observed by the detector is measured as a function of the displacement or retardation of the movable mirror, s, and is known as the interferogram, $I(s)$. To obtain a spectrum $I(\bar{v})$ from a measured interferogram, it is necessary to take the Fourier transform of the fluctuating portion of the interferogram, $\bar{I}(s) = [I(s) - 0.5I(0)]$. Thus,

$$I(\bar{v}) = \int_{-\infty}^{\infty} \bar{I}(s)\cos(2\pi\bar{v}s)\,ds \qquad (24)$$

An illustration of the relationship between $\bar{I}(s)$ and $I(\bar{v})$ for a spectrum consisting of three narrow lines is shown in Figure 17a.

In practice, the Fourier integral indicated in equation (24) can be determined experimentally only over a finite range $(-s_{max} \leq s \leq s_{max})$. As a consequence, the integration can be performed only over a finite range. Figure 17b shows that truncation of the limits of integration results in a broadening of the spectral peaks.

An additional consequence of finite retardation is the appearance of secondary extrema or "wings" on either side of the primary features. The

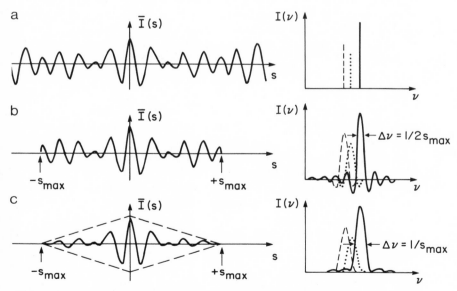

Figure 17. Appearance of the spectrum obtained by Fourier transformation of (a), an infinite interferogram; (b), a finite interferogram; (c), a finite interferogram with triangular apodization.

presence of these features is disadvantageous, especially when it is desired to observe a weak absorbance in proximity to a strong one. To diminish this problem the interferogram is usually multiplied by a triangular apodization function, which forces the product to approach zero continuously for $s = \pm s_{max}$. Fourier transformation of the apodized interferograms produces a spectrum such as that shown in Figure 17c.

Since the acquisition of an interferogram and its subsequent transformation to produce a spectrum requires a large amount of data handling and computation, these tasks are normally carried out with the aid of a dedicated minicomputer. The computer is also used to operate the spectrometer and to display spectra.

6.3. Comparison of Fourier-Transform and Dispersive Spectrometers

6.3.1. Resolving Power

The resolving power of a spectrometer is defined by the quantity

$$R = \bar{v}/\Delta\bar{v} \tag{25}$$

and is determined by either the characteristics of the resolving element in the spectrometer or the finite dimensions of the radiation source as it appears at the resolving element. The resolving power of a grating spectrometer is constrained by the width of the monochromator slits, w, and is given by

$$R_G = \frac{2f}{w}\tan\theta \tag{26}$$

where f is the focal length of the collimating mirror in the monochromator. For a fixed slit width, R_G decreases with increasing wave number (i.e., increasing θ) and as a consequence the resolution decreases. This pattern can be compensated by decreasing the slit width but only at the expense of decreasing the energy flux to the monochromator.

In most instances the resolving power of a Fourier-transform spectrometer is determined by the maximum retardation of the movable mirror in the interferometer so that

$$R_I = \bar{v}s_{max} \tag{27}$$

from equations (24) and (27) it follows that $\Delta\bar{v} = 1/s_{max}$. Consequently, the resolution of a Fourier-transform spectrometer is fixed by the maximum retardation, but the resolving power increases with increasing wave number.

6.3.2. Signal-to-Noise Ratio

The signal-to-noise ratio for both interferometers and monochromators can be determined in an identical fashion. The signal power, S, received at the detector is given by

$$S = B(T, \bar{v})\theta\eta\Delta\bar{v} \qquad (28)$$

where $B(T, \bar{v})$ is the source brightness, which is a function of both the source temperature and the wavenumber $\bar{v}$; θ is the optical throughput; η is the optical efficiency; and $\Delta\bar{v}$ is the resolution. Assuming that the detector itself is the dominant noise source, the noise power, N, is then expressed as

$$N = \text{NEP}/t_m^{1/2} \qquad (29)$$

where NEP is the noise equivalent power of the detector and t_m is time for observation of a single spectral element of width $\Delta\bar{v}$. Dividing equation (28) by equation (29) results in the following expression for the signal-to-noise ratio, S/N:

$$S/N = B(T, \bar{v})\theta\eta\Delta\bar{v}t_m^{1/2}/\text{NEP} \qquad (30)$$

Equation (30) indicates that high S/N's are favored by use of a bright source, high throughput optics, low resolution (i.e., a large value of $\Delta\bar{v}$), long spectral collection times, and a low noise detector.

The advantages of Fourier-transform over dispersive spectrometers can be discussed in terms of equation (30).[37-39] For this purpose, advantage is defined either by the ratio of S/N's for a fixed data-collection time, or by the ratio of data-collection times for a fixed S/N. One of the principal advantages of Fourier-transform over dispersive spectrometers, known as Fellgett's advantage, results from the fact that a Fourier-transform spectrometer observes all spectral elements simultaneously, whereas a monochromator spends only a fraction of the total data-collection time observing each resolution element. Table 2 illustrates several examples of Fellgett's advantage. It is apparent that the multiplexing capabilities of Fourier-transform spectrometers offer, in principle, very significant advantages, particularly in the speed of spectral acquisition. Griffiths et al.[38] have shown that Fellgett's advantage is never fully realized in practice and that the comparison of Fourier-transform and dispersive instruments must take into consideration all of the factors entering into equation (30). When this is done, a Fourier-transform spectrometer still outperforms dispersive spectrometers by a substantial margin. Thus, for a fixed resolution, Fourier-transform instruments offer S/N advantages of 10–10^2 (assuming a fixed data-collection time) and

TABLE 2. Tabulated Values of Fellgett's Advantage

Resolution $\Delta \bar{v}(\text{cm}^{-1})$	Range $\bar{v}_R(\text{cm}^{-1})$	t_m Advantage[a] M^c	S/N Advantage[b] $M^{1/2}$
8	3600	450	21
2	3600	1800	42
0.5	3600	7200	85
0.5	400	800	28

[a] For a fixed S/N, the ratio of the time required to obtain a spectrum by scanning through each optical element to the time required to obtain a spectrum by observing all optical elements simultaneously.
[b] For a fixed time of observation, the ratio of the S/N for a spectrum obtained by observing all spectral elements simultaneously to that obtained by scanning through each spectral element.
[c] $M = \bar{v}_R/\Delta\bar{v}$.

data-collection-time advantages of 10^2–10^3 (assuming a fixed S/N). It must be recognized, of course, that these advantages are determined on the assumption that the full midinfrared range of frequencies (4000–400 cm^{-1}) is of interest. For situations in which information is required from a narrower spectral range, the advantages of Fourier-transform spectroscopy diminish as either M or $M^{1/2}$, where $M = \bar{v}_R/\Delta\bar{v}$.

6.3.3. Data Acquisition, Storage, and Display

Fourier-transform spectrometers require access to an on-line computer in order to permit rapid transformation of interferograms to spectra, and, in fact, the development of commercial instruments was in large part made possible by the introduction of minicomputers in the mid-1960s. The availability of a computer provides a number of additional advantages including data storage on either magnetic tape or disks, comparison and subtraction of spectra, and great flexibility in displaying spectra. moreover, the computer can be used to smooth data, to subtract base lines, and to compute integral absorbances. It should be noted, however, that the same data acquisition and handling functions can be achieved by interfacing dispersive spectrometers to a minicomputer or microprocessor, and that at present there are several commercial grating spectrometers that are fully integrated with such data-handling systems. As a result, the presence of a computer as a part of a Fourier-transform spectrometer cannot be regarded as a feature providing such an instrument with an advantage over dispersive spectrometers.

7. Concluding Remarks

Infrared spectroscopy is a versatile and readily applicable technique for characterizing the structures of high-surface-area catalysts and species adsorbed on such catalysts. Recent developments in cell design permit *in situ* recording of transmission spectra at pressures ranging from 10^{-4} Pa to 30 MPa and temperatures from subambient to 773 K. Cell designs have also been developed for *in situ* studies by means of either diffuse reflectance or photoacoustic spectroscopy. The advent of commercial Fourier-transform spectrometers has made it possible to record spectra in a small fraction of the time required formerly using grating spectrometers. Since Fourier-transform infrared spectra with good S/N can be acquired in 0.1–1 s, it becomes possible to study many processes under dynamic conditions. Such investigations are particularly exciting and hold promise for providing data on the kinetics of adsorption, desorption, decomposition, and reaction of various gases. The dynamics of solid-state reactions can also be studied using similar techniques.

Acknowledgment

This work was supported by the Division of Chemical Sciences, Office of Basic Energy Sciences, U.S. Department of Energy under contract No. DE-ACO3-76SF-00098.

References

1. L. H. Little, *Infrared Spectra of Adsorbed Species,* Academic, New York (1966).
2. M. L. Hair, *Infrared Spectroscopy in Surface Chemistry,* Marcel Dekker, New York (1967).
3. A. V. Kiselev and V. I. Lygin, *Infrared Spectra of Surface Compounds,* Wiley, New York (1975).
4. A. T. Bell and M. L. Hair, eds., *Vibrational Spectroscopies for Adsorbed Species,* ACS Symposium Series 137, American Chemical Society, Washington, D.C. (1980).
5. R. P. Eischens and W. A. Pliskin, *Adv. Catal.* **10,** 1 (1958).
6. G. Blyholder, in: *Experimental Methods in Catalytic Research* (R. B. Anderson, ed.), pp. 323–360, Academic, New York (1968).
7. M. R. Basila, *Appl. Spectrosc. Rev.* **1,** 289 (1968).
8. J. Pritchard and T. Catterick, in: *Experimental Methods in Catalytic Research* (R. B. Anderson and P. T. Dawson, eds.) Vol. III, pp. 281–318, Academic, New York (1976).
9. G. L. Haller, in: *Spectroscopy in Heterogeneous Catalysis* (W. N. Delgass, G. L.

Haller, R. Kellerman, and J. H. Lunsford, eds.), pp. 19–57, Academic, New York (1979).

10. G. L. Haller, *Catal. Rev.-Sci. Eng.* **23**, 477 (1981).
11. J. Peri, in: *Catalysis Science and Technology* (J. R. Anderson and M. Boudart, eds.) Vol. 5, pp. 171–220, Springer-Verlag, New York (1984).
12. P. R. Griffiths and M. P. Fuller, in: *Advances in Infrared Spectroscopy* (R. E. Hester and R. J. H. Clark, eds.) Vol. 9, pp. 63–129, Heyden, London (1981).
13. P. Winslow and A. T. Bell, *J. Catal.* **86**, 158 (1984).
14. G. Duyckaerts, *Analyst* **84**, 201 (1969).
15. R. Kellerman, in: *Spectroscopy in Heterogeneous Catalysis* (W. N. Delgass, G. L. Haller, R. Kellerman, and J. H. Lunsford, eds.), pp. 86–131, Academic, New York, 1979.
16. K. Klier, in: *Vibrational Spectroscopies for Adsorbed Species* (A. T. Bell and M. L. Hair, eds.), ACS Symposium Series 137, pp. 141–162, American Chemical Society, Washington, D.C. (1980).
17. P. Kubelka and F. Munk, *Z. Tech. Phys.* **12**, 593 (1931).
18. M. J. D. Low and G. A. Parodi, *Appl. Spectrosc.* **34**, 76 (1980).
19. Y.-H. Pao, ed., *Optoacoustic Spectroscopy and Detection*, Academic, New York (1977).
20. A. Rosencwaig, *Photoacoustics and Photoacoustic Spectroscopy*, Wiley, New York (1980).
21. A. Rosencwaig and A. Gersho, *J. Appl. Phys.* **47**, 64 (1976).
22. Y. C. Teng and B. S. H. Royce, *Appl. Opt.* **21**, 77 (1982).
23. J. R. Anderson, *Structure of Metallic Catalysts*, Academic, New York (1975).
24. J. T. Yates, Jr., T. M. Duncan, S. D. Worley, and R. W. Vaughn, *J. Chem. Phys.* **70**, 1219 (1979).
25. P. B. Weisz and C. D. Prater, *Adv. Catal.* **6**, 143 (1954).
26. J. Harrod, R. W. Roberts, and E. F. Rissman, *J. Phys. Chem.* **71**, 343 (1967).
27. N. D. Parkyns and J. B. Patrick, *J. Sci. Instrum.* **43**, 695 (1966).
28. G. Blyholder and L. D. Neff, *J. Phys. Chem.* **66**, 1464 (1962).
29. T. R. Hughes and H. M. White, *J. Phys. Chem.* **71**, 2192 (1967).
30. G. L. Griffin and J. T. Yates, Jr., *J. Catal.* **73**, 396 (1982).
31. D. L. King, *J. Catal.* **61**, 77 (1980).
32. R. F. Hicks, C. S. Kellner, B. J. Savatsky, W. C. Hecker, and A. T. Bell, *J. Catal.* **71**, 216 (1981).
33. J. F. Edwards and G. L. Schraeder, *Appl. Spectrosc.* **35**, 559 (1981).
34. J. M. L. Penninger, *J. Catal.* **56**, 287 (1979).
35. I. M. Hamadeh, D. King, and P. R. Griffiths, *J. Catal.* **88**, 264 (1984).
36. S. J. McGovern, B. S. H. Royce, and J. B. Benziger, *Appl. Surf. Sci.* **18**, 401 (1984).
37. P. R. Griffiths, *Chemical Infrared Fourier Transform Spectroscopy*, Wiley, New York (1975).
38. P. R. Griffiths, H. J. Sloane, and R. W. Hannah, *Appl. Spectrosc.* **31**, 485 (1977).
39. A. T. Bell, in: *Vibrational Spectroscopies for Adsorbed Species* (A. T. Bell and M. L. Hair, eds.), ACS Symposium Series 137, pp. 13–36, American Chemical Society, Washington, D.C. (1980).

4

Inelastic Electron Tunneling Spectroscopy

Paul K. Hansma

1. Principles of the Method

1.1. A Water Analogy

Inelastic electron tunneling spectroscopy,[1,2] also known as IETS or tunneling spectroscopy, is a sensitive technique for measuring the vibrational spectra of molecules. At present, it is particularly well suited to measuring the vibrational spectra of a monolayer or submonolayer of molecules adsorbed on aluminum or magnesium oxide or on metal particles or metal complexes that are themselves supported on aluminum or magnesium oxide. In the future, it may be possible to extend it to a much wider range of systems with the use of mechanically adjusted tunnel junctions.

A water analogy[3] is helpful in understanding how tunneling spectroscopy works. Consider the tube shown in Figure 1. It has two open channels separated by height h. If we measure the steady-state flow as a function of pressure, we will find two components: (1) a steadily increasing flow through the bottom channel and (2) a flow through the top channel, which has a threshold pressure $\rho g h$, that steadily increases thereafter.

Thus, the total flow F has a kink at the characteristic pressure

Paul K. Hansma • Department of Physics, University of California, Santa Barbara, California 93106.

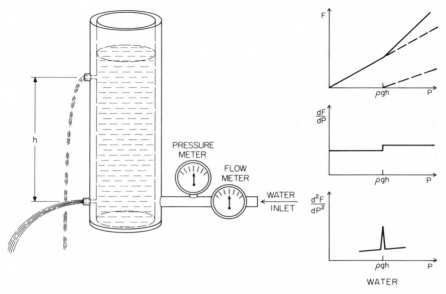

Figure 1. The basic physics behind tunneling spectroscopy can be understood with the help of a water analogy. The flow-versus-pressure curve has a kink when it becomes possible for water to flow out of the upper channel. This kink becomes a step in the first derivative and a peak in the second derivative.

$P = \rho g h$ as shown in Figure 1. This kink might be difficult to observe. It would be easier to see the step in dF/dP and even easier to see the peak in d^2F/dP^2 if we could find a way to measure them.

A key point is that, even if the tube were covered by a basket, we could determine the height of the opening by looking at the d^2F/dP^2 versus P curve: an opening at height h is revealed by a peak at pressure $P = \rho g h$.

Figure 2 shows an idealized view of a tunnel junction. Here, we measure the current as a function of voltage. Again, we will find two components: (1) a steadily increasing current due to elastic electron tunneling and (2) a current, which has a threshold voltage $h\nu/e$ that increases steadily thereafter, due to inelastic electron tunneling. This threshold is set by the requirement that the electrons must give up an energy $h\nu$ to excite the molecular vibration. Since their tunneling energy is eV, we must have $eV \gtrsim h\nu$.

Figure 2 also shows the total current I, which is the sum of the current through the elastic and inelastic tunneling channels. It has a kink at $V = h\nu/e$ which becomes a step in dI/dV versus V and a peak in

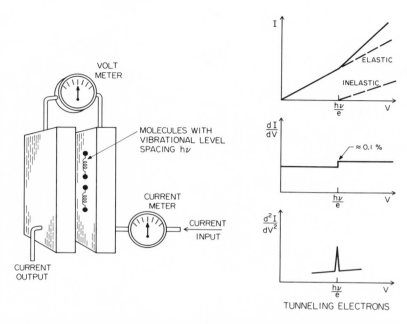

Figure 2. In this schematic view of a tunneling junction, the molecules with vibrational level spacing $h\nu$ are sandwiched between two metal electrodes. In an actual tunnel junction the metal electrodes would be separated by less than 100 Å. Current versus voltage has a kink when the inelastic electron tunneling channel opens up. This kink becomes a step in the first derivative and a peak in the second derivative.

d^2I/dV^2, just as for the water analogy. A plot of d^2I/dV^2 versus V is called a tunneling spectrum. [More generally, d^2V/dI^2 is actually measured. It has peaks in the same locations since $|d^2V/dI^2| = |d^2I/dV^2(dV/dI)^3|$ and $(dV/dI)^3$ is a smooth, relatively slowly varying junction.]

Thus, a tunneling spectrum reveals the vibrational energies of molecules included between the metal electrodes, since a vibrational energy of $h\nu$ results in a peak at $V = h\nu/e$. A real tunneling spectrum has many peaks since the typical molecules that are studied have many vibrational modes. For example, Figure 3 shows the tunneling spectrum of a monolayer of formate ions on alumina[4] Each of the sharp peaks in this spectrum corresponds to the opening of an inelastic electron tunneling channel. Each channel corresponds to a particular molecular vibration.

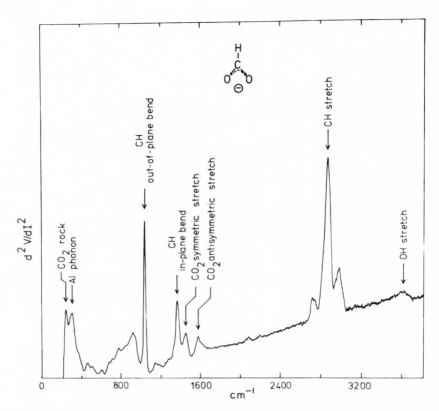

Figure 3. Tunneling spectra of Al–I–Pb sandwich doped with formic acid (see Ref. 4). The x axis is labeled with cm^{-1} for convenience in making comparisons with optical spectra. The conversion factor from the measured voltage is $1\,meV = 8.065\,cm^{-1}$.

1.2. Spectral Range, Sensitivity, Resolution, and Selection Rules

The spectral range of tunneling spectroscopy includes all molecular vibrations. It extends from 0 to beyond $4000\,cm^{-1}$. The easiest part of this range to study is the part from 400 to $4000\,cm^{-1}$. This is consequently the most common range for published tunneling spectra: it is the spectral range covered in most of the spectra in this chapter. With some difficulty, the range from 0 to $400\,cm^{-1}$ can be explored, as shown in Figure 4.

The sensitivity of tunneling spectroscopy is sufficient to see a fraction of a monolayer of material over a junction area less than $1\,mm$ on a side.[5–7] In fact, almost all of the tunneling spectra in this chapter are

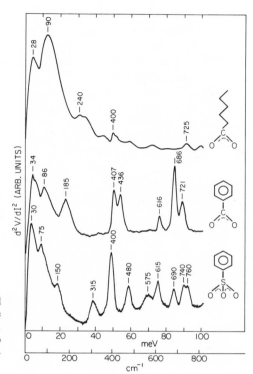

Figure 4. The low-energy vibrational modes of chemisorbed species can be studied with tunneling spectroscopy. These modes, which are inaccessible to most vibrational spectroscopies, remain largely unexplored.

taken from approximate monolayer coverage; one monolayer is just about the right amount of adsorbate for a tunneling spectrum.

The resolution in typical tunneling spectra ranges between 1 and 4 MeV (10 and 32 cm^{-1}). This is not, however, the maximum resolution that can be achieved. Professor Walmsley and co-workers,[8] by using a helium dilution refrigerator and small modulation voltages, have been able to set the current world record in the resolution of tunneling spectroscopy at roughly 2 cm^{-1}. This high resolution is not commonly used (even by Professor Walmsley) because the relative trace time—that is, the time to obtain a spectrum with a same signal-to-noise ratio—rises rapidly as the resolution improves. Thus, most spectra are taken at the poorest resolution sufficient to obtain the information desired. Fortunately, since the vibrational modes of surface adsorbed molecules typically have an intrinsic width of 1 meV or greater, the ultimate in resolution is not required.

Basically there are no strong selection rules in tunneling spectroscopy. Selection rules are the general consequence of symmetry. In the case of optical spectroscopy, selection rules arise because the wavelength of light is very long relative to the size of molecules. Thus, the electric field of the light is uniform over molecular dimensions. This is not true for tunneling spectroscopy: both Raman active and infrared active modes appear in tunneling spectra.[9-11]

There is some dependence of intensity on orientation[12,13] (favoring orientation perpendicular to the surface), but this is considerably less pronounced than for either electron energy loss or optical spectroscopy. Note, for example, that the CO_2 symmetric and antisymmetric stretching modes in Figure 3 occur with comparable intensity.

2. Description of Typical Apparatus

2.1. Apparatus for Sample Preparation

Sample preparation is the most critical part of tunneling spectroscopy research. After the apparatus for tunneling spectroscopy is assembled, researchers typically find most of their experimental time and energy is spent in sample preparation. Though measuring the spectrum involves cryogenics and low-noise electronics, it is relatively straightforward once the necessary equipment is assembled. Sample preparation remains a challenge. Therefore, these steps will be covered in some detail here. Further information can be found in books[14,15] and review articles[16-21] on tunneling spectroscopy and in review articles on electron tunneling.[22-24]

Many types of smooth insulating substrates have been used for tunneling spectroscopy. These include fused alumina substrates, sapphire substrates, silicon substrates, ordinary glass microscope slides, and cover glasses. It really makes no difference to the experimental results what substrate is used. Therefore, choice of substrates is based on convenience to the experimenter.

Cleaning substrates is an important part of sample preparation. Good junctions for tunneling spectroscopy cannot be obtained on dirty substrates. Many failures to obtain good tunneling junctions can be traced to dirty substrates.

Though there are roughly as many procedures as there are experimentalists, a few observations can be made:

1. Mechanical scrubbing with a brush, high-pressure water, or fingers using a detergent solution seems essential.
2. Thorough rinsing with water followed by deionized water is necessary and takes longer than people first expect.

3. Various organic solvent rinses are used by some groups but not by others.
4. The final rinse should be blown or spun off the substrates. The quantity of solvent that evaporates to completion should be minimized.

The clean substrates are then put into a vacuum evaporator. Many types of vacuum evaporators have been successfully used to make electron tunnel junctions for tunneling spectroscopy. The basic requirement is that it be relatively clean from organic contaminants. Experience has shown, that even inexpensive oil diffusion pumped systems with mechanical pumps as backing pumps can be maintained at the desired degree of cleanliness. Most groups have, however, found that common facilities cannot be maintained at the desired degree of cleanliness. In other words, regarding the vacuum evaporator, "Be it ever so simple, it should be your own."

Figure 5 shows a schematic view of junction fabrication inside the vacuum evaporator. First, an aluminum strip is evaporated onto the substrate through a mask that defines its shape. (The source of aluminum is typically aluminum wire wound around a stranded tungsten filament.) Next, the aluminum is oxidized in air or in an oxygen glow discharge. Typical parameters would be 1 min of oxidation in clean laboratory air or 10 min of oxygen glow discharge oxidation[25,26] at a pressure of 50 mTorr with currents of 10 mA.

Next, the junction is doped with organic molecules. There are four basic techniques for doing this: (1) gas-phase doping in which the oxidized aluminum is exposed to a vapor or molecular beam of the desired dopant,[1,2,27] (2) liquid-phase doping in which a drop of solution containing the desired dopant is put onto the substrate and the excess spun or blown off,[28–31] (3) infusion doping in which the dopant is infused through the electrode of a completed junction,[32–34] and (4) picomole doping in which a small drop of dilute solution is evaporated to dryness on the junction.[35] Figure 6 shows spectra from tunnel junctions doped with a compound from the vapor and from several liquid solutions. Though there are subtle and intriguing differences, the similarities are clear.

Finally, the junction is completed with an evaporated lead electrode. Though more than 90% of all inelastic tunnel junctions are made with an aluminum first electrode and lead second electrode, it should be emphasized that other materials have been used with varying degrees of success. Magnesium works fairly well as a first electrode since it also oxidizes to give a good tunneling barrier.[36,37] Other metals that will

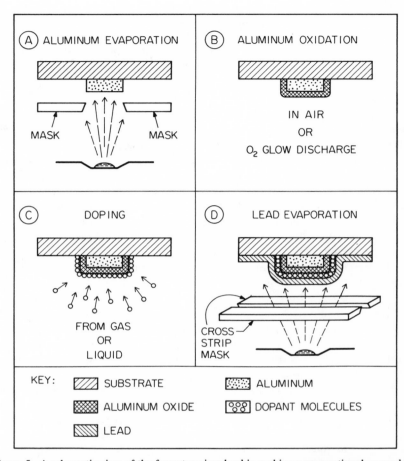

Figure 5. A schematic view of the four steps involved in making a conventional crossed film junction for tunneling spectroscopy. The masks are shown away from the substrate for clarity. In practice, the aluminum mask touches the substrate and the lead mask is held of order 0.1 mm from the substrate by dimpling it with a center punch.

work for the second electrode include thallium, indium, tin, silver, and gold, roughly in that order of difficulty. Figure 7 shows a slide with five aluminum strips on one side and a slide with five completed junctions that are formed at the intersections of the crossed strips on the other side. A schematic view of a single one of these junctions is shown in Figure 8.

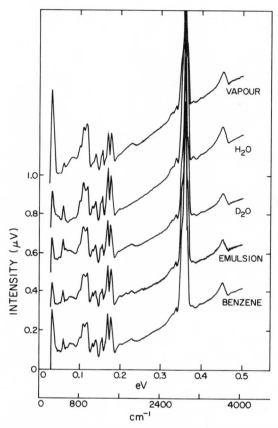

Figure 6. A comparison of 1,3-butanediol doped onto oxidized aluminum with various techniques. Note the similarity of the vapor-doped junction with those doped from various liquid solutions. The similarity of the liquid-doped junctions suggests that the solvent neither modifies the chemisorbed layer nor remains after doping (see Ref. 31).

2.2. Apparatus for Measuring Spectra

For good resolution, tunneling spectroscopy must be done at liquid helium temperatures.[38] After the junctions are inserted into liquid helium, it is important to measure the $I-V$ characteristics within a few millivolts of zero voltage to establish that the junctions are good tunnel junctions. This is done by looking at the structure due to the superconducting lead energy gap. (The origin of this structure is discussed in detail

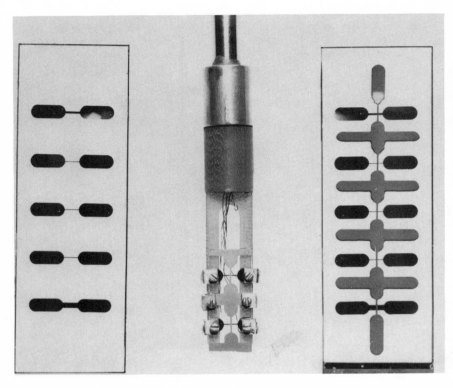

Figure 7. On the left is a 1 × 3-in. glass microscope slide with five aluminum strips ready for oxidation and doping. On the right five completed junctions are formed at the intersections of the aluminum bottom metal electrode and the lead top metal electrode. After measuring the resistances with an ohmmeter we cut out the pair that we most want to run and mount it with brass screw clamps on a probe as shown in the center. The probe is then inserted down the neck of a helium storage Dewar for measurements.

by, for example, Giaever[22] and McMillan and Rowell.[23]) For our purposes, it is only important to note that the conductance at zero voltage should be approximately $\frac{1}{7}$ the conductance far away from zero voltage. Thus, the slope of the $I-V$ curve should be about $\frac{1}{7}$ as large at zero voltage as it is for voltages far from zero voltage. If the ratio is closer to unity, it indicates current flow in channels other than tunneling channels or excessive noise that is smearing the characteristic.

After the $I-V$ characteristics are measured, it is time to measure tunneling spectra. Though it is possible, and for some experiments desirable, to build refined measuring circuits[39] including bridge

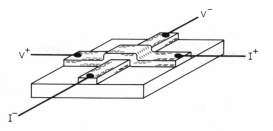

Figure 8. A schematic view of the
four-terminal measurement used in
tunneling spectroscopy to minimize
the effect of strip resistance.

circuits,[40–43] a very simple measurement apparatus will suffice for most
experiments—if high-quality components are used.

As shown in Figure 9, all that is really necessary is to apply a slowly
varying current to the junction with a slowly varying voltage in series with
a resistor and to apply an ac modulation current from a low-distortion
oscillator in series with a large resistor. It is, of course, important to have
only one ground in the circuit. Two common choices for where to put the
ground are (1) at one side of the slowly varying voltage source or (2) at
one side of the voltage measurement apparatus. Either way, the ac

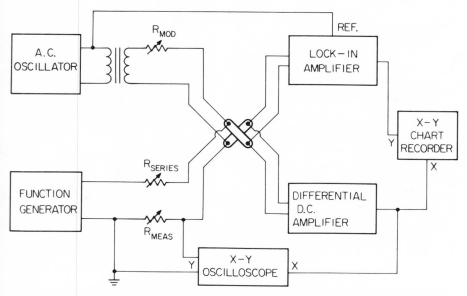

Figure 9. If high-quality components are used, tunneling spectra can be taken with a
minimal amount of noncommercial equipment: just resistors and an isolation transformer.
Careful attention to shielding and grounding is necessary, however, because of the low-level
signals.

Figure 10. A tunneling spectrum of the junction in the liquid helium storage Dewar is being plotted on the $x-y$ chart recorder. The apparatus is inside an optional, but highly desirable, screened room.

source must be isolated from ground. This can be done with a transformer.

Though the resistors are shown as variable resistors, our experience is that a decade switch with metal film resistors is preferable to a potentiometer. A potentiometer has inductance and becomes electrically noisy with time.

The dc and ac voltages across the sample must be measured. The ac voltage is measured with a lock-in amplifier. A lock-in amplifier with a notch filter at the first harmonic frequency (or bandpass filter at the second harmonic frequency) can measure the second harmonic voltage across tunnel junctions with no external filtering. However, resonant circuits,[39] such as used originally by Lambe and Jaklevic,[2] can give an appreciable boost in signal-to-noise ratio. The price one pays is uncertainty in the absolute magnitude and phase of the second harmonic signal.

The dc voltage should be applied through a buffer amplifier to the x axis of a chart recorder. We have found it important to use a buffer amplifier since inputs of chart recorders are usually very noisy electrically and this noise is directly across the input to the lock-in amplifier.

Figure 10 shows electronics hooked-up to measure a tunneling spectrum which is being traced out on the x–y chart recorder. Though the circuit of Figure 9 is conceptually simple, the necessity for good shielding and grounding complicates its execution (though, as the careful reader can see, not all of the boxes in the figure are in use). The electronics is connected to the top end of a sample holder, which is immersed in a liquid helium Dewar direct from the vendor. The bottom end of this sample holder with the junction attached was shown in Figure 7.

3. Calibration, Artifacts, and Miscellaneous Problems

3.1. Calibration

Fortunately, the calibration of tunneling spectra is trivial. The peak position is a dc voltage in the range 0–$\frac{1}{2}$ V that can be conveniently and accurately measured by readily available digital voltmeters. Most commonly, the spectrum is plotted with an x–y chart recorder. One or more calibration marks can be made on this plot by stopping the trace at a particular voltage, moving the pin up and down, and noting the value the voltage as measured with the digital voltmeter.

The relative peak intensity can, of course, be seen directly from the trace. Less commonly, the absolute peak intensity is of interest. If it is, it can be obtained from the following formula for the fractional change in the conductance due to the opening up of a particular inelastic tunneling channel:

$$|\Delta(dI/dV)_k/(dI/dV)_r| = 2\sqrt{2}V_{\omega,r}\int_{\text{over peak}} V_{2\omega}(1/V_\omega)^3\,dV$$

where $\Delta(dI/dV)_k$ is the change in the conductance due to the opening of an inelastic tunneling channel, dI/dV_r is the conductance at some reference voltage, $V_{\omega,r}$ is the rms modulation voltage at the reference voltage, $V_{2\omega}$ is the rms second harmonic voltage (measured from the background level), V_ω is the rms first harmonic voltage, and the integral is over the area of a peak. In practice V is a slowly varying function of voltage and so approximate values of the fractional conductance change due to the opening of an inelastic tunneling channel can be obtained by simply measuring peak areas. A more detailed description of calibration in digitized systems is contained in a book chapter by Adler.[44]

3.2. Artifacts

Tunneling spectroscopy is relatively free of artifacts but there are two that can creep in. The first consists of a noisy patch in the spectrum: a patch in which the apparent background noise level increases dramatically and then decreases again. It looks like a broadening of the base line noise. The peaks following this broadened region have reduced resolution. The broadened region is believed to be due to the superconducting to normal transition of the top lead electrode due to excess heat dissipation in the junction. It can be moved to beyond the region of interest in a tunneling spectrum, beyond $\frac{1}{2}$ V, by increasing the junction resistance. In practice, a junction resistance of order $100\,\Omega$ for a 1×1 mm junction is sufficient to move this superconducting to normal transition beyond the region of interest.

The second artifact is less well understood. It consists of a peak followed by a dip that is roughly equal in area to the peak. It is commonly called a microshort and can be observed to move with temperature. Though these are not well understood, they can be easily distinguished by their shape. In some cases, they can be "burned out" by cycling the junction to high voltage before running a trace. In practice, they seldom occur for junctions with good superconducting energy gap structure.

3.3. Junction Geometry

The most serious limitation of tunneling spectroscopy is that it can only be performed with tunnel junctions! Figure 11 shows that the unknown layer, the layer that can be identified with tunneling spectroscopy, must be sandwiched between metal electrodes together with an excellent quality tunneling barrier. The only systems that can be studied are systems that can be modeled within this constraint. To date, the only tunneling barriers that have proved to be satisfactory for routine use are aluminum and magnesium oxides. Thus, almost all tunneling spectra recorded to date are of molecules adsorbed directly on these oxides or molecules adsorbed on cluster compounds or small metal particles that are themselves supported by these oxides.

There is hope for the future, however. Mechanically adjusted tunnel junctions[45-52] in which the two electrodes are spaced mechanically rather than by an oxide will vastly broaden the applicability of tunneling spectroscopy. The key technological barrier, which is being chipped away at rapidly, is obtaining the necessary stability in the spacing of the two electrodes. This will be discussed in more detail in Section 6 of this chapter.

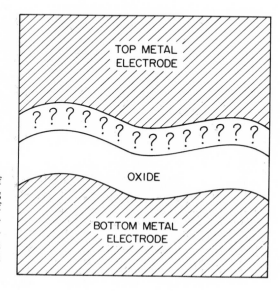

Figure 11. This schematic view of the barrier region of a tunneling junction illustrates a constraint of tunneling spectroscopy. The unknown layer, symbolized by question marks, must be part of an excellent tunneling barrier and covered by top metal electrode.

3.4. Top Metal Electrode

It is somewhat surprising that tunneling spectra can be observed at all. As Figure 11 shows, the unknown layer is covered with an evaporated top metal electrode. Why does the top metal electrode not destroy the unknown layer? It turns out that many top metal electrodes will indeed destroy the unknown layer. For example, junctions with a top electrode of aluminum, chromium, or most of the transition metals have tunneling spectra that bear little or no resemblance to spectra of the original molecules that were doped into the junction. Presumably the top metal electrode has damaged the layer of molecules. Fortunately, this does seem to be the case for lead, thallium, and, under some conditions, indium, tin, silver, and gold. For these metals, tunneling spectra are obtained that closely resemble infrared and Raman spectra of surface adsorbed molecules without a top electrode. There are, however, some shifts due to the top metal electrode.[53-57]

These shifts have been the topic of several investigations. The magnitudes of the shifts are correlated with the magnitude of the

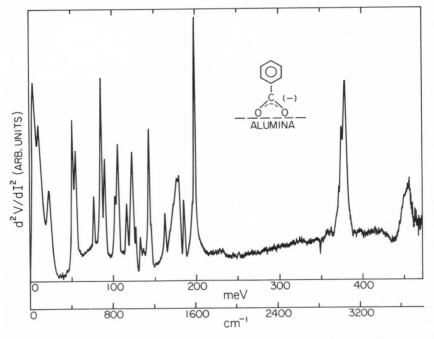

Figure 12. A differential tunneling spectrum[43] of benzoate ions on alumina. The top of the sharp peak near 198 meV is shown in the next figure.

oscillating dipole moment involved in a vibration.[53,54] The shifts are the largest for the largest oscillating dipole modes. Thus, somewhat ironically, tunneling spectroscopy has the most trouble with the vibrations that are the most easily seen with infrared and electron energy loss spectroscopy: in particular, the vibrations of carbon monoxide on supported metals. For hydrocarbons, the shifts are relatively small: less than 1% in general.[53,54] For example, Figure 12 shows a differential tunneling spectrum of benzoate ions adsorbed on aluminum oxide. Kirtley *et al.* examined the peak shifts of various peaks with different top metal electrodes[54] and found results like those displayed in Figure 13. This shows the top of the sharpest peak in the spectrum with the voltage axis much expanded. Note that the tunneling results with a lead top electrode are shifted by only a few wave numbers from the published optical results

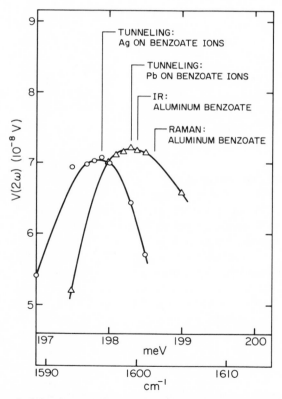

Figure 13. The top of the sharp peak near 198 meV from the spectrum shown in the last figure must be expanded to see the small frequency shifts due to the top metal electrode. Similar results were found for other peaks in the spectrum (see Ref. 54).

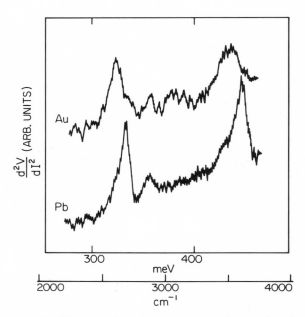

Figure 14. Larger peak shifts, of order a few percent, can be seen for vibrational modes with larger oscillating dipole derivatives such as the OD and OH stretching modes near 330 and 450 cm^{-1} (see Ref. 53).

(which should be the same, illustrating calibration difficulties with optical results). Note that with a silver top electrode there are larger shifts from the optical results but still less than one percent.

Figure 14 shows the larger shifts for a vibrational mode with larger oscillating dipole moment: an OH group.[53] Here we see shifts down to of order 446 meV for a lead top metal electrode and to 433 meV for a gold top metal electrode from the value of roughly 455 meV that could be expected with no top metal electrode.

Finally, Figure 15 shows the large and presently not understood changes in the spectrum of a molecule with very large oscillating dipole moment: carbon monoxide. Note that as the top metal electrode is changed not only are there large shifts but the spectrum changes its character.[55,56] As another example of pathology associated with large oscillating dipole moments, the N≡C stretching mode of methyl iso-cyanide adsorbed on alumina-supported rhodium particles appears as a *dip* rather than a peak in tunneling spectra.[55] It would seem that other spectroscopies would be preferable for these systems.

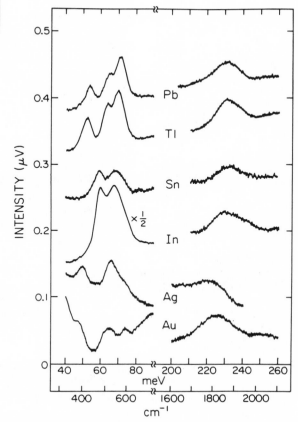

Figure 15. The effect of the top metal electrode can be much more dramatic in exotic systems. For CO, which has a very large dipole derivative, on small iron particles on alumina, the low-frequency vibrations change their character dramatically in going between pairs of metals with similar atomic radii (see Ref. 55). Only for Pb have the vibrations been analyzed with isotopic shift measurements. For Pb the modes near 436 and 519 cm^{-1} are primarily bending; the modes near 569 and 1856 cm^{-1} are primarily stretching, the lower one being primarily metal vs. CO stretching, and the upper one being primarily C vs. O stretching (see Ref. 95).

3.5. Cryogenic Temperatures

Cryogenic temperatures are necessary to obtain good resolution in tunneling spectra.[38] This can be a fundamental limitation. For example, tunneling spectra cannot be obtained under reaction conditions for most catalytic reactions. The reaction must be carried out, and whatever is on the surface frozen to low temperatures before a spectrum can be obtained. It is also a day-to-day nuisance to order and preserve liquid helium. This day-to-day nuisance can, however, be minimized with a type of sample probe displayed in Figures 7 and 10 since helium does not have to be transferred from the storage Dewar in which it is received from the vendor. The samples can be simply slipped down its neck.

4. Data Interpretation, Theory

4.1. Peak Position

There is, of course, a rich lore of theory that has been developed to interpret vibrational mode energies. About the only thing necessary for tunneling spectroscopy is to note that 1 meV is equivalent to 8.0655 cm^{-1}. One detail is that peaks are shifted up in energy by a fraction of the superconducting energy gap if one of the electrodes is superconducting. This shift is well understood and can be easily calculated.[54] For example, at 4.2 K with 2 mV modulation voltage, the shift is 0.7 meV. Shifts for other conditions are tabulated in Ref. 58. The downshifts due to the top metal electrode have been discussed previously in Section 3.4.

4.2. Peak Widths

In tunneling spectroscopy, there are two contributions to vibrational peak widths in addition to the natural width: modulation voltage broadening and thermal broadening.[17] As shown in Figure 16, the modulation voltage broadening has a full width at half maximum of 1.7 times the rms modulation voltage. The full width at half maximum of thermal broadening is 5.4 kT/e. In practice, these two broadening contributions both occur. Their combined effect[58] in most published tunneling spectra ranges from 3.9 meV for a modulation voltage of 2 mV at a temperature of 4.2 K down to 1.3 meV for a modulation voltage of 0.7 mV at 1 K. Thus, the peak widths due to these two broadening mechanisms range from roughly 10 to 30 cm^{-1}.

Not much work has been done on the interpretation of the natural peak width observed in tunneling spectroscopy. In part, this is due to uncertainties introduced by the top metal electrode.[53–57]

4.3. Peak Intensities

There have been a number of theories for the intensity of peaks in tunneling spectra. The two most widely used theories are due to

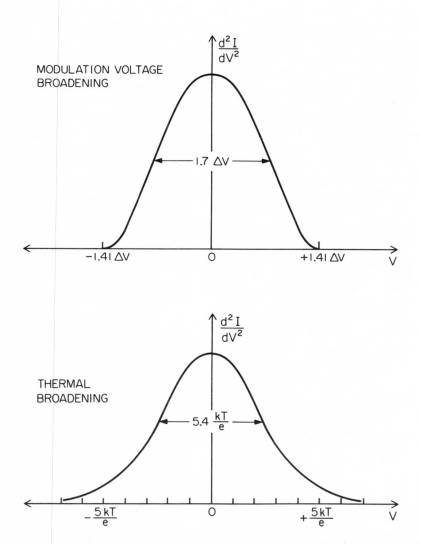

Figure 16. If there were a vibrational mode with negligible intrinsic width in a tunneling system, it would appear to have width from two contributions: (1) modulation voltage broadening that results from the modulation voltage applied to measure the derivative and (2) thermal broadening that results from the smearing of the electrons around the Fermi energy at nonzero temperature.

Scalapino and Marcus[59] and to Kirtley *et al.*[60] The theory of Scalapino and Marcus treats a molecular vibration as an oscillating dipole in the barrier and calculates the modification in the tunneling current due to the modification in the effective tunneling barrier caused by the presence of the molecule. This theory was later extended by Lambe and Jaklevic[2] to include the interaction of tunneling electrons with Raman active molecular vibrations. The theory of Scalapino and Marcus as extended by Lambe and Jaklevic was successful in that it could be used to calculate the order of magnitude of peaks seen in a typical tunneling spectrum. It has, however, several difficulties: (1) It assumes that the transverse electron momentum is conserved. In fact, inelastic scattering by an impurity usually changes the transverse electron momentum. The inclusion of the "off-axis scattering" substantially modifies selection rules. (2) It does not include the effect of the electron energy lost in exciting a molecular vibration on the tunneling probability of that electron. (3) It depends on a cutoff r_0 which is difficult to determine. How would we actually calculate relative intensities for the various vibrational modes of a relatively complex molecule? Is r_0 the same for each mode? (4) Finally, it implicitly assumes localized tunneling electrons by first computing $I_i(r_\perp)$ and then integrating over $r_\perp$.

It was natural to turn to the transfer-Hamiltonian formalism originated by Bardeen[61] to formulate a "second generation theory". The transfer-Hamiltonian formalism has been successfully used to explain many tunneling phenomena. Cohen, Falicov, and Phillips[62] used it to explain the presence of energy gaps in tunnel junctions with a superconducting electrode. Josephson won the Nobel Prize for using it to predict the Josephson effect. It can be used for inelastic electron tunneling.

The transfer-Hamiltonian formalism begins with the assumption that the Hamiltonian for a metal–insulator–metal junction can be written as a sum of three terms:

$$H = H_L + H_R + H_T$$

where H_L and H_R are the Hamiltonians that describe the electrons in the left and right metal electrodes. H_T is the transfer Hamiltonian that describes the tunneling of electrons from one electrode to the other. The wave functions for the electrons in each metal electrode are solutions to H_L and H_R.

This procedure was used by Kirtley *et al.* in developing the theory[60] that has been most extensively used for actually predicting or interpreting peak intensities. It was, for example, used to fit the intensities of methanesulfonate ions on alumina.[13] From the fit, it was concluded that this molecule bonds with its C–S bond normal to the surface. Recently,

Hipps and Knochenmuss have proposed modifications to the theory to better represent the oscillating charges present in a vibrating molecule.[63] It can be hoped that these modifications, which do not increase the complexity of calculation, will allow better theoretical understanding of peak intensities.

It is, however, still difficult to calculate peak intensities and there are still some theoretical uncertainties in fitting parameters that must be used. Thus, as for infrared, Raman, and electron energy loss spectroscopy, peak intensities are not commonly calculated from first principles.

Peak intensities can be used to get an indication of surface concentration. Figure 17 shows the results of an experimental test of the linearity of peak position on surface concentration. In this experiment, the surface concentration of benzoate ions was varied by dropping a

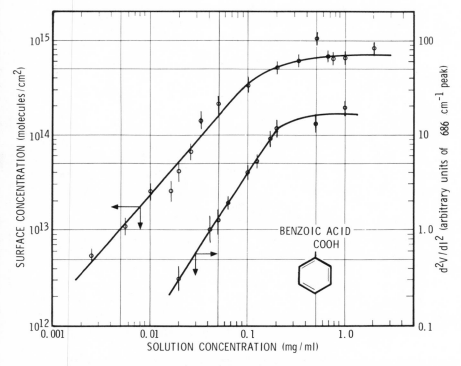

Figure 17. Intensity of the 686-cm^{-1} peak of the spectrum shown in Figure 12 is shown as a function of a solution concentration. Also shown is the surface concentration as a function of solution concentration. Note that the tunneling peak height increases and then saturates following the increase and then saturation of the surface concentration at approximately one molecule per 15 Å^2: monolayer coverage (see Ref. 64).

solution of radioactively labeled benzoic acid onto the oxidized aluminum. As the concentration of the solution was varied, the concentration
of benzoate ions deposited on the alumina varied. It could be experimentally measured by counting the radioactive molecules in a scintillation
counter. Note that the surface concentration rose roughly linearly with
solution concentration until it saturated at monolayer coverage for
solutions of order 0.2 mg/ml. The intensity of a peak in the tunneling
spectrum was also measured as a function of solution concentration. Note
that it also rose with solution concentration and that it also leveled off for
solution concentrations above 0.2 mg/ml. The curve was not, however,
parallel to the curve for surface concentration for smaller values of
solution concentration. In particular, the tunneling peak height fell more
rapidly than the surface concentration. Though there is a simple
theory[64] for this effect based on the tendency of tunneling electrons to
flow preferentially through undoped regions of the junction, the simple
fact is that tunneling peak intensity cannot be depended upon to be a
linear measure of surface concentration.

5. Examples of the Method's Use

5.1. Overview of Applications

Much of the recent interest in tunneling spectroscopy can be traced
to the pioneering work of R. V. Coleman. He showed that not
just a few special molecules but a wide range of organic molecules
including large molecules of biological interest could be studied with
tunneling spectroscopy. For example, Figure 18 shows the spectra
obtained by his group for the pyrimidine base uracil and for uracil
monophosphate.[65,66] His group has obtained spectra of many amino
acids,[67] the other pyrimidine and purine bases,[65,66] and nucleotides and
nucleosides.[65,66,68]

Spectra of inorganic ions have been obtained by Hipps and Mazur.
For example, Figure 19 shows their results for the ferrocyanide ion[69]
obtained by liquid doping of an Al–AlO$_x$–Pb junction with Fe(CN)$_6^{-4}$.
The figure also shows infrared and Raman spectra of comparable species.
Note that the tunneling spectrum has more bands than the infrared
structure and has significant differences from the Raman spectra. Hipps
and Mazur went on to dope their junctions with isotopically substituted
Fe(CN)$_6^{-4}$ ions to pin down the identification of the vibrational
modes.[69,70] One particularly interesting feature of the assignments that
they proposed as a result of these studies is that the metal–ligand modes

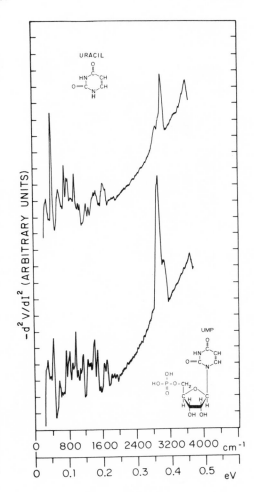

Figure 18. IET spectra of 5'-UMP (695 Ω) (lower spectrum) and uracil (56 Ω) (upper curve). Both spectra were recorded at 4.2 K from Al–AlO$_x$–Pb junctions doped from H$_2$O solutions (see Refs. 65 and 66).

observed are primarily metal–ligand bending modes rather than metal–ligand stretching modes.

Tunneling spectroscopy has also been used for studying radiation chemistry. The strength of tunneling spectroscopy for these studies is the ability to see changes in a monolayer of molecules under irradiation. Experiments with ultraviolet irradiation were performed by Coleman and co-workers.[66] Experiments with electron beam irradiation have been performed by Parikh and co-workers.[71–73] For example, Figure 20 shows their results for electron beam exposure of a monolayer of β-D-fructose.[72] Note that some peaks decrease in intensity during exposure

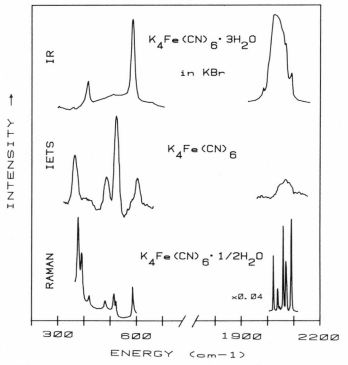

Figure 19. Raman, tunneling, and ir spectra of the ferrocyanide ion (see Ref. 69).

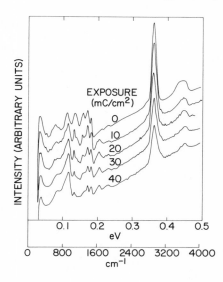

Figure 20. Tunneling spectra of junctions containing β-D-fructose irradiated by 30-keV electrons for different incident electron exposures (fluences) in a scanning electron microscope. Each curve was traced in approximately 100 min with a lock-in sensitivity of $2\,\mu V$ (full scale) and a time constant of 3 s. (See Ref. 72.)

while other peaks grow. Analysis of the spectrum suggested that COH and CH, CH_2 groups are destroyed by the action of energetic electrons and that double bonds are formed between carbon atoms. The transformation of the symmetrical peak at $900 \, \text{cm}^{-1}$ into an asymmetrical peak around $940 \, \text{cm}^{-1}$ leads to the speculation of the persistence of a ring stretching mode even at the high exposure values.

Proton beam damage has been studied by Saemann-Ischenko, Behrle, and co-workers in Germany.[74] Figure 21 shows their results on proton beam irradiation of a benzoate monolayer. Perhaps the most striking feature of these spectra is the disappearance of the aromatic C–H stretching mode and the corresponding growth of a nonaromatic C–H stretching mode when the experiments are performed at 4.2 K. In contrast, when the experiments are performed at room temperature, the growth of the strong nonaromatic peak at intermediate exposures is not seen. Thus, it seems clear that an unstable intermediate in the degradation process is being frozen at the low temperatures. The possibility of analyzing intermediates in the degradation process is exciting.

These selected examples perhaps give a flavor for the current applications of tunneling spectroscopy. A useful tabular review of tunneling spectroscopy with 283 references indexed by topic and by adsorbate has recently been published by Hipps.[75] In the next two subsections, two particular applications—the study of model catalysts and the study of corrosion—will be treated in somewhat more detail.

5.2. The Study of Model Catalysts

As mentioned earlier, the current limitation on tunneling spectroscopy is that it must be done with oxide barrier tunnel junctions. This precludes the possibility of taking an actual catalyst sample and studying the species on it with tunneling spectroscopy. What can be done is to study model catalysts. Fortunately, the two oxides that are commonly used for tunneling spectroscopy—aluminum oxide and magnesium oxide—are in themselves of some interest as catalysts and as catalyst supports.

Figure 22 shows the results of Walmsley, Nelson, and co-workers on phenol adsorbed on magnesium oxide and aluminum oxide.[76] Though the spectra are similar, the subtle differences that are pointed out with the arrows give clues to the subtle differences in structure and orientation of the adsorbed species on these two oxides. This same group has studied a wide range of organic acids,[76–79] alcohols,[80] and unsaturated hydrocarbons[81] on these two oxides. Their work combined with the work of other groups working worldwide has produced a library of

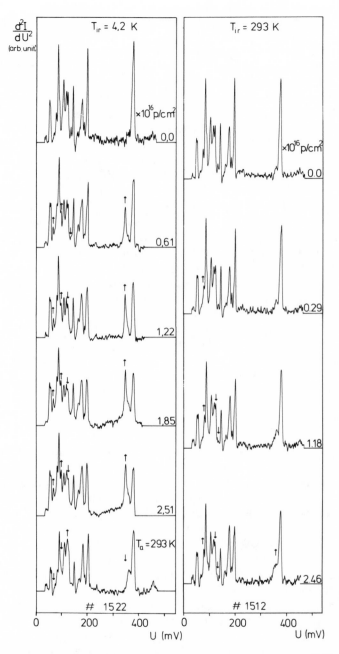

Figure 21. Proton beam irradiation of tunnel junctions doped with benzoate ions changes the spectrum. Comparison of room temperature and helium temperature results suggests that reaction intermediates were trapped at helium temperature (see Ref. 74).

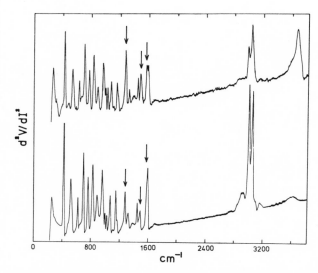

Figure 22. Upper curve: Tunneling spectrum of Mg–I–Pb sandwich doped with phenol. (I refers to "insulator," which is, in this case, magnesium oxide). Lower curve: Tunneling spectrum of Al–I–Pb sandwich doped with phenol (here the insulator is aluminum oxide; see Ref. 76).

hundreds of spectra of compounds adsorbed on alumina. This library is currently being assembled into a book,[82] and, as mentioned above, a tabular review is already available.[75]

The application of tunneling spectroscopy to the study of homogeneous cluster compounds supported on oxide surfaces is illustrated by Figure 23. In it, Weinberg and co-workers[83–85] have deposited $Zr(BH_4)_4$ on Al_2O_3 at 300 K. Their detailed analysis of this spectrum and other spectra obtained at various temperatures gives important clues to the nature of the supported complex.

Not only can the supported clustered compounds themselves be studied, but molecules can be adsorbed on them.[85–88] For example, Figure 24 shows the result of exposing the supported $Zr(BH_4)_4$ catalyst to acetylene at 300 and 400 K. A particularly interesting feature of these experiments by the Weinberg group was that at higher temperatures the concentration of hydrocarbon species on the surface increased rapidly with no apparent saturation limit. Their experimental results were consistent with the view that a polymer was formed on the surface at a few active centers, probably Zr–H sites.

Rhodium and ruthenium complexes were also studied.[83,89,90] For example, Figure 25 shows the results of exposing the oxidized aluminum

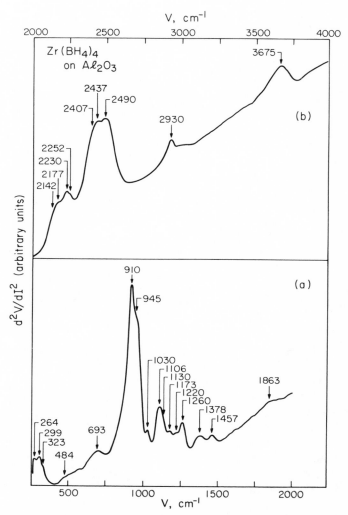

Figure 23. Tunneling spectrum for $Zr(BH_4)_4$ adsorbed on Al_2O_3 at 300 K over the spectral ranges (a) 240–2000 cm^{-1} and (b) 2000–4000 cm^{-1} (see Refs. 83 and 84).

to a rhodium complex from the gas phase.[83,99] Analysis of this spectrum suggested that the rhodium decomposed, losing chlorine via HCl evolution leaving a rhodium–CO species bound to the surface. This is similar to what is believed to happen when actual supported catalysts are prepared by doping large surface area alumina from the liquid phase.

An important challenge is to dope tunnel junctions with cluster

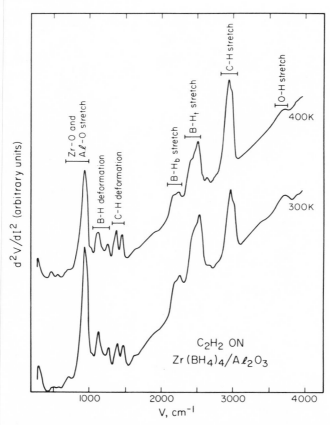

Figure 24. Tunneling spectra of acetylene (300 Torr s exposure) on $Zr(BH_4)_4/Al_2O_3$ at 300 and 400 K (see Refs. 83 and 87).

compounds from the liquid phase. The technological problem involved is that the species formed are quite often air sensitive. This is particularly a problem for tunnel junctions since the adsorbed species on a flat surface are completely exposed (rather than hidden in pores). Thus, what is needed is a controlled atmosphere liquid-phase doping apparatus.

Hipps and Mazur and co-workers have constructed such an apparatus consisting of an evaporator enclosed in a high-technology glove box. Figure 26 shows some of their results for a cobalt complex liquid doped onto aluminum oxide from a benzene solution under a nitrogen atmosphere.[91] On the basis of comparison of infrared and tunneling frequencies the surface species has both terminal and bridging CO

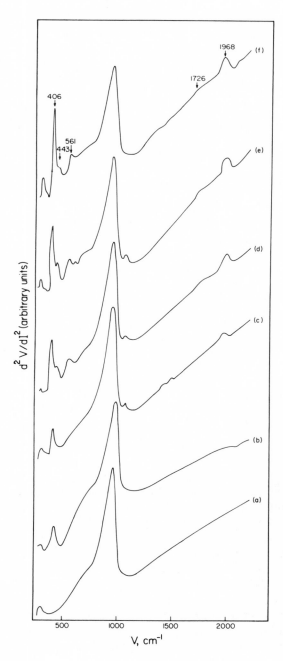

Figure 25. Tunneling spectra of alumina cooled to various temperatures and exposed to $[RhCl(CO)_2]_2$ from the gas phase. A clean surface spectrum is shown in (a). The other spectra represent the surface exposed to the complex at (b) room temperature (295 K), (c) 240 K, (d) 210 K, (e) 190 K, and (f) 180 K. The lead was evaporated on the cold surface. The lower two modes identified in spectrum (f) were assigned as M–C–O bending, the upper two as C–O stretching, and the middle one ($561\ cm^{-1}$) as either M–C–O bending or M–CO stretching. (See refs. 83 and 89.)

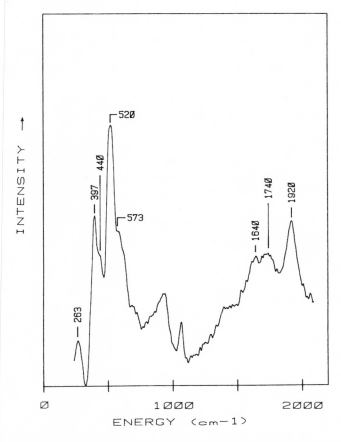

Figure 26. Tunneling spectrum of a transition metal complex liquid doped onto alumina in a controlled atmosphere apparatus (see Ref. 91).

ligands. In contrast, when this compound is doped onto the oxide in air no CO ligands at all remain.

A number of tunneling studies have also been done by Kroeker and co-workers on small metal particles which are supported on aluminum oxide.[92] The general experimental scheme was to evaporate a small quantity of transition metals such as rhodium,[93] nickel,[94] or iron onto the oxidized aluminum. The evaporated metal films formed islands with dimensions of order 20 Å on a side. These islands of metal can serve as adsorption sites for carbon monoxide and, as in the work of Bayman and co-workers, hydrocarbons[96] One exciting result of this research was the observation that carbon monoxide adsorbed on alumina supported rhodium particles could be converted into hydrocarbons[97] by heat treating with hydrogen gas at high pressures. The carbon monoxide could be converted to formate ions by heating without hydrogen gas. In this case the source of the hydrogen atoms may have been surface OH groups on the alumina.

5.3. The Study of Corrosion

Corrosion is, of course, of enormous economic significance. Tunneling spectroscopy can be used to study corrosion and corrosion inhibition of aluminum. Of particular importance in the aerospace electronic and chemical industries is the problem of corrosion of aluminum by chlorin-

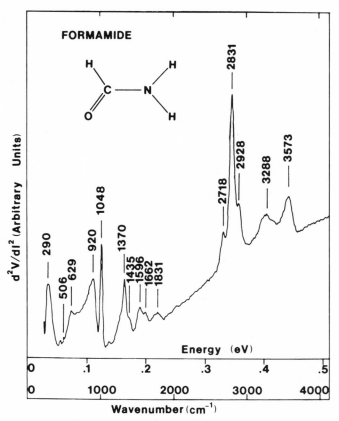

Figure 27. Tunneling spectrum for formamide HCONH$_2$. The wave numbers of significant peaks are labeled in units of cm^{-1} (see Refs. 100 and 101).

ated solvents. Figure 27 illustrates the work of White and co-workers[100,101] on this important problem. It shows the spectrum of formamide, which is known to inhibit the corrosion of aluminum by carbon tetrachloride. The spectrum was obtained by liquid doping the oxidized aluminum with a solution of formamide in carbon tetrachloride. The analysis of this spectrum and other spectra made with isotopically labeled molecules led to conclusions about the nature of the bonding of the formamide to the oxidized aluminum surface. This in turn led to conclusions about the nature of the inhibition mechanism.

Figure 28 shows the work of Shu and co-workers on the tunneling spectra of some molecules that were found to inhibit the corrosion of

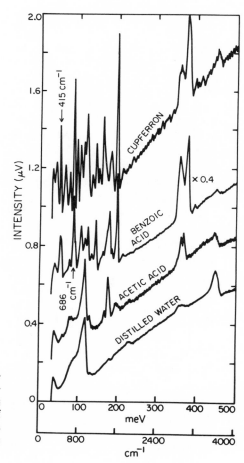

Figure 28. Tunneling spectra of cupferron, benzoic acid, acetic acid, and distilled water liquid doped onto the oxide of an Al–oxide–Pb junction. Analysis of these spectra reveals that the two acids bond as bidentate symmetric carboxylate ions (see Ref. 102).

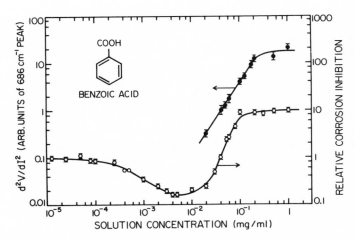

Figure 29. Surface concentration (left scale) and relative corrosion inhibition parameter (right scale) versus the solution concentration for benzoic acid in water (see Ref. 102).

aluminum by pure distilled water.[102] From the spectra it can be determined that the benzoic acid and acetic acid bond as symmetric bidentate carboxylate anions with the hydrocarbon groups oriented away from the surface.[102] The bonding of the cupferron has not yet been determined. For example, Figure 29 shows the relative corrosion inhibition for benzoic acid in water as a function of the solution concentration. (The relative corrosion inhibition is defined as the ratio of the corrosion rate without the inhibitor to the corrosion rate with the inhibitor. Thus, a relative corrosion inhibition of one indicates that the inhibitor has no net effect on the corrosion rate.) Note that at high solution concentrations, benzoic acid inhibited the corrosion by an order of magnitude. This saturation of corrosion inhibition can be correlated with the saturation in surface coverage at one monolayer.

Another interesting feature of these data is that at lower solution concentrations the benzoic acid actually accelerated corrosion. A possible explanation is that two competing effects are going on. Dilute solutions of benzoic acid do corrode aluminum more rapidly than pure water, as might be intuitively expected, but, as the concentration is raised, the corrosion inhibition provided by the adsorbed layer of carboxylate ions more than compensates and gives a net corrosion inhibition.

6. Comparison with Other Techniques

A detailed and authoritative comparison of tunneling spectroscopy with other surface analytical techniques has been made by Dubois.[103] A simple comparison of tunneling spectra such as Figures 4, 12, and 22 with infrared, Raman, or electron energy loss spectra of a monolayer of hydrocarbons adsorbed on a support can quickly convince one that tunneling spectroscopy is unexcelled in its combination of spectral range, sensitivity, and resolution. For example, Figure 30 shows a comparison of a tunneling for benzaldehyde adsorbed on alumina with infrared and Raman spectra for the same system.[17] Though infrared spectroscopy does exceedingly well over a narrow wavelength range, it is prevented from seeing the vibrations below $1000 \, cm^{-1}$. Raman spectroscopy picks out a few more peaks, but, again, it is best over a limited spectral range.

Comparisons like this one are, however, somewhat misleading. Though tunneling spectroscopy does very well on the systems that can be studied with it, the systems that can be studied are severely limited. The technique, at present at least, does not have the versatility of the other spectroscopies. In particular, it cannot be used in the same ultrahigh vacuum chamber with other surface analysis equipment to get vibrational spectra of molecules adsorbed on well characterized surfaces. This limitation in its applicability has limited its usefulness.

As mentioned previously in the chapter, because of vibrational mode shifts due to the top metal electrode, tunneling spectroscopy does the poorest on the molecules for which infrared and electron energy loss spectroscopy do the best job, specifically, on molecules such as carbon monoxide with a very high oscillating dipole derivative. Figure 31 shows a comparison of tunneling spectroscopy[104] on this worst case system with infrared[105] and electron energy loss[106] spectroscopies.[104] Note the dramatic downshift and broadening in the CO stretching peaks as observed with inelastic electron tunneling spectroscopy. Tunneling spectroscopy does a somewhat better job with the low-energy vibrations, but here one must be cautious because of the top metal electrode effects such as presented in Figure 15 of this chapter.

Will it ever be possible to make tunnel junctions without evaporating a top metal electrode directly on the molecules? Will it ever be possible to make tunnel junctions to well characterized single-crystal surfaces in ultrahigh vacuum? Recent experiments on mechanically adjusted tunneling junctions give some hope in this regard.[51,52] The goal would be to hold an electrode mechanically a very small distance, of order 10 Å, away from a well characterized surface and do tunneling spectroscopy between the well characterized surface and the mechanically held electrode.

Pioneering research on mechanically adjusted junctions was done by Young et al.,[45,46] Teague,[48] and Thompson and Hanrahan.[47] Great excitement has been generated recently by the beautiful surface topography done by Binnig and Rohrer with a point held above a surface.[49,50] In these experiments, a stability of order 0.2 Å was achieved.

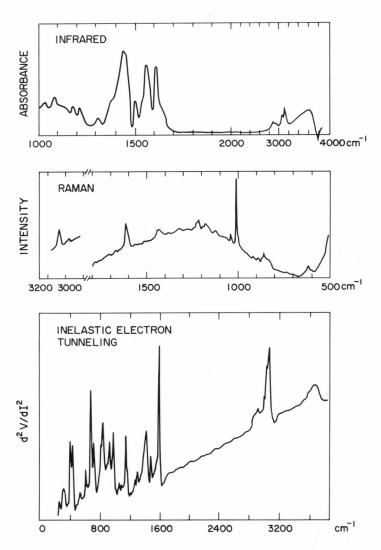

Figure 30. A comparison of infrared, Raman, and tunneling results for the same system: benzaldehyde adsorbed on alumina. Note that the tunneling spectrum has as good resolution as the infrared in the region where the infrared is the best, 1300–1700 cm^{-1}, but extends this resolution over a much wider range. The agreement between peak positions determined by the three spectroscopies is excellent in the regions where they overlap (see Ref. 17).

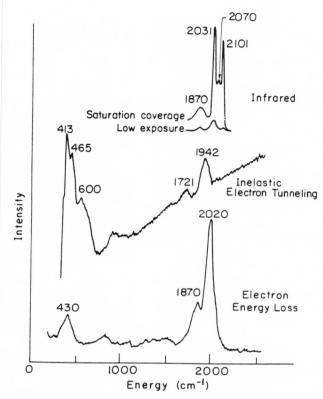

Figure 31. Vibrational spectra taken by three different techniques for carbon monoxide adsorbed on rhodium particles supported on alumina. The infrared spectra (upper traces) are from the work of Yates et al.[105] The high resolution of ir spectroscopy is evident. The inelastic electron tunneling spectrum (middle trace), taken from the work of Kroeker et al.[104] shows the downshift in the CO stretching vibrations that are characteristic of tunneling spectroscopy and the relatively strong low-frequency modes. The electron energy loss spectrum (see Ref. 106) (lower trace) approximates the low-frequency (400–1000 cm^{-1}) tunneling spectrum and the high-frequency (1000–2500 cm^{-1}) infrared spectra.

What kind of stability is necessary to do tunneling spectroscopy? Since typical tunneling peaks amount to conductance changes of order 0.1%, the spacing must be stable enough that the conductance not vary by more than 0.1%, at least over short times, say 10 s. In order for the conductance not to vary by more than 0.1%, the spacing must be stable to of order 0.001 Å since the tunneling conductance is an exponentially sensitive function of the spacing with a characteristic length scale of order 1 Å.

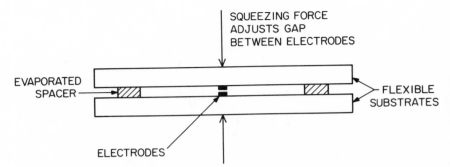

Figure 32. A schematic illustration of a squeezable electron tunneling junction (SET junction) showing two electrodes supported by flexible substrates separated by thin film spacers (see Refs. 51 and 52).

An approach to mechanically adjusted tunnel junctions that does at least approach this stability is illustrated in Figure 32. In this case, the two electrodes of the tunnel junction are evaporated on two flexible substrates. The flexible substrates are deformed by squeezing them together. If the gap between the electrodes is of order 1000 Å with no

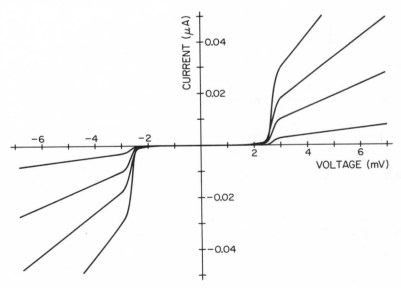

Figure 33. $I-V$ curves obtained at 1.2 K in liquid He for a SET junction consisting of superconducting Pb electrodes on glass slides separated by evaporated Pb spacers (see Refs. 51 and 52).

squeezing force and the squeezing force can be controlled to one part in 10^6, the the gap can be controlled to 0.001 Å.

Figure 33 shows results obtained with this geometry for a junction with two lead electrodes. The presence of the superconducting energy gap in the tunneling characteristic shows that the current flow is, indeed, by electron tunneling. Note that the conductance of the junction could be adjusted by changing the squeezing force. Clearly, the challenge is to produce junctions with adsorbed molecules and find a way to maintain this stability out to the higher bias voltages necessary to obtain tunneling spectra. It is a difficult but an exciting challenge.

7. Conclusions

1. Tunneling spectroscopy is a technique for measuring the vibrational spectrum of minute quantities of material.

2. The *spectral range* of tunneling spectroscopy includes all molecular vibrations.

3. Its *sensitivity* is sufficient to see a small fraction of a monolayer over an area of order 1 mm on a side.

4. Its *resolution* has no theoretical limits but is generally set at $10–40 \, cm^{-1}$ by practical considerations.

5. There are *no selection rules*. There is a slight selection preference for vibrations oscillating perpendicular to the plane of the junction's barrier.

6. *Sample preparation* occupies most of the time in tunneling spectroscopy research. Measuring spectra is straightforward once the necessary low-noise electronics has been assembled.

7. Problems include the limitations set by the *junction geometry*, the *cryogenic temperatures* necessary for good resolution, and the *top metal electrode*, which can perturb the molecular vibrations being measured.

8. *Mechanically adjusted junctions* have the potential to overcome the limitations currently imposed by junction geometry and the top metal electrode.

Acknowledgments

I thank Atiye Bayman, Bob Coleman, Jim Hall, Kerry Hipps, John Kirtley, Richard Kroeker, John Langan, Ursula Mazur, John Moreland, Mehir Parikh, Doug Scalapino, Qi Qing Shu, Richard Sonnenfeld,

George Walmsley, Henry White, and Henry Weinberg for discussions and data on which this chapter is based, and the National Science Foundation (current grant No. DMR86-13486) and the Office of Naval Research for their support. Special thanks is due Bill Kaska for a critical reading of this manuscript and his collaboration for many years.

References

1. R. C. Jaklevic and J. Lambe, Molecular vibration spectra by electron tunneling, *Phys. Rev. Lett.* **17**, 1139–1140 (1966).
2. J. Lambe and R. C. Jaklevic, Molecular vibration spectra by inelastic electron tunneling, *Phys. Rev.* **165**, 821–832 (1968).
3. The water analogy was developed by Atiye Bayman.
4. D. G. Walmsley and W. J. Nelson, in: *Tunneling Spectroscopy: Capabilities, Applications, and New Techniques* (P. K. Hansma, ed.), pp. 311–357, Plenum Press, New York (1982).
5. J. D. Langan and P. K. Hansma, Can the concentration of surface species be measured with inelastic electron tunneling?, *Surf. Sci.* **52**, 211–216 (1975).
6. A. A. Cederberg, Inelastic electron tunneling spectroscopy intensity as a function of surface coverage, *Surf. Sci.* **103**, 148–176 (1981).
7. R. M. Kroeker and P. K. Hansma, A measurement of the sensitivity of inelastic electron tunneling spectroscopy, *Surf. Sci.* **67**, 362–366 (1977).
8. D. G. Walmsley, R. B. Floyd, and S. F. J. Read, Inelastic electron tunneling spectra lineshapes below 100 mK, *J. Phys. C* **11**, L107–L110 (1978).
9. K. W. Hipps and Ursula Mazur, in: *Tunneling Spectroscopy: Capabilities, Applications, and New Techniques* (P. K. Hansma, ed.) pp. 229–269, Plenum Press, New York (1982).
10. N. I. Bogatina, Selection rules in tunnel spectroscopy for highly symmetrical molecules, *Opt. Spectrosc.* **38**, 43–44 (1975).
11. N. B. Bogatina, I. K. Yanson, B. I. Verkin, and A. G. Batrak, Tunnel spectra of organic solvents, *Sov. Phys.-JETP* **38**, 1162–1165 (1974).
12. J. Kirtley, D. J. Scalapino, and P. K. Hansma, Theory of vibrational mode intensities in inelastic electron tunneling spectroscopy, *Phys. Rev. B* **14**, 3177–3184 (1976).
13. J. Kirtley and J. T. Hall, Theory of intensities in inelastic-electron tunneling spectroscopy orientation of adsorbed molecules, *Phys. Rev. B* **22**, 848–856 (1980).
14. *Tunneling Spectroscopy: Capabilities, Applications, and New Techniques* (P. K. Hansma, ed.) Plenum Press, New York (1982).
15. *Inelastic Electron Tunneling Spectroscopy* (T. Wolfram, ed.) Springer, Berlin (1978).
16. S. K. Khanna and J. Lambe, Inelastic electron tunneling spectroscopy, *Science* **220**, 1345–1351 (1983).
17. P. K. Hansma, Inelastic electron tunneling, *Phys. Rep.* **30C**, 147–206 (1977).
18. W. H. Weinberg, Inelastic electron tunneling spectroscopy: A probe of the vibrational structure of surface species, *Ann. Rev. Phys. Chem.* **29**, 115–139 (1978).
19. P. K. Hansma and J. Kirtley, Recent advances in inelastic electron tunneling spectroscopy, *Accts. Chem. Res.* **11**, 440–445 (1978).
20. R. G. Keil, T. P. Graham, and K. P. Roenker, Inelastic electron tunneling spectroscopy: A review of an emerging analytical technique, *Appl. Spectrosc.* **30**, 1–18 (1976).

21. N. M. Brown and D. G. Walmsley, IETS—A new tool, *Chem. Br.* **12**, 92–94 (1976).
22. J. Giaever, Electron tunneling and superconductivity, *Rev. Mod. Phys.* **46**, 245–250 (1974) (his Nobel Prize acceptance speech).
23. W. L. McMilland J. Rowell, in: *Superconductivity* (R. D. Parks, ed.), p. 561, Marcel Dekker, New York (1969).
24. R. V. Coleman, R. C. Morris, and J. E. Christopher, *Methods of Experimental Physics VII. Solid State Physics* (R. V. Coleman, ed.), Academic, New York (1974).
25. J. L. Miles and P. H. Smith, The formation of metal oxide films using gaseous and solid electrolytes, *J. Electrochem. Soc.* **110**, 1240–1245 (1963).
26. R. Magno and J. G. Adler, Inelastic electron-tunneling study of barriers grown on aluminum, *Phys. Rev. B* **13**, 2262–2269 (1976).
27. M. G. Simonsen and R. V. Coleman, Inelastic-tunneling spectra of organic compounds, *Phys. Rev. B* **8**, 5875–5887 (1973).
28. P. K. Hansma and R. V. Coleman, Spectroscopy of biological compounds with inelastic electron tunneling, *Science* **184**, 1369–1371 (1974).
29. M. G. Simonsen, R. V. Coleman, and P. K. Hansma, High-resolution inelastic tunneling spectroscopy of macromolecules and adsorbed species with liquid-phase doping, *J. Chem. Phys.* **61**, 3789–3799 (1974).
30. Y. Skarlatos, R. C. Barker, G. L. Haller, and A. Yelon, Detection of dilute organic acids in water by inelastic tunneling spectroscopy, *Surf. Sci.* **43**, 353–368 (1974).
31. A. Bayman and P. K. Hansma, Inelastic electron tunneling spectroscopic study of lubrication, *Nature* **285**, 97–99 (1980).
32. R. C. Jaklevic and M. R. Gaerttner, Electron tunneling spectroscopy—External doping with organic molecules, *Appl. Phys. lett.* **30**, 646–648 (1977).
33. R. C. Jaklevic and M. R. Gaerttner, Inelastic electron tunneling spectroscopy. Experiments on external doping of tunnel junctions by an infusion technique, *Appl. Surf. Sci.* **1**, 479–502 (1978).
34. R. C. Jaklevic, in: *Tunneling Spectroscopy: Capabilities, Applications, and New Techniques* (P. K. Hansma, ed.), pp. 451–473, Plenum Press, New York (1982).
35. P. K. Hansma and H. G. Hansma, in: *Tunneling Spectroscopy: Capabilities, Applications, and New Techniques* (P. K. Hansma, ed.), pp. 475–481, Plenum Press, New York (1982).
36. D. G. Walmsley, W. J. Nelson, N. M. D. Brown, and R. B. Floyd, Development of inelastic electron tunneling spectroscopy. Comparison of adsorbates on aluminum and magnesium oxides, *Appl. Surf. Sci.* **5**, 107–120 (1980).
37. D. G. Walmsley and W. J. Nelson, in: *Tunneling Spectroscopy: Capabilities, Applications, and New Techniques* (P. K. Hansma, ed.), pp. 311–357, Plenum Press, New York (1982).
38. R. J. Jennings and J. R. Merrill, The temperature dependence of impurity-assisted tunneling, *J. Phys. Chem. Solids* **33**, 1261 (1972).
39. D. G. Walmsley, I. W. N. McMorris, W. E. Timms, W. J. Nelson, J. L. Tomlin, and T. J. Griffin, A sensitive robust circuit for inelastic electron tunneling spectroscopy, *J. Phys. E: Sci. Instrum.* **16**, 1052–1057 (1983).
40. D. E. Thomas and J. M. Rowell, Low-level second-harmonic detection system, *Rev. Sci. Instrum.* **36**, 1301–1306 (1965).
41. J. G. Adler and J. E. Jackson, System for observing small nonlinearities in tunnel junctions, *Rev. Sci. Instrum.* **37**, 1049–1054 (1966).
42. A. F. Hebard and P. W. Shumate, A new approach to high resolution measurements of structure in superconducting tunneling currents, *Rev. Sci. Instrum.* **45**, 529–533 (1974).

43. S. Colley and P. K. Hansma, Bridge for differential tunneling spectroscopy, *Rev. Sci. Instrum.* **48**, 1192–1195 (1977).
44. J. G. Adler, in: *Tunneling Spectroscopy: Capabilities, Applications, and New Techniques* (P. K. Hansma, ed.), pp. 423–450, Plenum Press, new York (1982).
45. R. Young, J. Ward, and F. Scire, Observation of metal–vacuum–metal tunneling, field emission, and the transition region, *Phys. Rev. Lett.* **27**, 922–924 (1971).
46. R. Young, J. Ward, and F. Scire, The topografiner: An instrument for measuring surface microtopography, *Rev. Sci. Instrum.* **43**, 999–1011 (1972).
47. W. A. Thompson and S. F. Hanrahan, Thermal drive apparatus for direct vacuum tunneling experiments, *Rev. Sci. Instrum.* **47**, 1303–1304 (1976).
48. C. Teague, thesis, available from University Microfilms as dissertation No. 78-24-678, North Texas State University, Denton, Texas 1978.
49. G. Binnig, H. Rohrer, Ch. Gerber, and E. Weibel, Tunneling through a controllable vacuum gap, *Appl. Phys. Lett.* **40**, 178–180 (1982).
50. G. Binnig, H. Rohrer, Ch. Gerber, and E. Weibel, Surface studies by scanning tunneling microscopy, *Phys. Rev. Lett.* **49**, 57–60 (1982).
51. J. Moreland, S. Alexander, M. Cox, R. Sonnenfeld, and P. K. Hansma, Squeezable electron tunneling junctions, *Appl. Phys. Lett.* **43**, 387–388 (1983).
52. J. Moreland and P. K. Hansma, An electromagnetic squeezer for compressing squeezable electron tunneling junctions, *Rev. Sci. Instrum.* **55**, 399–403 (1984).
53. J. R. Kirtley and P. K. Hansma, Effect of the second metal electrode on vibrational spectra in inelastic electron tunneling spectroscopy, *Phys. Rev. B* **12**, 531–536 (1975).
54. J. Kirtley and P. K. Hansma, Vibrational mode shifts in inelastic electron tunneling spectroscopy: Effects due to superconductivity and surface interactions, *Phys. Rev. B* **13**, 2910–2916 (1976).
55. A. Bayman, P. K. Hansma, and W. C. Kaska, Shifts and dips in inelastic electron tunneling spectra due to the tunnel junctions environment, *Phys. Rev. B* **25**, 2449–2455 (1981).
56. A. Bayman, P. K. Hansma, and W. C. Kaska, The effect of the top metal electrode on tunneling spectra, *Physica* **108B**, 1171–1172 (1981).
57. R. Magno, M. K. Konkin, and J. G. Adler, Effect of cover electrode metal on inelastic electron tunneling structure, *Surf. Sci.* **69**, 437–443 (1977).
58. P. K. Hansma, in: *Tunneling Spectroscopy: Capabilities, Applications, and New Techniques* (P. K. Hansma, ed.) pp. 1–41, Plenum Press, New York (1982).
59. D. J. Scalapino and S. M. Marcus, Theory of inelastic electron-molecule interactions in tunnel junctions, *Phys. Rev. Lett.* **18**, 459–461 (1967).
60. J. Kirtley, D. J. Scalapino, and P. K. Hansma, Theory of vibrational mode intensities in inelastic electron tunneling spectroscopy, *Phys. Rev. B* **14**, 3177–3184 (1976).
61. J. Bardeen, Tunnelling from a many-particle point of view, *Phys. Rev. Lett.* **6**, 57–59 (1961).
62. M. H. Cohen, L. M. Falicov, and J. C. Phillips, Superconductive tunneling, *Phys. Rev. Lett.* **8**, 316–318 (1962).
63. K. V. Hipps and R. Knochenmuss, Some proposed modifications in the theory of inelastic electron tunneling spectroscopy and the source of parameters utilized, *J. Phys. Chem.* **86**, 4477–4480 (1982).
64. J. D. Langan and P. K. Hansma, Can the concentration of surface species be measured with inelastic electron tunneling? *Surf. Sci.* **52**, 211–216 (1975).
65. R. V. Coleman, in: *Tunneling Spectroscopy: Capabilities, Applications, and New Techniques* (P. K. Hansma, ed.) pp. 201–227, Plenum Press, New York (1982).
66. J. M. Clark and R. V. Coleman, Inelastic electron tunneling study of uv radiation damage in surface adsorbed nucleotides, *J. Chem. Phys.* **73**, 2156–2178 (1980).

67. M. G. Simonsen and R. V. Coleman, Tunneling measurements of vibrational spectra of amino acids and related compounds, *Nature* **244**, 218–220 (1973).
68. J. M. Clark and R. V. Coleman, Inelastic electron tunnelling spectroscopy of nucleic acid derivatives, *Proc. Natl. Acad. Sci. USA* **73**, 1598–1602 (1976).
69. K. W. Hipps and U. Mazur, in: *Tunneling Spectroscopy: Capabilities, Applications, and New Techniques* (P. K. Hansma, ed.), pp. 229–269, Plenum Press, New York (1982).
70. K. W. Hipps and U. Mazur, An IETS study of some iron cyanide complexes, *J. Phys. Chem.* **84**, 3162–3172 (1980).
71. M. Parikh, in: *Tunneling Spectroscopy: Capabilities, Applications, and New Techniques* (P. K. Hansma, ed.), pp. 271–285, Plenum Press, New York (1982).
72. P. K. Hansma and M. Parikh, A tunneling spectroscopy study of molecular degradation due to electron irradiation, *Science* **188**, 1304–1305 (1975).
73. M. Parikh, P. K. Hansma, and J. Hall, Quantitative tunneling spectroscopy study of molecular structural changes due to electron irradiation, *Phys. Rev. A* **14**, 1437–1446 (1976).
74. R. Behrle, W. Rösner, H. Adrian, G. Saemann-Ischenko, F. Bömmel, and l. Söldner, Inelastic electron tunneling as vibrational spectroscopy of adsorbed organic molecules after 3 MeV proton irradiation at 4.2 and 293 K, Proceedings of 10th International Conference on Atomic Collisions in Solids (Bad Iburg, Germany, 1983).
75. K. W. Hipps, A tabular review of tunneling spectroscopy, *J. Electron Spectrosc. Relat. Phenom.* **30**, 275–285 (1983).
76. D. G. Walmsley and W. J. Nelson, in: *Tunneling Spectroscopy: Capabilities, Applications, and New Techniques* (P. K. Hansma, ed.) pp. 311–357, Plenum Press, New York (1982).
77. N. M. D. Brown, R. B. Floyd, and D. G. Walmsley, Inelastic electron tunneling spectroscopy (IETS) of carboxylic acids and related systems chemisorbed on plasma-grown aluminum oxide. Part 1. Formic acid (HCOOH and DCOOD), acetic acid (CH₃COOH, CH₃COOD and CD₃COOD), trifluoroacetic acid, acetic anhydride, acetaldehyde and acetylchloride, *J. Chem. Soc. Faraday Trans. 2* **75**, 17–31 (1979).
78. D. G. Walmsley, W. J. Nelson, N. M. D. Brown, S. de Cheveigne, S. Gauthier, J. Klein, and A. Leger. Evidence from inelastic electron tunneling spectroscopy for vibrational mode reassignments in simple aliphatic carboxylate ions, *Spectrochim Acta* **37A**, 1015–1019 (1981).
79. N. M. D. Brown, W. J. Nelson, and D. G. Walmsley, Inelastic electron tunneling spectroscopy (IETS) of carboxylic acids and related systems chemisorbed on plasma-grown aluminum oxide. Part 2. Propynoic acid, propenoic acid and 3-methyl-but-2-enoic acid, *J. Chem. Soc. Faraday Trans 2* **75**, 32–37 (1979).
80. N. M. D. Brown, R. B. Floyd, W. J. Nelson, and D. G. Walmsley, Inelastic electron tunneling spectroscopy of selected alcohols and amines on plasma-grown aluminum oxide, *J. Chem. Soc. Faraday Trans. 1* **76**, 2335–2346 (1980).
81. N. M. D. Brown, W. E. Timms, R. J. Turner, and D. G. Walmsley, Inelastic electron tunneling spectroscopy (IETS) of simple unsaturated hydrocarbons adsorbed on plasma-grown aluminum oxide, *J. Catal.* **64**, 101–109 (1980).
82. The book is being compiled by D. G. Walmsley and J. F. Tomlin.
83. W. H. Weinberg, in: *Tunneling Spectroscopy: Capabilities, Applications, and New Techniques* (P. K. Hansma, ed.), pp. 359–391, Plenum Press, New York (1982).
84. H. E. Evans and W. H. Weinberg, Inelastic electron tunneling spectroscopy of zirconium tetraborohydride supported on aluminum oxide, *J. Am. Chem. Soc.* **102**, 872–873 (1980).
85. W. H. Weinberg, W. M. Bowser, and H. E. Evans, Reduced metallic clusters and

homogeneous cluster compounds "supported" on aluminum oxide as studied by inelastic electron tunneling spectroscopy, *Surf. Sci.* **106,** 4720–4724 (1980).

86. H. E. Evans and W. H. Weinberg, A vibrational study of zirconium tetraborohydride supported on aluminum oxide. 1. Interactions with deuterium, deuterium oxide and water vapor, *J. Am. Chem. Soc.* **102,** 2548–2553 (1980).

87. H. E. Evans and W. H. Weinberg, A vibrational study of zirconium tetraborohydride supported on aluminum oxide. 2. interactions with ethylene, propylene and acetylene, *J. Am. Chem. Soc.* **102,** 2554–2558 (1980).

88. L. Forester and W. H. Weinberg, A vibrational study of $Zr(BH_4)_4$ supported on alumina: Interactions with cyclohexene, 1,3-cyclohexadiene and benzene, *J. Vac. Sci. Technol.* **18,** 600–601 (1981).

89. W. M. Bowser and W. H. Weinberg, An inelastic electron tunneling spectroscopic study of the interaction of $[RhCl(CO)_2]_2$ with an aluminum oxide surface, *J. Am. Chem. Soc.* **103,** 1453–1458 (1981).

90. W. M. Bowser and W. H. Weinberg, An inelastic electron tunneling spectroscopic study of $Ru_3(CO)_{12}$ adsorbed on an aluminium oxide surface, *J. Am. Chem. Soc.* **102,** 4720–4724 (1980).

91. K. Hipps and U. Mazur, Construction and application of a novel combination glove box deposition system to the study of air-sensitive materials by tunneling spectroscopy, *Rev. Sci. Instrum.* **55,** 1120 (1984).

92. R. M. Kroeker, in: *Tunneling Spectroscopy: Capabilities, Applications, and New Techniques* (P. K. Hansma, ed). pp. 393–421, Plenum Press, New York (1982).

93. R. M. Kroeker, W. C. Kaska, and P. K. Hansma, How carbon monoxide bonds to alumina-supported rhodium particles, *J. Catal.* **57,** 72–79 (1979).

94. R. M. Kroeker, W. C. Kaska, and P. K. Hansma, Vibrational spectra of carbon monoxide chemisorbed on alumina-supported nickel partices: A tunneling spectroscopy study, *J. Chem. Phys.* **74,** 732–736 (1981).

95. R. M. Kroeker, W. C. Kaska, and P. K. Hansma, Low-energy vibrational modes of carbon monoxide on iron, *J. Chem. Phys.* **72,** 4845–4852 (1980).

96. A. Bayman, P. K. Hansma, W. C. Kaska, and L. H. Dubois, Inelastic electron tunneling spectroscopic study of acetylene chemisorbed on alumina supported palladium particles, *Appl. Surf. Sci.* **14,** 194–208 (1982).

97. R. M. Kroeker, W. C. Kaska, and P. K. Hansma, Formation of hydrocarbons from carbon monoxide on rhodium/alumina model catalysts, *J. Catal.* **61,** 87–95 (1980).

98. R. M. Kroeker and P. K. Hansma, Tunneling spectroscopy for the study of adsorption and reactions on model catalysts, *Catal. Rev. Sci. Eng.* **23,** 553–603 (1981).

99. R. M. Kroeker, W. C. Kaska, and P. K. Hansma, Sulfur modifies the chemisorption of carbon monoxide on rhodium/alumina model catalysts, *J. Catal.,* **63,** 487–490 (1980).

100. H. W. White, in: *Tunneling Spectroscopy: Capabilities, Applications, and New Techniques* (P. K. Hansma, ed.), pp. 287–309, Plenum Press, New York (1982).

101. R. M. Ellialtioglu, H. W. White, L. M. Godwin, and T. Wolfram, Study of the corrosion inhibitor formamide in the aluminum-carbon tetrachloride system using IETS, *J. Chem. Phys.* **75,** 2432 (1981).

102. Q. Q. Shu, P. J. Love, A. Bayman, and P. K. Hansma, Aluminum corrosion: Correlations of corrosion rate with surface coverage and tunneling spectra of organic inhibitors, *Appl. Surf. Sci.* **13,** 374–388 (1982).

103. L. H. Dubois, in: *Tunneling Spectroscopy: Capabilities, Applications, and New Techniques* (P. K. Hansma, ed.), pp. 153–199, Plenum Press, New York (1982).

104. R. M. Kroeker, W. C. Kaska, and P. K. Hansma, How carbon monoxide bonds to

alumina-supported rhodium particles: Tunneling spectroscopy measurements with isotopes, *J. Catal.* **57,** 72–79 (1979).
105. J. T. Yates, Jr., T. M. Duncan, S. D. Worley, and R. W. Vaughn, Infrared spectra of chemisorbed CO on Rh, *J. Chem. Phys.* **70,** 1219–1224 (1979).
106. L. H. Dubois, P. K. Hansma, and G. A. Somorjai, The application of high resolution electron energy loss spectroscopy to the study of model supported metal catalysts, *Appl. Surf. Sci.* **6,** 173–184 (1980).

5

Incoherent Inelastic Neutron Scattering: Vibrational Spectroscopy of Adsorbed Molecules on Surfaces

R. R. Cavanagh, J. J. Rush, and R. D. Kelley

1. Introduction

Given the variety of laboratory-based, surface-sensitive vibrational spectroscopies, the need for an additional vibrational spectroscopy such as incoherent inelastic neutron scattering (IINS), which is based on a centralized user facility, may not be immediately obvious. In the absence of special capabilities or suitability for probing relevant classes of materials, it would be difficult to justify such experiments owing to both the expense and the limited accessibility of these facilities. However, the range of incident wavelengths (0.5–15.0 Å) and corresponding energies (0.8–300 meV) accessible with neutrons clearly distinguishes IINS from more conventional probes of molecular vibrations. In fact, incoherent inelastic neutron scattering embodies a number of attributes that make this reactor-based technique an often unique method for surface characterization and a powerful complement to other techniques.

R. R. Cavanagh, J. J. Rush, and R. D. Kelley • National Bureau of Standards, Gaithersburg, Maryland 20899.

Neutrons are not inherently surface sensitive probes. Unlike electrons or photons, which are of readily reflected, absorbed, or scattered by the electrons which bind samples, neutrons interact with the much smaller volume that is occupied by the nuclear cores. Statistically, then, neutrons reflect or scatter weakly from most materials, and a large percentage of those neutrons that impinge on most samples of thickness of ~1 cm will be transmitted. As a consequence, surface information is only accessible by employing samples where the surface/bulk atom ratio is large, and where the contributions to the scattering arising from the nuclei in the bulk can be separated from effects due to nuclei at the surface.

While a variety of experimental constraints must be met in order to conduct neutron scattering measurements of adsorbates, access to a nuclear reactor or high-intensity pulsed source represents the first major hurdle. Modern high-performance research reactors typically provide a flux of thermal neutrons in the range of 10^{14}–10^{15} neutrons/cm^2 per s. With such a flux, it is feasible to probe samples with as few as 10^{21} scatterers. However, flux alone does not adequately define the neutron beam requirements. The available beam size (cross-sectional area) is of comparable importance, since the probability of multiple scattering will increase exponentially with sample depth, but not at all with sample cross section. For the high sensitivity required in most neutron inelastic scattering studies of surface species, background is also critical.

As illustrated by the wide range of methods addressed in this book, vibrational spectroscopy provides a powerful probe of the nature of surface species. Recently, studies on well-characterized, single-crystal substrates have demonstrated the sensitivity of adsorbate vibrational modes to changes in surface topography and to the presence of chemical modifiers. However, there are few spectroscopic techniques that are directly applicable to the actual materials and conditions used in many practical chemical processes (e.g., the high-surface-area, optically opaque materials used as catalysts). It therefore becomes important to be able to compare the vibrational features observed on idealized single-crystal surfaces with the vibrational modes characteristic of those surfaces more commonly encountered in the real world. In addition, IINS is a technique that can probe the molecular species present at a solid–gas interface when the gas pressure is close to that encountered in practical chemical processes (i.e., atmospheres). A comparison can then be made of the molecular species present at the interface for high (i.e., atmospheric) and low (high-vacuum) gas pressures so that typical "surface science" experiments can be related to both the materials (catalysts) and operating pressures encountered in practical reaction systems.

Since the eigenvalues of individual vibrational modes are not affected by the nature of the scattering probe, the observed spectral features ought to be a common factor to all vibrational spectroscopies. Consequently, given data from comparable adsorption systems, the major distinction between the features observed with a technique like IINS and the features observed in other surface vibrational spectroscopies is simply a question of the operative selection rules and the magnitude of the matrix elements for excitation. In the case of IINS, it is straightforward to quantify and predict the relative intensities of various features in the vibrational spectrum. As will be seen in Section 2, the IINS signal is dominated by scattering from hydrogen. The relative intensity of each of the vibrational modes will be proportional to the root mean squared hydrogen displacement associated with each eigenvalue for all of the normal modes in a hydrogenous molecular species. Spectrometer design and sample preparation are considered in Section 3, while in Section 4, various aspects of such spectral analysis and subsequent spectroscopic assignments will be discussed in greater detail. Selected examples that illustrate the power of neutron scattering as a surface probe are presented in Section 5.

2. Fundamental Physics of Neutron Scattering

A number of introductory texts and reviews have been written on the subject of neutron scattering with emphasis on inelastic neutron scattering[1-9] Briefly summarized, the differential cross section for incoherent inelastic scattering in the single-phonon limit is given by

$$\frac{\partial^2 \sigma}{\partial \Omega\, \partial \omega} = \frac{k}{k_0} \frac{1}{N} \sum_{j,\,n}^{N} (b_n)^2 \exp(-2W_n) \exp(-\hbar\omega/2kT)$$
$$\times [\hbar(\bar{Q} \cdot \bar{C}_j^n)^2 / 4N\omega_j M_n] \mathrm{csch}(\hbar\omega_j/2kT) \delta(\omega_j - \omega) \qquad (1)$$

where b_n is the known scattering amplitude of the nth atom, M_n is the mass of atom n, N is the number of atoms in the molecule, ω_j and $\bar{C}_j^n$ are the frequency and displacement vectors for the nth atom in the jth normal mode, $\exp(-2W_n)$ is the Debye–Waller factor for the nth atom, δ is a delta function, $\bar{k}_0$ and $\bar{k}$ are the incident and scattered momenta, and $\bar{Q} = \bar{k} - \bar{k}_0$. The critical terms for understanding and interpreting inelastic neutron scattering are also the terms that distinguish it from the other forms of vibrational spectroscopy. First, thermal neutrons are more readily scattered by certain nuclei than by others. Tables 1 and 2 provide

Table 1. Summary of Important Terms and Units

$$E \text{ (meV)} = (0.286/\lambda(\text{Å}))^2 \times 1000$$
$$E \text{ (joules)} = 1.602189 \times 10^{-19} \times E \text{ (eV)}$$
$$m_{\text{neut}} = 1.008665 \text{ a.u.}$$
$$= 1.67495 \times 10^{-27} \text{ kg}$$
$$h = 6.62618 \times 10^{-34} \text{ J s}$$
$$k = 1.38066 \times 10^{-23} \text{ J}/K$$
$$b_H^{(+)} = 1.04 \times 10^{-14} \text{ m (triplet)}$$
$$b_H^{(-)} = -4.74 \times 10^{-14} \text{ m (singlet)}$$
$$\overline{|b_H|^2} = \tfrac{3}{4}|b_H^{(+)}|^2 + \tfrac{1}{4}|b_H^{(-)}|^2 = 6.49 \times 10^{-28} \text{ m}^2$$
$$\bar{b}_H = \tfrac{3}{4}b_H^{(+)} + \tfrac{1}{4}b_H^{(-)} = -0.375 \times 10^{-14} \text{ m}$$

Coherent cross section

$$\sigma_c = (4\pi \, |\bar{b}^2|)$$

Total cross section

$$\sigma = (4\pi \, \overline{|b|^2})$$

Incoherent cross section

$$\sigma_{\text{inc}} = \sigma - \sigma_c$$

Table 2. Selected Scattering and Capture Cross Sections[1,6,8] (in Units of 10^{-24} cm^2) and Scattering Lengths (in Units of 10^{-12} cm)

Isotope	Coherent cross section	Incoherent cross section	Capture cross section	b scattering length
^{3}He		1.2	5500	0.62
^{4}He	1.1	0.0	0	0.30
Ne	2.8	0.1	<2.8	0.46
Kr	7.3		31	0.74
H	1.8	79.7	0.19	-0.374
D	5.6	2.0	0.0003	0.667
^{12}C	5.5	0.0	0.003	0.663
^{14}N	11.1	0.3	1.1	0.94
^{16}O	4.24	0.0	0.0001	0.58
Ni(natural abundance)	13.2	5.2	4.6	1.024
^{58}Ni	26		4.2	1.47
^{60}Ni	1.1		2.5	0.28
^{62}Ni	9.5		15	-0.85
Pt	11.2	0.7	8.8	0.95
Al	1.5	0.01	0.13	0.35
Si	2.16	0.03	0.06	0.42
Mg	3.34	0.36	0.04	0.52
Fe(natural abundance)	11.4	0.44	1.4	0.96
SiO$_2$		9		
Al$_2$O$_3$		15		

definitions and examples of coherent, incoherent, and capture cross sections for a number of isotopes in addition to their scattering amplitudes. Hydrogen dominates all other isotopes with a total cross section of 82 barns (1 barn = 10^{-24} cm^{-2}); deuterium has a total cross section of only 7.6 barns, and most other nuclei scatter even more weakly. The source of these large differences is related to the change between real and virtual levels in scattering from H (due to the forbidden nature of the triplet state formed by the scattering of two odd-spin particles) compared to the real levels in D or other nuclei (i.e., no forbidden triplet states are involved). Table 1 illustrates the effect for hydrogen. Similar but less dramatic evidence of the isotopic specificity can be seen in the cross sections associated with various nickel isotopes, and also in the scattering cross section for nickel as it occurs in natural abundance (see Table 2). A major consequence of the variation in scattering cross section is that, by judicious choice of adsorbate (namely, hydrogen or hydrogen-containing species) and substrate (such as graphite, platinum, or nickel), an enhanced surface sensitivity can be attained.[10–13]

The second influence on the intensity of observed neutron scattering features arises from the term $(\bar{Q} \cdot \bar{C}_j^n)^2$. This term is similar to the dipole operator term which determines the intensity of all infrared and certain electron energy loss spectroscopy (EELS) features. In those more familiar experimental methods, the evaluation of the dipole matrix element which appears in the differential scattering cross section requires a detailed knowledge of both the dipole moment function [i.e., $\mu(r)$], and the vibronic wave functions which characterize the initial and final states. Evaluation of the $\bar{C}_j^n$ is much less complicated. To the extent that hydrogenous adsorbates are involved, the scattering intensity is dominated by $\bar{C}_j^n$ (in other words, the hydrogen atom displacements associated with each normal mode). Hydrogen atom displacements can be predicted using a force-constant function and a normal-mode analysis. Prior knowledge of the dipole moment function is not pertinent. Consequently, within the framework of a simple ball-and-spring model of the adsorbate, the spectral features and their relative intensities are, in principle, directly interpretable.

While the remainder of this chapter will be devoted to incoherent inelastic neutron scattering as a probe of surface vibrational modes, it should be acknowledged that neutron-scattering techniques are applicable to a significantly broader range to topics. The utility of neutrons in probing crystallographic structures and in extracting crystallite sizes is well known.[3] Recent efforts in the characterization of zeolite structures provide a notable example.[14,15] Neutrons can also probe the microstructures and particle or pore sizes in materials in the 10^1–10^4 Å regime by using small-angle neutron scattering (SANS).[16,17] Further, as one approaches lower and lower energies in characterizing the dynamics of adsorbed layers, it is no longer appropriate to consider adsorbate modes in terms of simple harmonic oscillators. For instance, large-amplitude, anharmonic, adsorbate torsional modes will exist and may have discrete, resolvable energy levels. Neutron scattering is capable of quantifying the relevant parameters such as the spectroscopy of the torsional modes and tunnel splittings, thus characterizing the relevant potential energy surface. Furthermore, the mechanistic details for diffusional motion of adsorbates become accessible using high-resolution (<0.1 meV) quasielastic neutron scattering. From quasielastic scattering (analogous to the Doppler shift due to the moving adsorbate), residence times for reorientational or translational diffusion and, most particularly, molecular-scale diffusion mechanisms can be determined. In part, the beauty of neutron scattering for such experiments is that the measurement is totally independent of the symmetries of the initial and final states and does not demand a spatial variation in the moments of the local charge distribution. Thus even large amplitude torsional modes are readily accessible in IINS.

3. Experimental

Historically, most neutron scattering research has been accomplished at high-performance research reactors where intense beams of neutrons are produced by uranium fission. More recently "pulsed" neutron sources have been developed which produce bursts of neutrons by using the collisions of high-energy protons or electrons with heavy-metal targets. Our discussion will focus on results using reactor-based instrumentation, but the interested reader is referred elsewhere[18] for a detailed discussion of the emerging pulsed neutron sources for studies of materials structure and dynamics.

In all of the experiments to be considered here, a beam of either thermal neutrons or cold neutrons is employed.[19] Since all of the neutrons are originally created with high kinetic energy in the core of the reactor, some form of moderator must be employed in order to shift the flux distribution into an energy range of interest. For a thermal beam, one typically uses heavy water. The temperature of the moderator characterizes the approximately Maxwellian distribution of energies of the emerging neutron gas, even though the kinetic energy distribution has a persistent high-energy tail due to incomplete equilibration within the moderator. Thus, a typical "thermal" neutron beam for a 330 K moderator will have a most probable kinetic energy of about 30 meV, while a beam emerging from a cold moderator at 30 K will have a most probable kinetic energy of 3 meV. The characteristic temperature of the reactor moderator is important for determining the feasibility of different experiments. The higher the mean kinetic energy, the greater the flux of neutrons with sufficient energy to excite the high-frequency modes of the sample. Frequently, these high-energy neutrons are accompanied by a higher background of unmoderate ("fast") neutrons which contribute to the experimental background. In order to minimize such contributions, a variety of filters and shields are employed to prevent the "fast background" from reaching the counter. The technique used in both major types of neutron spectrometers (Sections 3.1 and 3.2) for dealing with the fast background is the same. Namely, the detectors, filters, and relevant flight paths are surrounded with paraffin combined with a neutron absorbing material (e.g., B, Cd, or Li). The high density of hydrocarbon modes present in the paraffin results in an efficient slowing of any "fast"incident neutron. Thus, the "fast background" which is characteristic of the reactor environs is greatly reduced by slowing down and then absorbing stray neutrons before they reach the detector.

3.1. Modified Triple-Axis Spectrometer

Figure 1 depicts a typical triple-axis spectrometer. The moderated neutron beam emerges from a port in the biological shield of the reactor and is immediately collimated by a set of cadmium-coated blades (Soller slits). The cadmium coating serves to absorb neutrons whose trajectories lie outside the acceptable angular divergence. Typical angular divergences are in the range of 20–60 min. This collimated beam then strikes a crystal monochromator. The monochromator can vary from metal single crystals (copper, beryllium, germanium) to pyrolytic graphite. The monochromator of choice is determined by the interplanar spacing needed to provide (by Bragg reflection) the appropriate energy or wavelength resolution and the energy range of interest. Great care is taken to optimize the reflectivity of the monochromator crystal, since neutron scattering is an inherently signal limited technique. The Bragg-scattered neutrons of interest pass through a port ($\sim$10–30 cm^2) in the spectrometer shield which rotates in relation to the monochromator crystal and which houses a second set of collimators. This collimated and monochromatized neutron beam subsequently impinges on a sample.

The neutrons that pass through the sample without being scattered

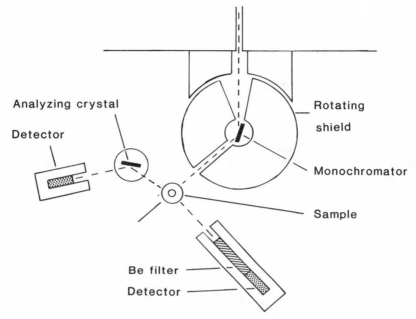

Figure 1. Triple-axis spectrometer with monochromator and analyzing crystal and modified triple-axis spectrometer with monochromator and Be filter.

are intercepted by a beam stop. The signal of interest is contained in those neutrons that have lost part of their initial kinetic energy as a result of exciting an internal mode of the sample. The scattered neutron energy can be measured in one of two ways. In a conventional triple-axis spectrometer, a well-shielded crystal monochromator and detector assembly is placed after the scattering sample, so that the scattered neutron energies can be selected by adjusting the Bragg angle between the monochromator and detector (see Figure 1). Such a detection system, which is widely used for ordered specimens, is not generally appropriate for the rather low intensities observed in spectroscopic studies on chemically active surfaces. In this case, a less sophisticated energy analyzer is used to increase greatly the solid angle scanned, and thus, the observed spectral intensity. A low-energy bandpass filter is located between the sample and the final detector (a ^{3}He gas-filled proportional counter). The filter consists of a block of beryllium which is cooled to liquid nitrogen temperature. The beryllium scatters all neutrons other than those between 0 and 5 meV due to the 4-Å Bragg cutoff wavelength of beryllium. (By cooling the filter from 300 to 77 K, the transmission of the beryllium in the range between 0 and 5 meV is increased by a factor of 2 to about 90%.) This filter is housed in a polyethylene-boron carbide shield which absorbs those neutrons Bragg-scattered by the cold beryllium. It is this detector configuration that comprises the modifications made to the conventional triple axis spectrometer (Figure 1).

The energy resolution of IINS experiments using a modified triple-axis spectrometer is determined by two factors. In the low-energy region (<100 meV), the resolution is fixed by the 5-meV bandpass of the beryllium filter. The Cu(220) monochromator resolution (using 40-min collimation both in the pile and between the monochromator and the sample) is $\Delta E/E \approx 5\%$. Thus, above 100 meV, the resolution is no longer limited by the detection system but rather by the performance of the monochromator.

Energy scans of the incident neutron beam require rotation of the monochromator crystal, the shield, the sample, the detector, and the beam stop. The mass of these components, coupled with the large displacements associated with spectral scans, can be dealt with by using naval gun mount bearings or their equivalent to support the instrumentation. Scanning is typically achieved using computer control of the moving components in the spectrometer. Since the incident flux varies non-monotonically with the beam energy (due to the initial distribution of energies in the beam combined with the Bragg diffraction of the monochromator crystal) some normalization must be made in order to compare the signals observed at different energies. This is accomplished by placing a beam monitor between the downstream side of the final collimator and the sample. The monitor consists of a neutron-absorbing foil (99.99% transmitting), which releases charged particles that are subsequently detected. The computer advances the instrument to the next energy of interest when the predetermined number of neutrons has passed through the monitor. In this fashion, the absolute number of neutrons impinging on a sample at a given incident energy can be held fixed during a run, and thus, signals obtained under varying experimental conditions can be compared directly. Furthermore, in cases where blank subtraction procedures are unnecessary, it is still possible to correct the observed signal for contributions from the fast-neutron background since such noise will scale linearly with the sampling time at each incident energy.

Clearly, the resolution minimum of 5 meV presented by the beryllium filter presents an unacceptable limit for certain measurements. Higher resolution can be obtained by use of the more conventional triple-axis-spectrometer configuration discussed above. However, such measurements greatly sacrifice the count rate in order to attain the somewhat improved energy resolution along with high (but unnecessary) momentum resolution. Alternatively, energy resolution can be improved somewhat without paying the price of high momentum resolution by using tighter collimation and bandpass filters with longer-wavelength (lower energy) cutoffs, such as graphite or beryllium oxide.

3.2. Time-of-Flight Spectrometry

An alternative approach to inelastic neutron scattering exists. In this second methodology, control and measurement of neutron velocities is critical. Thus, rather than using a crystal monochromator combined with a low-energy bandpass filter or a crystal analyzer to monitor the inelastically scattered neutrons, one can use time-of-flight (TOF) techniques. Since a typical thermal neutron (wavelength ~ 2 Å) has a velocity of 2000 m s^{-1} it is apparent that an inelastically scattered neutron will have a measurably different velocity compared to an elastically scattered neutron, and that this velocity and related energy can be measured directly by determining the time-of-flight of a neutron over a finite flight path ($\sim$meters).

Figure 2 depicts a schematic of one type of TOF spectrometer for such neutron scattering experiments at a steady-state (reactor) neutron source. In this case, a set of parallel crystal monochromators select an incident neutron energy and direction. A neutron chopper is then used to create a discrete burst of neutrons (FWHM $\sim 10 \text{ μs}$). [At a pulsed neutron source, the bursts of neutrons are provided by the periodic collisions (~ 10–10^2 s^{-1}) of charged particles with a heavy metal target.] The elastically scattered neutrons will generally provide an intense peak at a time corresponding to no velocity change (energy transfer) in the scattering process. However, an inelastic scattering event corresponding to neutron energy loss (gain) at the sample will result in late (early) arrival of that neutron at the detector. As in the modified triple-axis

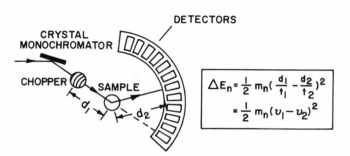

Figure 2. Time-of-flight spectrometer.

experiments, hydrogenous surface species will be most efficiently probed because of their large inelastic scattering cross section.

In the case of TOF experiments, the signals are usually measured as a function of arrival time after the formation of the neutron burst from a neutron chopper or pulsed source. In this case, in order to relate the time-of-flight spectrum to the vibrational or rotational energy levels in the system, it must be transformed to an energy scale by

$$\frac{\partial^2 \sigma}{\partial \Omega \, \partial \omega} = \frac{h\tau^3}{m_{neut}} \frac{\partial^2 \sigma}{\partial \Omega \, \partial \tau} \tag{2}$$

where τ is the absolute value of the difference in the time of flights of the elastically and inelastically scattered neutrons.

Additional information is often available by taking data as a function of Q. As can be seem from equation (1), the sensitivity to inelastic scattering is optimized by measuring energy transfers at large momentum transfer (Q). Unfortunately, the contribution of multiphonon scattering processes to the measured spectrum also increases with Q (Q^4 dependence). Hence, it is possible to distinguish multiphonon scattering from single-phonon processes by examining the Q dependence of the IINS spectrum. A primary advantage of the TOF technique is that it becomes possible to detect the inelastic scattering spectrum simultaneously over a wide range of energy and wavevector transfers. This possibility exists for a multidetector TOF spectrometer (as compared to the single detector crystal spectrometer), since the wide range of scattering angles (and therefore, Q's) are sampled simultaneously and neutrons of all energies are detected after each chopper (or source) burst ($\sim 100 \, s^{-1}$). On the other hand, the TOF spectrometer is much less useful for measuring coherent scattering concentrated around particular energies or momenta such as in studies of individual phonons in single crystals. In addition, the required beam pulsing means that the beam is "on" only 0.1%–1% of the time. Typical TOF instruments consist of 10–100 detectors on an arc with sample scattering angles between 10° and 120°. A timing resolution of 10 μs at 5 meV results in an energy resolution of ~ 0.1 meV. Additional experimental details may be found elsewhere.[19]

3.3. Sample Cells

Various experimental factors must be considered in the design of cells for inelastic neutron scattering studies of high-surface-area materials. The size and shape of the cell will depend on the characteristics of the probing neutron beam (e.g., the geometric dimensions and intensity) and the surface-to-bulk ratio of the sample. In general, sample cells are cylindrical, but wafer-shaped cells are also employed to accommodate large cross section ($30 \, cm^2$) neutron beams.

Materials used to construct sample cells must transmit neutrons with high efficiency. They must, therefore, be free of elements with large neutron absorption cross sections (such as cobalt and cadmium). For inelastic measurements, aluminum is the most frequently used cell material owing to its low scattering cross section and its ease of fabrication, although thin-walled cells of stainless steel and quartz have also been utilized successfully. For neutron diffraction experiments, vandium is widely used and is the material of choice owing to its very low coherent cross section. Typically, all extraneous portions of the sample cell (e.g., valves, heaters, thermocouples, etc.) are masked with thin sheets of cadmium metal to reduce background contributions to the scattering intensity.

Cryogenic to elevated temperatures ($\sim$4–700 K) are of interest in IINS studies of adsorbates. Most reported adsorbate IINS spectra have been measured at 77–100 K due to the reduction of bulk multiphonon scattering from the sample material (in general, the largest single source of background scattering intensity) at these low temperatures. Low-temperature measurement improves the ratio of desired signal-to-background scattering intensities and thus decreases the necessary measurement time. From a surface chemical viewpoint, low temperatures are needed to probe weakly bound surface species (such as physisorbed molecules) and the low-frequency modes (such as hindered rotational modes) of strongly chemisorbed species. Higher temperatures enable the study of both thermally induced changes in the adsorbed layer, such as adsorbate rearrangement or decomposition,[20] and catalytic reactions in which reactant gases are converted to products via interaction with the surface.[21]

Cryogenic temperatures are produced and maintained by mounting the sample cell in a cryostat. The cell is attached to a thermally conductive block through which either liquid nitrogen or liquid helium is passed. The desired block temperature is maintained both by controlling the flow of the cryogenic fluid and by a resistive heater located in the mounting block. Alternatively, Joule–Thomson refrigerators can be used to produce low temperatures.

Elevated temperatures can be produced in much the same way—sample heating by conduction from a controlled temperature block. A heater and thermocouple imbedded in the block, together with a standard temperature-control device, can provide a constant temperature block. As in cryogenic applications, a vacuum jacket (aluminum) is used to thermally isolate the sample and heater block from the atmosphere. It should be noted that, for temperatures above ~400 K, shielding of extraneous material near the sample by contact with cadmium cannot be employed because of the high vapor pressure of cadmium.

A schematic drawing of a wafer-shaped (7.5 × 7.5 × 1.2 cm with a 0.04 cm wall thickness) stainless steel sample cell used at the National Bureau of Standards is depicted in Figure 3.[22] The border of the cell is in intimate contact with copper bars to facilitate heat transfer and minimize temperature gradients across the sample. The sample is loaded into the cell through a flanged port at the top. Bakable, all-metal bellows-sealed valves are used for introduction and evacuation of gases. The gas inlet tube is perforated to achieve more uniform adsorbate dosing.

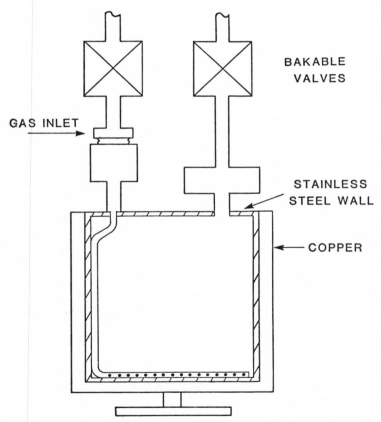

Figure 3. All-metal cell used for neutron inelastic scattering.[22]

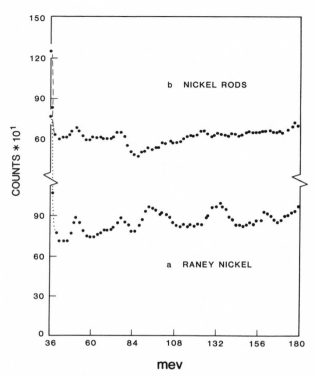

Figure 4. IINS spectra from nickel, corrected for contributions from fast neutrons: (a) Raney nickel blank and (b) nickel rods.[22]

Figure 4 shows raw data (corrected for contributions from fast background) obtained with this cell for inelastic scattering from a "clean" Raney nickel (W. R. Grace Co., type Raney 4100[23]) sample (a) compared with the scattering observed from the same quantity of bulk nickel (b).[22] The sample of Raney nickel (a nickel powder with crystallites of average diameter <70 Å) had a mass of 90 g. The intense feature near 36 meV in both spectra is due to scattering from bulk nickel phonons. In fact, these data demonstrate that the cell contributes little to the IINS signal compared to the contributions from the specimen, and that the clean Raney nickel spectrum is dominated by scattering from nickel. Comparison of the spectra indicates that the high-surface-area Raney nickel also shows discernible scattering features in the region between 80 and 160 meV indicative of bound hydrogen. The intensity of these features is small in comparison to that observed at saturation hydrogen coverage.

3.4. Sample Preparation

The preparation of clean surfaces is particularly important in chemisorption studies where surface impurities can drastically alter adsorption behavior. The materials traditonally utilized in neutron scattering studies are frequently of catalytic interest—small metal particles (supported and unsupported), oxides, zeolites, etc. Extensive techniques and procedures have been developed with regard to the preparation and characterization of these catalyst materials for adsorption and reactivity studies.[24]

There are few published reports[25] of inelastic scattering from adsorbates on supported metal catalysts owing undoubtedly to the presence of large quantities of hydrogen on the oxide support (mainly in the form of OH). It should be possible to use IINS to investigate the surface chemistry of these catalysts provided that a support with a low OH concentration is used. However, most neutron scattering work on chemisorption to date has involved unsupported metal powders—Raney nickel,[26] Raney platinum,[27] platinum black,[28] and palladium black.[29] These powders sinter rapidly at high temperatures and pretreatments are restricted to lower temperatures where the metal surface area is not significantly reduced. Treatment in hydrogen for extended periods followed by helium purging and evacuation at these lower temperatures was found necessary to produce a metal surface relatively free of hydrogenous impurities (such as hydrogen, hydrocarbons, and hydroxyls) as determined by inelastic neutron scattering.[22] The amount of such impurities remaining on the "clean" surface can be estimated from their contribution to the total IINS intensity or from a gas-phase isotopic mass balance following equilibrated exchange between a known quantity of gas-phase deuterium and the impurity hydrogen present.[30]

Because of its low incoherent neutron scattering cross section, deuterium can also be used to mask hydrogenous-impurity vibrations.[22] For example, in the preparation of Raney nickel, the last 5% of the hydrogen removed was characterized by a vibrational mode at 95 meV. Replacement of this hydrogen with deuterium, followed by additional H_2 adsorption at 150 K, gave no evidence of the 95 meV peak. The absence of H–D exchange at 150 K strongly suggested that the 95-meV mode was not associated with chemisorption on the nickel surface sites. This vibrational mode was thought to be due to OH species associated with a small concentration of alumina present in the Raney nickel. For supported metal catalysts, there is potential for similar masking by deuterium exchange with hydrogen present on the support.

The total surface area of these materials can be estimated by the

BET method.[31] With metal powders, the BET area can be compared with the area calculated from quantitative hydrogen adsorption as a diagnostic. For example, for platinum black, the measured BET/H_2 area ratio is close to unity;[32] for Raney nickel, this ratio is about 2.[22] The high ratio for Raney nickel can be attributed to the presence of both unreacted Ni–Al alloy and high-surface-area alumina, which do not possess sites for hydrogen adsorption.

Two gas phase analytical techniques are useful for following reactions on these high-surface-area materials. First, a quadrupole mass spectrometer (QMS) has been used to measure the products (H_2, C_2H_6, C_2H_4) desorbed into the gas phase during warming after a low-temperature chemisorption of C_2H_4 on Raney nickel.[33] The QMS was also used to determine the residual H atom coverage on platinum black by measuring the $H_2/HD/D_2$ distribution following exchange of chemisorbed H and gas phase D_2.[32] Second, a gas chromatograph (GC) has been used to monitor the cleaning by hydrogenation of Raney nickel.[22] The disappearance of CH_4 from the flowing hydrogen during the hydrogenation correlated with the disappearance of a broad vibrational peak centered at 135 meV (see Figure 4a), which is therefore to be associated with hydrogen bound to surface carbon. The long time required for the removal of this surface carbon may be due to diffusion of carbon from the bulk to the nickel surface. The two techniques provide direct methods for monitoring reaction rates during *in situ* IINS measurements on catalytically active samples.

Characterization by high-resolution neutron powder diffraction and small-angle neutron scattering can yield the distribution of sizes and shapes of the particles and pores in these materials. For example, an analysis of the widths and shapes of diffraction peaks can reveal the distribution of crystallite sizes. SANS data, on the other hand, can be used to characterize the size and shape of the particles and pores themselves. An inherent advantage of using neutron scattering to measure vibrational spectra is that these other neutron techniques are available to characterize the samples further. These techniques are convenient and complemetary to incoherent neutron scattering studies; they can be applied to the identical sample and cell used in the vibrational experiments. This is especially useful for highly air-sensitive, reduced-metal catalysts.

3.5. Data Acquisition

It should be evident at this point that a variety of factors influence the rate at which useful data can be acquired. In addition to source-dependent terms such as flux per unit band width, the beam cross section, and the level of background counts, the performance of the instrumentation required for different measurements needs to be considered. Furthermore, the duration of a series of experimental measurements will be influenced by the size of the samples since a significant amount of time may be required in order to achieve thermal equilibrium. At a 20-MW reactor typical scan times will range from several hours up to a full day.

4. Analysis of Data

As indicated earlier, neutrons are not preferentially scattered at surfaces. The major experimental consequence of this is that the neutrons directly measure the vibrational spectrum of the underlying substrate, even in the absence of adsorbates. In order to deal with scattering from the bulk, a spectrum of the adsorbate-free material is recorded. Since neither the energy calibration nor the incident neutron flux is affected by the chemisorption process, a quantitative subtraction is possible. *In situ* dosing of the clean sample subsequent to measurement of the blank spectrum is recommended wherever possible. Alternatively, exactly reproducible mounting of the sample must be assured.

The magnitude and spectral distribution of the substrate-dependent background is a function of temperature, since both single-phonon and multiphonon scattering processes vary somewhat with temperature.[1] Thus, for samples with major contributions from the substrate, it will be necessary to record spectra of the blank at the same temperature. In this way, multiphonon corrections will be automatically accounted for upon subtraction of the blank spectrum.

Preparation of hydrogen-free adsorbents has been informative. Since neutron scattering is, in fact, equally sensitive to surface and subsurface hydrogen, it is possible spectroscopically to distinguish residual surface hydrogen from hydrogen in the bulk as a consequence of their differing force constants, bond lengths, and site symmetries. In addition, the capacity for the remaining hydrogen (surface or bulk) to react subsequently during the course of the experiment can be ascertained by a series of experiments using selective deuteration.

4.1. Force Models and Vibrational Dynamics

The simplest system to consider in vibrational spectroscopy is a diatomic harmonic oscillator. Within this framework, given the masses and bond strength, the vibrational energy levels of the system can be directly related to a Hooke's law force constant, k, by $\omega = (k/m)^{1/2}$. For more complex species (more than two atoms), there are a variety of ways to formulate the problem within a ball-and-spring approach. In this section, the well-known treatment of Wilson[34] will be recalled. Although originally developed for spectroscopies that involved optical selection rules (where a knowledge of the symmetries of the vibrational states is useful), the Wilson FG matrix technique is equally well suited to IINS. Since a variety of texts exist that fully develop the FG technique[34-37] a simple heuristic approach will be adopted here.

If one writes down the equations of motion of the N atoms in some arbitrary molecule, one will have $3N$ equations. Each atom will have associated with it both a kinetic energy and a potential energy. However, a key step in solving any vibrational problem lies in judicious choice of the coordinate system. Typically, internal coordinates that reflect the bonding due to the valence electrons are chosen. In the Wilson FG method, the relationship between a set of internal coordinates (S) based on the Cartesian displacement coordinates (dx) is established. In this new coordinate system, the kinetic energy (T) is

$$2T = \sum_{t, t'} G_{tt'}^{-1} \dot{S}_t \dot{S}_{t'} \tag{3}$$

where

$$G_{tt'} = \sum_{i=1}^{3N} m_i^{-1} B_{ti} B_{t'i} \qquad t, t' = 1, 2, \ldots, 3N - 6 \tag{4}$$

and the B_{ti} are constants determined by the geometry of the molecule, which transform the Cartesian displacement coordinates (dx) to the internal coordinates (S). Explicitly, the transformation to internal coordinates is given by the $\bar{B}$ matrix

$$S_t = \sum_{i=1}^{3N} B_{ti} \, dx_i, \qquad t = 1, 2, \ldots, 3N - 6 \tag{5}$$

The $\bar{G}$ matrix is the mass-weighted coordinate transformation. The potential energy (V) is similarly written as

$$2V = \sum_{t, t'} F_{tt'} S_t S_{t'} \tag{6}$$

The elements of the $\bar{F}$ matrix are simply defined as

$$F_{ij} = (d^2 V / dS_i \, dS_j) \tag{7}$$

or the force constants in terms of the internal coordinates. Since it is apparent from equations (3) and (6) that the kinetic and potential energy of the system can be expressed in the internal coordinate system (S),

Newton's equations of motion can be cast in the form

$$\frac{d}{dt}\frac{\partial T}{\partial \dot{S}_i} + \frac{\partial V}{\partial S_i} = 0, \qquad i = 1, 2, \ldots, 3N - 6 \tag{8}$$

Since equation (3) is only a function of $\dot{S}$, and equation (6) is only a function of the internal coordinates, it is apparent that the solutions of equation (8) must be of the form

$$[\bar{F} - \bar{G}^{-1}\lambda] = 0 \tag{9}$$

or

$$[\bar{F}\bar{G} - \bar{E}\lambda] = 0 \tag{10}$$

where the eigenvalue λ is equal to ω^2, and $\bar{E}$ is the unit matrix.

The advantages of this formulation are that the coordinates have simple physical meanings (e.g., bond angles and bond distances), and the force constants are associated with easily visualized motions. For the purposes of incoherent inelastic neutron scattering experiments, it is most convenient to formulate the problem in valence coordinates, but to then solve the eigenvalue problem in Cartesian coordinates. Thus by always expressing the force field in valence coordinates it is possible to avoid introducing symmetry-dependent terms into the problem. By solving the problem in Cartesian coordinates the introduction of redundant coordinates is avoided. The transformation matrix (B) from Cartesian (X) to valence (S) coordinates is simply

$$\bar{S} = \bar{B} \cdot \bar{X} \tag{11}$$

Such valence coordinates convey a chemical-bond sense of the vibrational modes. Generally, the coordinates can be classified as (i) bond stretches, (ii) angle bends, (iii) out-of-plane wagging, (iv) perpendicular pairs of linear angle bends, and (v) torsions. Force constants expressed in terms of these valence coordinates F_{ij} can then be transformed back to force constants Φ_{ab} in the original Cartesian coordinate scheme by

$$\Phi_{ab} = \sum_{i,j} \frac{B_{ai}F_{ij}B_{bj}}{(M_a M_b)^{1/2}} \tag{12}$$

The eigenvectors [the $\bar{C}_j^n$ of equation (1)] obtained in the Cartesian coordinate representation are then proportional to the extent of displacement of each atom in that mode. By taking the square of the H-atom coefficients, thus forming $(\bar{Q} \cdot \bar{C}_j^H)^2$ in equation (1), one obtains the relative IINS signal.

4.2. Neutron Spectra

The utility of combining such force field methods with IINS data can be illustrated by recent work on the inorganic complex benzenetricarbonylchromium(0), $(\eta^6\text{-}C_6H_6)Cr(CO)_3$.[38] Figure 5 shows the IINS spectrum and the calculated spectrum from a force field model which has taken into account the optical data, the IINS spectrum, and x-ray diffraction data. In this work, it was possible to distinguish between several existing force fields in the region above $800\ cm^{-1}$, and to provide a better fit to the IINS data at low frequencies than was provided by any of the existing models. In general, the existence of force constants from other molecular systems provides a convenient starting point for IINS analysis.[39–43]

Such neutron scattering data provide a measure of the density of vibrational states $g(\omega)$. Equation (1) is frequently written in terms of an effective density of vibrational states:

$$\frac{\partial^2 \sigma}{\partial \Omega\, \partial \omega} = g_{\text{eff}}(\omega) Q^2 \frac{k}{k_0} \frac{1}{\omega} [\exp(\hbar\omega/kT) - 1]^{-1} \tag{13}$$

where $g_{\text{eff}}(\omega) = b_H^2 (C_j^H)^2 \exp(-2W) g(\omega)/2M_H$. Thus, the IINS spectrum is the density of vibrational states weighted by the incoherent scattering cross section and the rms displacement of each atom.

4.3. Diffusion and Reorientation

While not strictly related to quantized surface vibrational energies, the nature of the diffusion or reorientation of surface species is often acknowledged to be closely coupled with low-frequency adsorbate modes. In addition, there is a direct correlation between rotational diffusion and adsorbate torsional modes and the rotational potential. Therefore, it is worth noting the special capabilities that neutron scattering brings to studies of surface rotational modes and dynamics.

The cross section for neutron scattering can be written in ways that better display its sensitivity to diffuse or reorientational motion. For instance, equation (1) can also be expressed in terms of the intermediate scattering law $S(Q, \omega)$[1]:

$$\frac{\partial^2 \sigma_{\text{inc}}}{\partial \Omega\, \partial \omega} = \frac{k}{k_0} b_{\text{inc}}^2 \exp(-\hbar\omega/2kT) \bar{S}_{\text{inc}}(Q, W) \tag{14}$$

where $\bar{S}_{\text{inc}}(Q, \omega)$ is the symmetrized form of $S(Q, \omega)$. Since any spectrum can be described as the Fourier transform of a correlation function which describes the time delay of fluctuations of the property

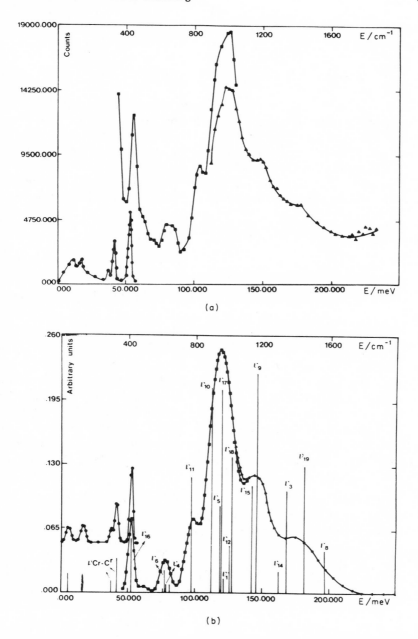

Figure 5. (a) Observed neutron spectrum of $(C_6H_6)\ Cr(CO)_3$ (note that the spectrum is not corrected for the beryllium filter transmission function) and (b) calculated spectrum.[38]

coupling the system to the exciting radiation, the scattering law can be written as

$$S_{inc}(Q, \omega) = 1/(2\pi\hbar) \int \int G_s(r, t) \exp(i[Q \cdot r - \omega \cdot t]) \, dr \, dt \quad (15)$$

where $G_s(r, t)$ is the van Hove correlation function. Note that $G_s(r, t)$ is the probability that, given a particle at the origin at $t = 0$, the same particle will be found at r at time t. In general, a form of $G_s(r, t)$ is chosen and Fourier-transformed to give $S(Q, \omega)$, which is then compared to the observed spectra.

Consider, first, classical diffusion in three dimensions. Solving the classical diffusion equation for the translational correlation function, one obtains

$$\partial G_s(r, t)/\partial t = Dv^2 G_s(r, t) \quad (16)$$

where $G_s(r, 0) \equiv \delta(r)$, v is the velocity, and D is the macroscopic diffusion coefficient. This equation is readily solved to give

$$G_s(r, t) = (4\pi D \, |t|)^{-3/2} \exp[-r^2/(4D \, |t|)] \quad (17)$$

Transforming this correlation function into a scattering law gives

$$S(Q, \omega) = (1/\pi)DQ^2/[(DQ^2)^2 + \omega^2] \quad (18)$$

a Lorentzian function. However, as one moves to larger values of Q, the corresponding velocity of the diffusing particle will exceed the thermally accessible regime, $(2kT/m)^{1/2}$. In this large-Q limit, it is necessary to account for the site-to-site details of the atomic (or molecular) diffusion process. Such a jump-diffusion model was originally developed by Chudley and Elliot.[44] While more realistic models have now been developed, they generally make certain assumptions about the diffusion process:

1. diffusion jumps are not correlated with thermal vibrations,
2. successive jumps are uncorrelated, and
3. interactions between diffusing particles are negligible.

Such models yield a scattering law of the form[45]

$$S(Q, \omega) = (1/\pi)F(Q)\Gamma/(\omega^2 + [F(Q)\Gamma]^2) \quad (19)$$

where

$$F(Q) = (1/n)\sum_{i=1}^{n} (1 - \exp(-iQ \cdot L_i)]$$

In this expression the intermediate scattering law has been taken in the form $I(Q, t) = \exp[-F(Q)t\Gamma]$, where Γ is the inverse of the time between jumps and L_i is nearest neighbor distance. In the small $Q^2\langle L^2\rangle$ limit (typically $Q < 0.3$ Å^{-1}), the scattering law becomes $(1/\pi)\Gamma/[\omega^2 + \Gamma^2]$, such that Γ is associated with the classical diffusion constant (D); in the limit of large $Q^2\langle L^2\rangle$, Γ is associated with the inverse of the time between jumps. The width Γ produces elastic-peak broadening in the neutron spectrum which is directly related to the residence time and jump distance for individual diffusing species. Thus, just as Doppler spectroscopy can provide data that measure the translational energy distribution of species, the width of elastically scattered neutrons will also reflect the nature of the nondiscrete levels of the system. It is this sample-dependent broadening of the elastic peak that is referred to as quasielastic scattering.

Next let us consider an adsorbed species undergoing rotational motion relative to a surface. Ethylidyne ($\equiv$C—CH$_3$), ammonia (—NH$_3$), or hydroxyl (—OH) species might be examples of such adsorbates. For these species, even if the adsorbate were not undergoing translational diffusion, rotational diffusion (reorientation) would contribute to the inelastic scattering by introducing a broadened component to the elastic peak, and the width and intensity of this quasielastic peak provides a direct probe or the reorientation dynamics. These widths are best measured using high-resolution time-of-flight or crystal spectrometers (if possible using near back-reflection). Such neutron studies in bulk materials have yielded information on molecular scale diffusion or reorientation processes in the time regime from 10^{-8} to 10^{-13} s. In this case, the scattering law can be approximated by

$$S(Q, \omega) = \left\{\left[1 + 2\frac{\sin(Q \cdot L)}{Q \cdot L}\right]\delta\omega + 2\Gamma/\pi(\Gamma^2 + \omega^2)\right.$$
$$\left. \times \left[1 - \frac{\sin(Q \cdot L)}{Q \cdot L}\right]\right\}\exp(-2W_n) \qquad (20)$$

The first term is the elastic coherent structure factor (due to fixed adsorbates) and the second factor is due to rotational diffusion. Taking $\tau_R = \tau_0 \exp(-E_R/RT)$ and extracting the temperature dependence of τ_R, one obtains the activation energy for rotational reorientation (E_R). Thus, measurement of $S(Q, \omega)$ at a single temperature will provide a measure of the classical diffusion rate, the periodicity of the potential, and the diffusion mechanism. Measuring $S(Q, \omega)$ as a function of temperature will then permit determination of the activation energy for such diffusion processes. Combining these quasielastic scattering measurements with inelastic measurements of the bound torsional modes can provide an exceptionally detailed picture of the local potential experienced by the adsorbate.

5. Examples

In this section we present several illustrative examples of applications of neutron spectroscopy in studies of chemisorbed speices on high-surface-area catalysts. These examples are in no sense intended as a comprehensive review of the field. For such a perspective the reader is urged to consult the many references at the end of the chapter which trace the body of work that has emerged over the past decade.

5.1. Hydrogen/Raney Nickel

The vibrational spectroscopy of hydrogen on Raney nickel[46-49] represented the first use of neutrons to study the surfaces of transition metals. The early work was performed in Europe and clearly demonstrated the utility of the technique. This hydrogen/nickel system illustrates a number of the unique capabilities of neutron scattering for studying transition metal surfaces.[25,28,29,50-52] For instance, Figure 6 shows the IINS spectra as a function of hydrogen uptake on a Raney nickel sample at 80 K.[22] It is clear that the spectrum develops uniformly with uptake, which indicates that the surface is dominated by one type of binding site. It is also worth noting that the integrated area in this type of experiment is not subject to changes in matrix elements or cross sections due to adsorbate reorientation as a function of coverage as it can occur in electron energy loss spectroscopy (EELS) or reflection absorption infrared (RAIR).

The assignment of these hydrogen vibrations is facilitated by two further pieces of experimental data. Figure 7 illustrates the inelastic neutron scattering spectra for hydrogen at 10% of saturation coverage. In the lower spectrum, the hydrogen is diluted in the ratio of 2:1 by deuterium. In the upper spectrum, the dilution in deuterium is 7:1. Such data provide an immediate check on the predicted isotopic shift due to the observation of new modes at 86 and 104 meV associated with the presence of deuterium. In addition, the 117- and 141-meV modes associated with saturation hydrogen coverage reflect a shift (to 122 and 138 meV) in addition to substantial narrowing of the line shapes. These effects are different from those observed for CO in RAIR upon isotopic dilution. In the optical experiments, the observed shift (increasing frequency with increasing mole fraction) was of opposite sign to the shift

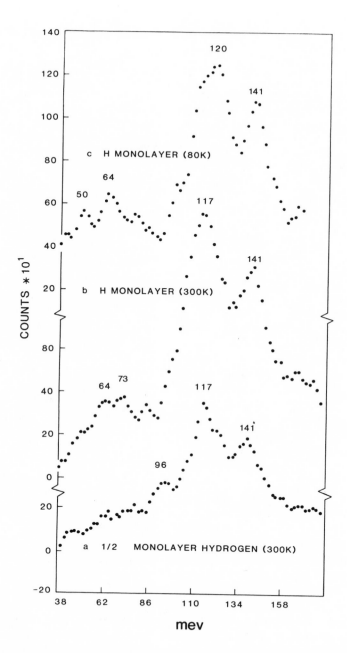

Figure 6. IINS spectrum due to hydrogen on Raney nickel, recorded at 80 K: (a) half saturation hydrogen coverage with sample annealed at 300 K prior to recording spectrum, (b) saturation hydrogen coverage with sample annealed at 300 K prior to recording spectrum, and (c) saturation hydrogen exposure with sample held at 80 K.[22]

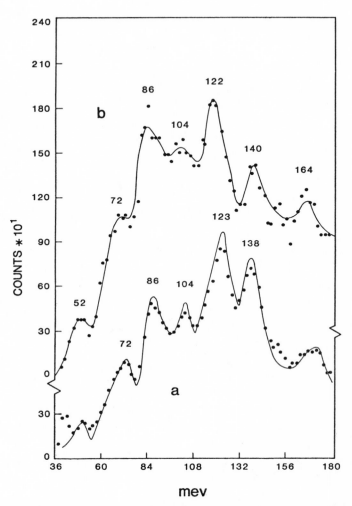

Figure 7. IINS spectra of hydrogen diluted in deuterium on Raney nickel annealed at 300 K (spectra recorded at 80 K, solid lines are drawn to guide the eye)[22] (a) $\theta_H = 0.1$, $\theta_D = 0.2$ of saturation; (b) $\theta_H = 0.1$, $\theta_D = 0.7$ of saturation.

observed in the neutron work. Since the infrared observation appears to be a result of dipole–dipole coupling, and since the optical transition moment of the Ni–H system is much weaker, a different coupling mechanism was required.[22] Dynamical coupling through the hydrogen atom rms displacements was capable of accounting for the shifts. (It is worth noting that the instrumental resolution was identical for the data in

Figures 6 and 7, yet the presence of deuterium adsorbed in the hydrogen overlayer results in much narrower line shapes.)

Comparison of the integrated areas in Figure 6 shows that the intensities associated with the dominant features at 117 and 141 meV are in the ratio of 2:1. The areas and positions of these peaks provide the basis of a force field model of the adsorbed hydrogen. Four general binding sites need to be considered: atop, twofold, threefold, and fourfold. It is rather straightforward to eliminate the atop site, since hydrogen adsorbed in such a site would yield a nondegenerate vibrational mode corresponding to hydrogen motion perpendicular to the substrate with large scattering intensity above 200 meV.[53] However, no feature of sufficient intensity is observed in this region of the IINS spectrum. With regard to the fourfold site, EELS has been used to measure the hydrogen vibrational spectrum on Ni(100) and found that such fourfold sites result in Ni–H modes near 74 meV.[54] In fact, the less intense shoulder in the neutron spectrum at ~70 meV has been tentatively assigned to a low concentration of such fourfold sites on the Raney nickel crystallites. It therefore remains to distinguish between two- and threefold sites. Symmetry arguments indicate that hydrogen adsorbed in a twofold site should yield three distinct modes, while hydrogen in a threefold site should yield a single mode perpendicular to the surface and a pair of degenerate modes parallel to the surface. Thus, the relative integrated intensities (2:1) of the 117- and 141-meV spectral features are suggestive of hydrogen in a threefold binding site.

To test this assignment further, one can perform a series of calculations to model the vibrational spectrum using the force constant formalism discussed in Section 4 assuming a range of metal–hydrogen bond lengths while also varying the force constants. Figure 8 indicates the results of such calculations; only a relatively narrow range of bond lengths and force constants are compatible with the IINS results. Indeed, the Ni–H bond length emerging from this analysis (1.9 Å) is in good agreement with the bond length obtained by LEED for H on Ni(111).[55]

5.2. Hydrogen + Carbon Monoxide/Raney Nickel

Figure 8 demonstrates the sensitivity of the vibrational spectrum to the binding site of the chemisorbed H atom and to the Ni–H bond length and force constant. This sensitivity suggests that any subtle changes in the adsorption of hydrogen on nickel induced by coadsorption of another molecule would be observable. The coadsorption of H_2 and CO on Raney nickel has been studied by incoherent inelastic neutron

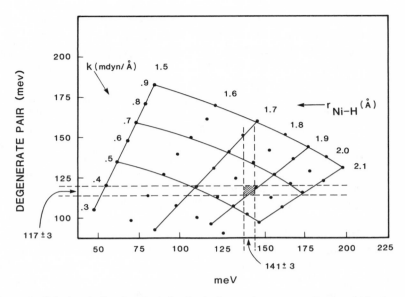

Figure 8. Calculated vibrational modes for hydrogen in a symmetric threefold adsorption site as a function of distance and force constant. The ordinate indicates the energy of the degenerate pair of vibrational modes, while the abscissa gives the energy of the nondegenerate mode. The shaded region indicates the range of bond distances and force constants consistent with the observed frequencies and relative intensities.[22]

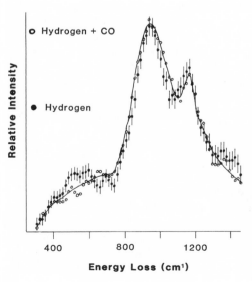

Figure 9. IINS intensity as a function of energy loss recorded at 80 K. The solid points indicate the spectrum of hydrogen adsorbed at room temperature on Raney nickel. The solid line (open points) indicates the spectrum observed for CO adsorbed at 80 K on the hydrogen-covered sample. Both spectra have been corrected for scattering due to bulk nickel.[21]

scattering.[21] In these experiments, CO was chemisorbed at 80 K on a hydrogen-saturated Raney nickel surface such that the surface contained one adsorbed CO molecule for every adsorbed H atom. Figure 9 shows the superposition of the pure hydrogen and coadsorbed hydrogen-CO spectra. Error bars [± (total counts)$^{1/2}$] are indicated for the CO free spectrum. Aside from a slight broadening of the 118-meV peak, there is little difference between the two spectra in the scattering associated with hydrogen adsorbed in threefold sites. Hence it appears that, at 80 K, adsorbed CO does not displace H from the threefold binding site, and has only a small effect on the Ni–H vibrational force constant. As the sample temperature is raised, significant changes are observed in the spectra which are a direct result of changes in the chemical binding of the surface hydrogen. These changes include an increase in the inelastic scattering intensity over the entire spectral region, a redistribution of the scattering intensity in the region of the chemisorbed H–Ni vibrations, and the appearance of new vibrations at low frequencies ($<800\ cm^{-1}$).[21] These observations indicate substantial interaction and possibly chemical reaction between the coadsorbed H and CO at temperatures greater than 200 K.

5.3. Hydrocarbons on Platinum Black

The adsorption and reaction of unsaturated hydrocarbons on transition metal surfaces have served as the focus of a variety of experiments in surface science. Complementary to EELS studies on single-crystal surfaces under ultrahigh vacuum conditions, neutron scattering has been useful for probing the low-frequency modes of adsorbed hydrocarbons on high-surface-area metal powders (Pt,[20] Ni[21]). In addition, it has been possible to discuss the adsorbed hydrocarbon species in terms of their local force fields. In Figures 10 and 11 are shown neutron scattering data with error bars of one standard deviation for C_2H_2 and C_2H_4 molecularly adsorbed on platinum black.[20] The results of simple force field models for the adsorbed species are also indicated. In the force field calculations for C_2H_2, the adsorbate had to be distorted from its linear, gas-phase configuration in order to account for the observed IINS spectrum; for C_2H_4, it was necessary to bend the hydrogen atoms out of the ethylene plane in addition to counter-rotating (staggering) the two CH_2 groups. Such structures had been proposed on the basis of EELS experiments on single-crystal metal surfaces,[56] but the neutron data provide a method for quantifying the extent of the molecular distortions in addition to establishing the molecular nature and geometry of adsorbed hydrocarbons on catalyst particles. The molecular geometries derived from the

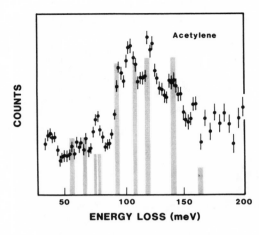

Figure 10. Acetylene adsorbed on platinum black at 120 K (data points) and IINS modes and intensities as determined from a force field model.[20]

model fits to the neutron data are shown schematically in Figure 12. The resulting force constants and bond angles obtained for C_2H_2 are listed in Table 3.

At this point it should be clear that the IINS technique provides a powerful method for comparing adsorbate conformations and adsorbate chemistry as one moves from well-defined single-crystal surfaces to the polycrystalline surfaces more commonly encountered in high-surface-area materials. While these results on platinum black indicate a strong

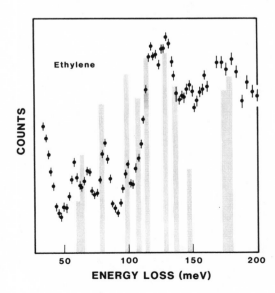

Figure 11. Ethylene adsorbed on platinum black at 120 K (data points) and IINS modes and intensities (shaded verticle bars) as determined from a force field model.[20]

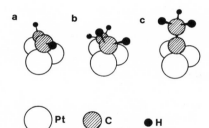

Figure 12. Model geometries for adsorbed hydrogenous species on platinum: (a) acetylene, (b) ethylene, and (c) ethylidyne.[20]

correlation between single-crystal surfaces and high-surface-area materials, IINS data on Raney nickel indicate that such good agreement is not always observed.[57,58] Specifically, IINS experiments have demonstrated that Raney nickel is far more active in terms of hydrocarbon reactivity than would have been anticipated from studies on single-crystal nickel surfaces. The ability to compare the properties of adsorbates on high-surface-area substrates to those on well-defined single-crystal surfaces represents an exciting aspect of neutron scattering in the study of surface processes.

5.4. Hydrocarbons in Zeolites

High-surface-area materials can achieve their characteristic ratio of surface to bulk atoms in two ways. In one limit, very small solid particles are formed. Another manner in which a material can achieve increased surface area is by means of internal voids or cages which are accessible

Table 3. Bond Lengths and Force Constants Used to Fit the IINS Spectral Features of C_2H_2 on Platinum Black.[a]

Bonding mode	Bond	Length (Å) or angle	Force constant (mdyne/Å)
Stretch	Pt–Pt	2.77	0.88
	Pt–C	2.05	2.61
	C–H	1.08	4.9
	C–C	1.39	5.5
Bend	H–C–C	36.5	0.35
	H–C–Pt		0.31
			0.30
Torsion	Pt–Pt–C–C		0.01
	H–C–C–H		0.015

[a] The platinum surface is represented by a symmetric Pt cluster.

for adsorption. Thus, the material presents an open lattice rather than a solid wall to approaching molecules. A particularly practical example of such an internal surface is provided by zeolites. Here, too, neutrons have been found to be a useful probe. Interesting examples are provided by the adsorption of ethylene[59] and acetylene[60] on silver and sodium exchanged type X zeolites. These experiments exploited both TOF and Be filter spectrometers in order to probe the spectral region which is inaccessible with optical techniques. Based on work with model compounds, Howard et al. were able to distinguish and assign the hindered translations (t_x, t_y, and t_z) and the hindered rotations (i.e., torsions).[60] For both C_2H_2 and C_2H_4, the lowest-frequency torisonal mode on silver exchanged X was assigned to a hindered rotation parallel to the surface. This torsional mode was split by steric interactions with the zeolite resulting in features at 3.0 and 5.7 meV (C_2H_2) and 5.0 and 10 meV (C_2H_4). A simple model for the barrier to this rotation indicated a potential that was shallower by a factor of 10 compared to the torsional potential found in model compounds.

5.5. Diffusive Motions

Study of the diffusive motion of adsorbates on surfaces remains an area ripe for further work.[13] The motions of hydrogen and hydrocarbons have already been probed on a number of surfaces[61-64] (particularly in physisorbed systems on pyrolytic graphite). Among the chemically more relevant systems are aluminosilicates. For example, the mean time between jumps for translational and rotational diffusion have been measured (1.3×10^{-10} s and 6×10^{-12} s, respectively) for acetylene on sodium 13 X by Wright and Riekel.[61] Similarly, in a calcium exchanged Montmorillonite, Tuck et al.[65] found the diffusion coefficient of water at 300 K to be 50% smaller than for bulk water, with a residence time between jumps of $\sim 10^{-11}$ s and an activation energy of 11 kJ mol^{-1}.

Rather than presenting the data from any of the surface diffusion studies, a single example from solid state chemistry will be presented. This approach serves to better illustrate the range of detail pertaining to diffusive and rotational dynamics which is available from neutron scattering. Figure 13 is the quasielastic scattering spectrum of solid CH_3NO_2 (nitromethane) as a function of Q.[66] Included in the figure is a theoretical fit of the quasielastic-peak broadening induced by the internal rotation of the CH_3 part of the molecule. Note that the data are capable of distinguishing between reorientations through 60° and 120° of the methyl group about the C–N bond. By recording the spectrum as a function of sample temperature, it is possible to determine the activation

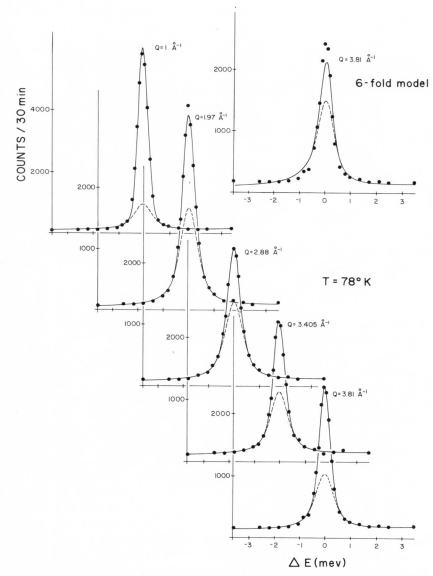

Figure 13. Quasielastic neutron scattering spectra for several values of momentum transfer of CH_3NO_2 at 78 K. The solid lines are the results of fits to the data by a threefold jump reorientation model. The spectrum in the upper right-hand corner of the figure is the best fit to the data with a sixfold jump reorientation model for $Q = 3.81 \, \text{Å}^{-1}$. The dashed lines represent the Lorentzian contribution from reorientation after accounting for the instrumental resolution.[66]

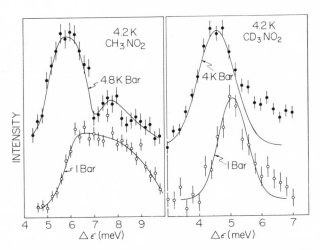

Figure 14. The inelastic neutron scattering spectra of CH_3NO_2 and CD_3NO_2 at two pressures and 4.2 K. This is a measurement of the pressure dependence of the first excited torsional level of both isotopic species.[67]

energy of the diffusive or rotational motion. For example, in this case, the barrier to internal rotation was determined to be 234 cal/mol.[66] The positions of the bound torsional levels of this system have been determined experimentally as shown in Figure 14.[67] The modes observed between 4 and 7 meV correspond to transitions between states with increasing torsional motion about a local CH_3 potential minimum, whereas the broadening in Figure 13 corresponds to the onset of free rotation of the CH_3 group. In this case, the dependence on pressure and deuterium substitution permit reliable spectral assignments. The observation of the ground-state tunnel splitting as shown in Figure 15 completed the picture of the local torsional potential.[68] Note that the observation of modes that are shifted less than 2 μev (~0.015 cm^{-1}) away from the elastic beam is critical for the deuterium check on these tunnel-split states. These results can be summarized in the potential energy diagram of Figure 16 (see Ref. 67). The torsional barrier, tunnel splittings, and torsional modes are indicated. While this level of detail would be much more difficult to achieve in studying diffusing or reorienting species on a surface, this example does illustrate a unique ability to map out the details of such motions and related molecular potentials on an atomic or molecular scale over a wide time/energy regime (10^{-8}–10^{-13} s, 10^{-6}–10^{-1} eV).

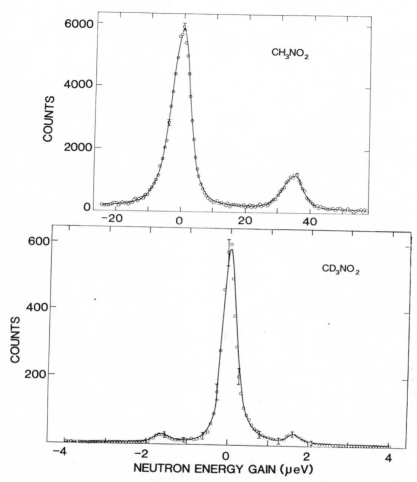

Figure 15. The ground state tunnel splitting of CH_3NO_2 and CD_3NO_2 at 5.0 K.[68]

6. Summary

Incoherent inelastic neutron scattering is proving to be a valuable, versatile, and sometimes unique probe of the dynamics of hydrogenous species chemisorbed on high-surface-area catalytic substrates. The well-understood neutron–nuclear scattering cross sections make it possible to interpret (and predict) inelastic scattering intensities due to various

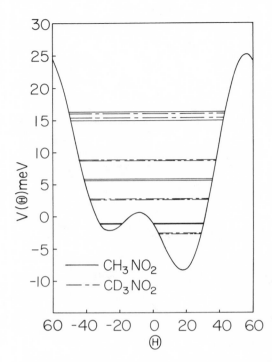

Figure 16. The rotational potential as a function of methyl rotation. Also presented are the calculated levels for CH_3NO_2 (solid lines) and CD_3NO_2 (dashed lines). The level splittings are schematic for the lower levels and drawn to scale only for the highest level.[67]

normal modes for hydrogenous adsorbates and to relate these intensities to force models and geometric parameters for the chemisorbed species. In addition, it is possible to obtain neutron spectra within reaction vessels over a wide range of temperature and pressure. Thus neutron spectroscopy provides a complement to techniuqes such as EELS, which measure the vibration spectrum of adsorbed species on well-characterized single-crystal surfaces at high vacuum. Neutron methods can cover an energy range from 0.5 μeV to 300 meV and thus offer the promise for unique studies of the diffusion and rotational dynamics and related potentials of surface molecular fragments in a time range of 10^{-8}–10^{-13} s. Such applications are in their early stages, but should yield exciting new insights into molecular bonding and processes in and on catalysts in the coming decade.

Acknowledgments

The authors would like to thank Drs. S. Trevino, J. M. Rowe, T. Udovic, and M. Wax for their critical comments on this chapter. Some of the work presented was supported, in part, by the Department of Energy, Office of Basic Energy Sciences.

References

1. W. Marshall and S. W. Lovesey, *Theory of Thermal Neutron Scattering,* Oxford University Press, Oxford (1971).
2. P. A. Egelstaff, *Thermal Neutron Scattering,* Academic, New York (1965).
3. G. E. Bacon, *Neutron Scattering in Chemistry,* Butterworths, London (1977).
4. J. J. Rush, in: *Critical Evaluation of Chemical and Physical Structure Information,* pp. 369–385, National Academy of Sciences, Washington, D.C. (1974).
5. J. A. Janik, in: *The Hydrogen Bond III* (P. Schuster, G. Zundel, and C. Sandorfy, eds.) pp. 891–936, North-Holland, Amsterdam (1976).
6. T. Springer, in: *Dynamics of Solids and Liquids by Neutron Scattering* (S. W. Lovesey and T. Springer, eds.) Topics in Current Physics, Vol. 3, 255–300, Springer-Verlag, New York (1977).
7. J. W. White, in: *Dynamics of Solids and Liquids by Neutron Scattering* (S. W. Lovesey and T. Springer, eds.) Topics in Current Physics, Vol. 3, pp. 197–254, Springer-Verlag, New York (1977).
8. D. K. Ross and P. L. Hall, in: *Advanced Chemical Methods for Soil and Clay Minerals Research* (J. W. Stucki and W. L. Banwart, eds.), pp. 93–168, D. Reidel, Dordrecht, (1980).
9. P. L. Hall, in: *Advanced Techniques for Clay Mineral Analysis* (J. J. Fripiat, ed.), pp. 51–75, Elsevier, New York (1982).
10. J. Howard and T. C. Waddington, The observation of the normal modes of vibration of adsorbed species by inelastic neutron scattering spectroscopy, *Appl. Surf. Sci.* **2,** 102–104 (1978).
11. J. W. White, R. K. Thomas, T. Trewern, I. Marlow, and G. Bomchil, Neutron diffraction and inelastic scattering from adsorbed molecules, *Surf. Sci.* **76,** 13–49 (1978).
12. P. G. Hall and C. J. Wright, Neutron scattering from adsorbed molecules, surfaces, and intercalates, *Chem. Phys. Solids Surf.* **7,** 89–117 (1978).
13. R. K. Thomas, Neutron scattering from adsorbed systems, *Prog. Solid State Chem.* **14,** 1–93 (1982).
14. J. B. Parise and E. Prince, The structure of cesium-exchanged zeolite-rho at 293 K and 493 K determined from high resolution neutron powder data, *Mater. Res. Bull.* **18,** 841–852 (1983).
15. J. B. Parise, L. Abrams, T. E. Gier, D. R. Corbin, J. D. Jorgensen, and E. Prince, Flexibility of the framework of zeolite rho. Structural variation from 11 to 573 K. A study using neutron powder diffraction data, *J. Phys. Chem.* **88,** 2303–2307 (1984).
16. A. H. Baston, J. A. Potton, M. V. Twigg, and C. J. Wright, Determination of particle-size distributions of heterogeneous catalysts on high-electron-density supports by neutron small-angle scattering: Dispersed nickel oxide on α-alumina, *J. Catal.* **71,** 426–429 (1981).
17. G. Kostorz, in: *Threatise on Materials Science and Technology* (G. Kostorz, ed.), Vol. 15, pp. 227–286, Academic, New York (1979).
18. J. W. White and C. G. Windsor, Neutron scattering—modern techniques and their scientific impact, *Rep. Prog. Phys.* **47,** 707–765 (1984).

19. R. Pynn, Neutron scattering instrumentation at reactor based installations, *Rev. Sci. Instrum.* **55**, 837–848 (1984).
20. R. R. Cavanagh, J. J. Rush, R. D. Kelley, and T. J. Udovic, Adsorption and decomposition of hydrocarbons on platinum black: Vibrational modes from NIS, *J. Chem. Phys.* **80**, 3478–3484 (1984).
21. R. D. Kelley, R. R. Cavanagh, and J. J. Rush, Coadsorption and reaction of H_2 and CO on Raney nickel: Neutron vibrational spectroscopy, *J. Catal.* **83**, 464–468 (1983).
22. R. R. Cavanagh, R. D. Kelley, and J. J. Rush, Neutron vibrational spectroscopy of hydrogen and deuterium on Raney nickel, *J. Chem. Phys.* **77**, 1540–1547 (1982).
23. Manufacturers are identified in order to provide a complete description of experimental conditions; this identification is not intended as an endorsement by the National Bureau of Standards.
24. A. P. Bolton and R. L. Moss, in: *Experimental Methods in Catalytic Research* (R. B. Anderson and P. T. Dawson, eds.), Vol. II, pp. 1–91, Academic, New York (1976).
25. A. Renouprez, P. Fouilloux, and B. Moraweck, in: *Growth and Properties of Metal Clusters* (J. Bourdon, ed.), pp. 421–434, Elsevier, Amsterdam (1980).
26. P. Fouilloux, The nature of Raney nickel, its adsorbed hydrogen and its catalytic activity for hydrogenation reactions (Review). *Appl. Catal.* **8**, 1–42 (1983).
27. H. Jobic and A. Renouprez, Neutron inelastic spectroscopy of benzene chemisorbed on Raney platinum, *Surf. Sci.* **111**, 53–62 (1981).
28. J. Howard, T. C. Waddington, and C. J. Wright, Low frequency dynamics of hydrogen adsorped upon a platinum surface, *J. Chem. Phys.* **64**, 3897–3898 (1976).
29. J. Howard, T. C. Waddington, and C. J. Wright, The vibrational spectrum of hydrogen adsorbed on palladium black using inelastic neutron scattering spectroscopy, *Chem. Phys. Lett.* **56**, 258–262 (1978).
30. J. J. Rush, R. R. Cavanagh, R. D. Kelley, and J. M. Rowe, Interaction of vibrating H atoms on the surface of platinum particles by isotope dilution neutron spectroscopy *J. Chem. Phys.* **83**, 5339–5341 (1985).
31. J. M. Thomas and W. J. Thomas, *Introduction to the Principles of Heterogeneous Catalysis,* Academic, London (1967).
32. M. A. Vannice, J. E. Benson, and M. Boudart, Determination of surface area by chemisorption: Unsupported platinum, *J. Catal.* **16**, 348–356 (1970).
33. R. D. Kelley, R. R. Cavanagh, J. J. Rush, and T. E. Madey, Neutron spectroscopic studies of the adsorption and decomposition of C_2H_2 and C_2H_4 on Raney nickel, *Surf. Sci.,* **155**, 480–498 (1985).
34. E. B. Wilson, J. C. Decius, and P. C. Cross, *Molecular Vibrations,* McGraw-Hill, New York (1965).
35. R. G. Snyder and J. H. Schnachtschneider, A valance force field for saturated hydrocarbons, *Spectrochim. Acta* **21**, 169–195 (1965).
36. A. Warshel and S. Lifson, Consistent force field calculations. II. Crystal structures, sublimation energies, molecular and lattice vibrations, molecular conformations and enthalpies of alkanes, *J. Chem. Phys.* **53**, 582–594 (1970).
37. O. Ermer and S. Lifson, Consistent force field calculations, III. Vibrations conformations, and heats of hydrogenation of nonconjugated olefins, *J. Am. Chem. Soc.* **95**, 4121–4132 (1973).
38. H. Jobic, J. Tomkinson, and A. Renouprez, Neutron inelastic scattering spectrum and valence force field for benzenetricarbonylchromium, *Mol. Phys.* **39**, 989–999 (1980).
39. J. Hiraishi, The vibrational spectra of several platinum-ethylene complexes: $K[PtCl_3(C_2H_4)]H_2O$ (Zeise's salt), $K[PtCl_3(C_2D_4)]H_2O$ and $[PtCl_2(C_2H_4)]_2$, *Spectrochim. Acta* **25A**, 749–760 (1969).

40. Y. Iwashita, Force constants in the acetylene molecule in a cobalt–carbonyl complex and in an excited electronic state., *Inorg. Chem.* **9,** 1178–1182 (1970).
41. M. W. Howard, S. F. Kettle, I. A. Oxton, D. B. Powell, N. Sheppard, and P. Skinner, Vibrational spectra and the force field of the $HCCo_3$ group in $HCCo_3(CO)_9$, *J. Chem. Soc., Faraday Trans.* 2 **77,** 397–404 (1981).
42. P. Skinner, M. W. Howard, I. A. Oxton, S. F. A. Kettle, D. B. Powell, and N. Sheppard, Vibrational spectra and the force field of ethylidyne tricobalt nonacarbonyl: Analogies with spectra from the chemisorption of ethylene upon the Pt(111) crystal face, *J. Chem. Soc. Faraday Trans.* 2 **77,** 1203–1215 (1981).
43. I. A. Oxton, D. B. Powell, N. Sheppard, K. Burgess, B. F. G. Johnson, and J. Lewis, The infrared vibrational assignment for the μ_2-bridging methylene ligand in metal cluster complexes and its comparison with frequencies assigned to CH_2 species chemisorbed on metal surfaces, *J. Chem. Soc. Chem. Commun.* 719–721 (1982).
44. C. T. Chudley and R. J. Elliot, Neutron scattering from a liquid in a jump diffusion model, *Proc. Phys. Soc.* (London) **77,** 353–361 (1961).
45. K. Sköld, in *Hydrogen in Metals I* (G. Alefeld and J. Volkl, eds.), Topics in Applied Physics Vol. 28, pp. 267–287, Springer-Verlag, New York (1978).
46. R. D. Kelley, J. J. Rush, and T. E. Madey, Vibrational spectroscopy of adsorbed species on nickel by neutron inelastic scattering, *Chem. Phys. Lett.* **66,** 159–164 (1979).
47. A. J. Renouprez, P. Fouilloux, J. P. Candy, and J. Tomkinson, Chemisorption of water on nickel surfaces, *Surf. Sci.* **83,** 285–295 (1979).
48. C. J. Wright, Alternative explanation of the inelastic neutron scattering from hydrogen adsorbed by Raney nickel, *J. Chem. Soc. Faraday Trans.* 2 **73,** 1497–1500 (1977).
49. A. J. Renouprez, P. Fouilloux, G. Coudurier, D. Tocchetti, and R. Stockmeyer, Different species of hydrogen chemisorbed on Raney nickel studied by neutron inelastic spectroscopy, *J. Chem. Soc. Faraday Trans.* 1 **73,** 1–10 (1977).
50. I. J., Braid, J. Howard, and J. Tomkinson, Inelastic neutron scattering study of hydrogen adsorbed on impure palladium black, *J. Chem. Soc. Faraday Trans.* 2 **79,** 253–262 (1983).
51. J. Howard, T. C. Waddington, and C. J. Wright, *Neutron Inelastic Scattering, 1977,* II, pp. 499–510, I.A.E.A., Vienna (1978).
52. J. J. Rush, R. R. Cavanagh, and R. D. Kelley, Neutron scattering from adsorbates on platinum black. *J. Vac. Sci. Technol. A* **1,** 1245–1246 (1983).
53. W. A. Pliskin and R. P. Eischens, Infrared spectra of hydrogen and deuterium chemisorbed on platinum, *Z. Phys. Chem.* **24,** 11–23 (1960).
54. S. Andersson, Vibrational excitations and structure of H_2, D_2 and HD adsorbed on Ni(100), *Chem. Phys. Lett.* **55,** 185–188 (1978).
55. K. Christmann, R. J. Behm, G. Ertl, M. A. Van Hove, and W. H. Weinberg, Chemisorption geometry of hydrogen on Ni(111): Order and disorder, *J. Chem. Phys.* **70,** 4168–4184 (1979).
56. T. E. Felter and W. H. Weinberg, A model of ethylene and acetylene adsorption on the (111) surface of platinum and nickel, *Surf. Sci.* **103,** 265–287 (1981).
57. R. D. Kelley, R. R. Cavanagh, J. J. Rush, and T. E. Madey, Neutron inelastic scattering study of C_2H_4 adsorbed on Raney nickel, *Surf. Sci.* **97,** L335–L338 (1980).
58. H. Jobic and A. Renouprez, Inelastic neutron spectrum of acetylene and ethylene chemisorbed on Raney nickel, Proceedings of the Fourth International Conf. on Solid Surfaces, Cannes, pp. 746–749 (1980).
59. J. Howard, T. C. Waddington, and C. J. Wright, Inelastic neutron scattering study of C_2H_4 adsorbed on type X zeolites, *J. Chem. Soc., Faraday Trans.* 2 **73,** 1768–1787 (1977).

60. J. Howard and T. C. Waddington, An inelastic neutron scattering study of C_2H_2 adsorbed on type 13X zeolites, *Surf. Sci.* **68**, 86–95 (1977).
61. C. J. Wright and C. Reikel, The uniaxial rotation of ethylene adsorbed by sodium 13X zeolite, *Mol. Phys.* **36**, 695–704 (1978).
62. A. J. Renouprez, R. Stockmeyer, and C. J. Wright, Diffusion of chemisorbed hydrogen in a platinum zeolite, *Trans. Faraday Soc. I*, **75**, 2473–2480 (1979).
63. E. Cohen de Lara and R. Kahn, Neutron and infrared study of the dynamical behaviour of methane in NaA zeolite, *J. Phys. (Paris)* **42**, 1029–1038 (1981).
64. H. Jobic, M. Bee, and A. Renouprez, Quasi-elastic neutron scattering of benzene in Na-mordenite, *Surf. Sci.* **140**, 307–320 (1984).
65. J. J. Tuck, P. L. Hall, M. H. B. Hayes, D. K. Ross, and C. Poinsignon, Quasi-elastic neutron-scattering studies of the dynamics of intercalated molecules in charge-deficient layer silicates, *J. Chem. Soc., Faraday Trans. 1* **80**, 309–324 (1984).
66. S. F. Trevino and W. H. Rymes, A study of methyl reorientation in solid nitromethane by neutron scattering, *J. Chem. Phys.* **73**, 3001–3006 (1980).
67. D. Cavagnat, A. Magerl, C. Vettier, I. S. Anderson, and S. F. Trevino, Anomalous pressure dependence of the torsional levels in solid nitromethane., *Phys. Rev. Lett.* **54**, 193–196 (1985).
68. B. Alefeld, I. S. Anderson, A. Heidemann, A. Mageral, and S. F. Trevino, The measurement of tunnel states in solid CH_3NO_2 and CD_3NO_2, *J. Chem. Phys.* **76**, 2758–2759 (1982).

6

Electron Energy Loss Spectroscopy

Neil R. Avery

1. Background

Before embarking on this chapter it is worth reflecting on the current status of the spectroscopic methods, and in particular electron spectroscopy, in modern surface science, and in so doing show why high-resolution electron energy loss spectroscopy (EELS) for the vibrational analysis of adsorbed molecules is already and will become increasingly the technique to open up a new understanding of surface science in general and surface chemistry in particular.

With a scattering cross section close to atomic dimensions, low-energy electrons ($<1 \, keV$) understandably have become the principal probe-particles of modern surface science. Indeed, even in the widely used ultraviolet and x-ray photoelectron spectroscopies, it is the intrinsic reactivity of the ejected electron that confers the surface sensitivity on these techniques. However, these spectroscopies are usually characterized by atomic and molecular excitation processes in which the energy transfer is greater than $1 \, eV$ or so. From the electron–photon energy correlation diagram (Figure 1) it may be seen that this corresponds to electronic excitations of either the valence or core levels. While in some instances it is possible to derive molecular structural information from excitations of this kind, these spectroscopies should be seen as better

Neil R. Avery • CSIRO Division of Materials Science, University of Melbourne, Parkville, Victoria 3052, Australia.

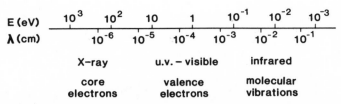

Figure 1. Electron–photon energy correlation diagram showing the approximate regions of molecular and atomic excitations.

suited to the study of atomic and the simpler molecular systems. For the larger molecules, chemists have long recognized that a powerful route to understanding the geometric structure of molecules was via their vibrational spectrum. Here, we may reflect on the surge of interest in molecular structural chemistry when infrared absorption spectrometers became readily available in the immediate postwar years.

In more recent years many techniques have been devised to determine the vibrational spectrum of molecules adsorbed on metal surfaces. The earliest efforts centered on compensating for the inherent insensitivity of infrared absorption spectroscopy (resulting directly from the low scattering cross section for photons) by using high-area adsorbates. Both oxides and oxide-supported metals have been studied in this way with the earlier work the subject of an extensive review by Little.[1] Aside from its limited sensitivity, this approach is complicated by strong phonon absorption by the support which "blacks-out" large and often important regions of the vibrational spectrum. While the more recently developed infrared reflection–absorption spectroscopy from well-defined single-crystal surfaces is not subject to this limitation, its wider application must still be limited by poor sensitivity. The realization of an inherently surface sensitive electron spectroscopy for vibrational analysis has been found over the past few years in high-resolution EELS. The viability of the technique was first demonstrated by an Urbana group[2] as early as 1967. However, the resolution of this pioneering experiment was poor ($\sim 400 \text{ cm}^{-1}$) and with the apparently intractable problems with its improvement to a useful level, the technique languished until the Julich group[3,4] developed and refined the technique to its current status in surface science. Since then, instruments of different designs have been developed in several laboratories, and as a result of these often independent efforts, different aspects of the design and operation are emphasized. Indeed, it is probable that the ideal configuration of the experiment has not yet been achieved. As a result of this, the comments expressed in this chapter must reflect more my own experience, and prejudices, in the development of a successful instrument. Similarly, no attempts will be made to give a full treatment of the electron optical principles that underpin the spectroscopy. Instead, the discussion will be pitched at a level designed to give the EELS experimentalist a feel for the manipulation of the electron beam. For more detailed discussions of the electron optical principles relevant to EELS the interested reader is directed to Refs. 5–9.

2. The Spectrometer

2.1. General Requirements

With the energy of molecular vibrations in the range of about 100–4000 cm^{-1} (1 meV $\equiv$ 8.066 cm^{-1}) the overriding demand on the instrumentation is for unprecedented electron monochromatization and energy analysis. Modern EEL spectrometers with electrostatic energy selectors of the general configuration shown in Figure 2 routinely achieve an overall resolution of 40–80 cm^{-1}. Briefly, for the moment, thermal electrons at relatively high energy are retarded and focused (with controlled angular dispersion) into a dispersive energy selector where only a small fraction in a narrow energy band are transmitted. The monochromatized beam is then accelerated to the desired energy and focused onto the surface of interest at a relatively large angle of

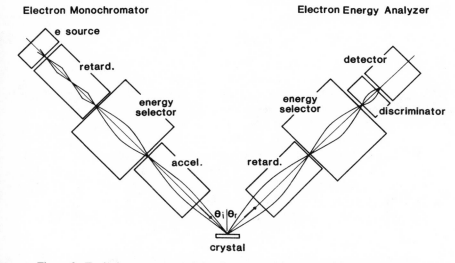

Figure 2. Typical arrangement of the electron optical components used in EELS.

incidence, θ_i, which is typically set at 45–70°. Single-crystal surfaces are preferred in work of this kind since the electron may be coherently reflected by the crystal potential and in so doing preserve the momentum information associated with the inelastic energy loss event. In this way, optical events in which essentially no momentum is transferred ($\Delta k \sim 0$) are confined to a small cone about the specularly reflected beam ($\theta_i = \theta_r$). The incoherent, angularly dispersed impact events ($\Delta k \gg 0$) may be distinguished from the optical events by either (i) rotating the crystal or (ii) rotating the analyzer. Unless the crystal mounting procedure guarantees that the rotation axis intersects the point of irradiation the latter method is preferable. High-energy resolution of electrons inelastically scattered from the surface is achieved by retarding and focusing the beam into a second dispersive energy selector, usually, but not necessarily, of the same type as that in the monochromator. Finally, electrons adventitiously scattered from the electron optical components and which would otherwise reach the electron detector should be suppressed.

In the design of a total instrument, it is preferable that each stage in the electron path be made as far as possible independent of the others. In this way, when one stage of the electron optics has been optimized, little or no change will occur during optimization of an adjacent stage. A second important design rule recognizes that mechanical and field perfection cannot be presumed with the low-electron kinetic energies used in EEL spectrometers and electrostatic trim deflectors should be incorporated between *all* windows and pupils.

2.2. The Energy Selectors

The heart of an EEL spectrometer lies in the electrostatic energy selectors, for it is here that the ultimate resolution of the spectroscopy is achieved. Energy selectors are required in both the monochromator and analyzer. Since the stringent electron optical requirements of the energy selector dictate many of the other design features it is appropriate to consider this component of the spectrometer first.

Possible geometrical arrangements that give the linear energy dispersion and better than first-order focusing required for this purpose have been tabulated by Roy and Carette.[9] An appraisal of the relative merits of these arrangements may be obtained from the general expression describing the maximum energy spread, or base resolution, ΔE_B:

$$\Delta E_B / \Delta E_p = AS + B\alpha^n + C\beta^2 \qquad (1)$$

where E_p is the pass energy in the selector, s is the mean source diameter

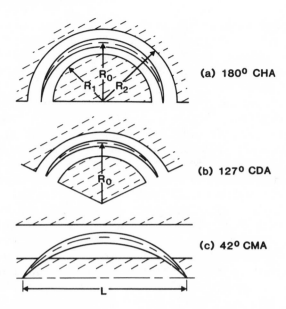

Figure 3. Geometrical arrangement of (a) 180° CHA, (b) 127° CDA, and (c) 42° CMA energy selectors.

at the input and output of the selector, and α and β are the semiangular divergences in and perpendicular to the plane of mean deflection, respectively. Energy selectors that have been successfully used in EEL spectrometers include the 42° cylindrical mirror analyzer (CMA), 127° cylindrical deflector analyzer (CDA), and 180° concentric hemispherical analyzer (CHA). The general geometrical arrangement of these selectors is shown in Figure 3 and the corresponding coefficients A, B, C, and n listed in Table 1. Additionally, energy selectors operating in tandem have been described for both the monochromator[12,17,18] and analyzer.[17]

Table 1. Coefficients A, B, C, and n for the Energy Selectors Used in Current EEL Spectrometers

Energy selector	A	B	C	n	References to use in EELS
42° CMA	$2.2L^{-1}$	5.55	0	3	10
127° CDA	$2R_0^{-1}$	1.33	1	2	3, 4, 11, 12, 17, 18
180° CHA	1	1	0	2	5, 13–16,[a] 18

[a] Reference 16 describes at 90° CHA spectrometer designed for dispersion compensation (Section 6.2).

Dual pass spectrometers of this kind have been claimed to give a superior (low count rate) background. Following the successful CDA-based instrument developed at Julich many instruments have followed this design. However, substitution of the coefficients A, B, C, and n in equation (1) shows that, for comparable input and output source sizes, angular dispersion, and characteristic physical dimensions the CHA is the preferred arrangement. Moreover, the CHA (along with the CMA) are point focusing, i.e., focusing occurs not only in the plane of mean deflection but also perpendicular to it. This is not the case for the CDA, where only line focusing is achieved. While this is the desired arrangement for energy analysis of the line source of ultraviolet photoelectrons from gases, for which the CDA design became popular,[19] it is less suitable for EELS work, where point illumination of the crystal target is preferable. With these advantages in mind, the CHA energy selectors were chosen for the EEL spectrometer developed in this laboratory. While the following discussion will center on this instrument, many of the design considerations and operating procedures described in this chapter will be relevant also to the CMA and CDA based spectrometers.

Turning now to a more detailed discussion of the CHA energy selector, substituting the parameters A, B, C, and n from Table 1 into equation (1) yields

$$\Delta E_B/\Delta E_p = s/R_0 + \alpha^2 \tag{2}$$

While ΔE_B is the more easily calculated resolution parameter, it is the full width at half maximum (FWHM) or ΔE_H that is the more useful measure of an energy selector. As discussed by Read et al.[7] ΔE_H is a complex function of not only the selector type but also its physical dimensions. For a CHA with a circular input and output sources these authors estimate

$$\Delta E_H/E_p = (1.6 \text{ to } 1.9)s/4R_0 + (0.17 \text{ to } 0.25)\alpha^2$$

which is more commonly rounded to[9,20]

$$\Delta E_H/E_p = s/2R_0 + \alpha^2/4 \tag{3}$$

For real systems with finite α, it is seen that ΔE_H is somewhat less than $0.5\Delta E_B$.

An R^{-2} electrostatic field is established in the concentric hemispherical selector when a potential difference, V, is applied between the inner (convex) and outer (concave) hemispheres. Electrons with kinetic energy $E_p = eV_p$, in electron volts, entering the selector normal to the equatorial plane and midway between the hemispheres are deflected through 180°, with unit magnification, if the potential difference, V, between the hemispheres is given by[5]

$$V = V_p[(R_2/R_1) - (R_1/R_2)] \tag{4}$$

and the potential on the inner and outer hemispheres are, respectively,

$$V_{in} = V_p[3 - 2(R_0/R_1)] \tag{5}$$

and

$$V_{out} = V_p[3 - 2(R_0/R_2)] \tag{6}$$

Under these conditions the energy selector will transmit a beam of electrons with angular divergence, α, in the energy range from $E_p - \Delta E_B/2$ to $E_p + \Delta E_B/2$.

An EEL spectrometer calls for energy selectors in both the monochromator and energy analyzer. Taking the FWHM in each of these elements as ΔE_M and ΔE_A, respectively, the overall FWHM resolution of the spectrometer, ΔE_s, becomes for Gaussian distributions, $\Delta E_s = (\Delta E_M^2 + \Delta E_A^2)^{1/2}$. Typically, the spectrometer is operated with $\Delta E_M \sim \Delta E_A$ so that an overall spectroscopic resolution in the range 5–10 meV (40–80 cm^{-1}) requires each selector to operate at 3.5–7 meV resolution. For CHA energy selectors this may be achieved by arranging for $s \sim 0.6$ mm and $R_0 = 30$ mm. Substitution in equation (3) results in a resolving power that is within 10% of 0.01 if α is contained to ~0.06. In this case the desired resolution may be expected for selector pass energies of 320–640 meV. The input and output source size on the equatorial plane of the hemisphere may be established with either a real window (aperture) or by imaging a window at higher potential. In the EEL spectrometer used in this laboratory (shown diagramatically in Figure 4; typical operating voltages are recorded in Table 2) the latter method is used for the monochromator input whereas the remaining three hemisphere windows are real. In either case, electron optics are required to retard the electron beam to the desired pass energy, E_p, with controlled source size, s, and divergence, α.

Before proceeding to the electron optics that achieve this it remains to comment on the methods that have been devised to correct for the deviation from the ideal field (R^{-2} over the 180° of the concentric hemispheres in the case of the CHA) at the selector input and output. In general, it is usual to minimize this distortion by choosing a spacing between the selector elements as small as possible, which of course is limited by the angular divergence (α) of the beam entering the selector. Additionally, various arrangements of correcting rings near the input and output have been empirically tested by computational methods.[19] However, the more common procedure is to use Herzog correctors in the

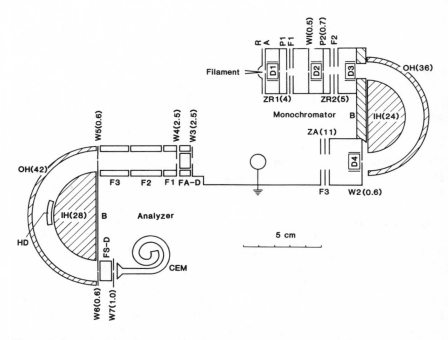

Figure 4. Layout of the electron monochromator and energy analyzer. Figures in brackets are the aperture diameters of the appropriate component (in millimeters).

Table 2. Typical Operating Voltages (Relative to Ground) for the Monochromator and Energy Analyzer shown in Figure 4

Monochromator		Analyzer	
Fil.	−3.5	FA	−2.1
R	−6.9	F1	+8.8
A	+6.5	F2	+27
F1	+10.4	F3	+27
R1	−2.3	B	−3.14
F2	+5.9	IH	−2.95
B	−3.25	OH	−3.23
IH	−3.13	HD	−3.4
OH	−3.35	FS	+1.5
F3	+7	S	+2.5

immediate vicinity of the input and output to the energy selector. The design of these correctors may be interpolated from the nomographs devised by Wollnik and Ewald[22] and more recently discussed by Roy and Carette.[9] Although the present instrument was designed accordingly, other CHA based spectrometers built in this laboratory, in which the effect of fringe fields was disregarded, did not show a noticeable deterioration in performance. Indeed, it is possible that with the iterative procedures used to tune an EEL spectrometer (Section 5.1), adequate compensation for the fringe fields may be achieved more subjectively.

2.3. The Electron Monochromator

2.3.1. Input and Output Electron Optics

As pointed out in Section 2.2, the resolution of the EELS experiment depends critically on the source energy, size, and divergence of the electron beam at the input of the energy selector. For practical reasons, the thermal electrons in the electron emitter assembly (Section 2.33) must be produced with energies (~ 10 eV) considerably greater than the 320–640 meV pass energy required for acceptable resolution. Although, as will be shown for the analyzer (Section 2.4), no obvious disadvantage is encountered with input and output windows (necessitating, of course, deflection within the hemisphere gap), the monochromator was designed to eliminate the need for deflectors within the hemisphere gap by using a real output window, W2 ($s = 0.6$ mm) and arranging for the input optics to produce a small source (0.6 mm) at the input to the selector. The beam may then be trimmed within the hemisphere gap with the external deflectors, D3. The third demand on the input optics is to restrict, with an appropriate window–pupil arrangement, the transverse momentum of the beam in order to keep the divergence, α, to an acceptable level.

In view of the importance of the input electron optical system to the proper operation of the monochromator, a brief description of the salient features of electrostatic optics will be given. Much of the terminology of electrostatic optics follows, with some license, that of conventional light optics. A geometrical representation showing the four cardinal points— viz., object focal point, F_O, and principal plane, P_O, and their image counterparts, F_I and P_I, respectively, are seen in Figure 5 for a strong asymmetric electrostatic lens in which object space is at a lower potential, V_1, than image space, V_3, i.e., acceleration of the beam occurs from left to right. However, as a result of time reversal of the electron trajectories the same cardinal points are appropriate for the present requirement of

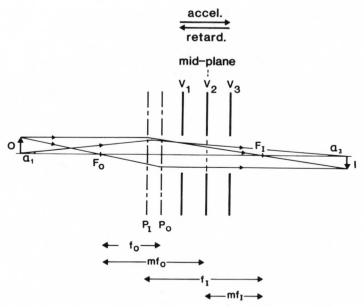

Figure 5. Diagrammatic representation of the cardinal points of a strong electrostatic lens. The object is at lower potential than its image so that electron acceleration occurs from left to right. The cardinal points and ray tracing shown here have been interpolated from the data of Harting and Read[21] for a three-aperture lens with a gap to aperture diameter ratio of 0.5. O and I are positioned 4 and 3 aperture diameters, respectively, from the midplane of the lens.

retardation from right to left. With the usual optical sign convention of left to right propagating rays and also that all object and image distances are positive if they are left and right, respectively, of the midplane of the lens, which for the present purpose is taken as the plane of symmetry of the lens normal to the center ray, the relationship between object and image distances from the mid plane, O and I, respectively, and linear magnification, m, may be written,

$$(O - mf_O)/f_O = f_I/(I - mf_I) \qquad (7)$$

and

$$M = f_O/(mf_O - O) = (mf_I - I)/f_I \qquad (8)$$

where mf_O and mf_I are the midfocal lengths measured from the midplane of the lens. Finally, the angular divergence of a beam entering and leaving an asymmetric lens (Figure 5) is related to the acceleration and magnification, M, by the law of Helmholz and Lagrange[7]:

$$\alpha_1^2 V_1 = \alpha_3^2 V_3 M^2 \tag{9}$$

Thus, from a knowledge of the cardinal points of a lens, all the parameters needed to predict the properties of a retarding lens suitable for injecting the electrons into the energy selector may be estimated. In principle, two, three, and higher element lenses may be used for this purpose. For fixed object and image positions two element lenses operate with an invariant ratio, V_1/V_3, and therefore lack the versatility required for an EEL spectrometer. On the other hand, three element lenses operating with fixed object and image positions may vary the retardation ratio (and magnification) by appropriate adjustment of the focus potential applied to the central element (V_2). Lenses of this type are referred to as zoom lens by analogy with their optical counterparts. Actually this analogy is not strictly pedantic since the optical zoom lens achieves the variable magnification over fixed working distances with the object and image in media of the same refractive index, whereas the three-element electrostatic lens zooms only with the object and image at different potentials. However, the term is widely used for electrostatic lenses of this kind and will be retained here. While greater control over the retardation ratio and magnification may be achieved with higher (>3) element lenses, the additional focus elements that must be added do not operate independently of one another and add greatly to the complexity of operating the lens. Indeed, as we will see, the three-element zoom lens may be made to operate over a wide range of retardation without excessive change in magnification, and for its predictive behavior and simplicity in operation it is ideally suited to EEL spectroscopy.

The cardinal points for a selection of different geometrical configuration of two- and three-elemental lenses have been computed by Harting and Read[23] and are a valuable guide to selecting a lens. For the three-element zoom lenses of present interest, these authors have presented comprehensive charts of the variations of focus voltage and magnification for variable object and image distances at a selection of acceleration ratios. For the present monochromator, constraints imposed by the limited size of the vacuum chamber dictated the choice of compact three-element aperture lenses with adjacent elements separated by half the aperture diameter. From the data provided by Harting and Read a zoom curve (i.e., the relationship between the acceleration ratio V_3/V_1 and focus potential V_2/V_1 for selected working distances) for a lens of this type was constructed for the object and image placed, respectively, four and three aperture diameters from the midplane of the lens (Figure 6). Since the data of Harting and Read[23] are presented for a left-to-right

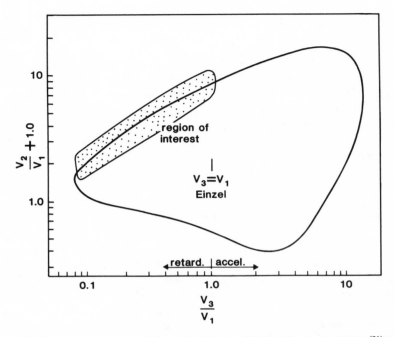

Figure 6. Zoom curve constructed from the data provided by Harting and Read[21] for the three-aperture lens shown in Figure 5 (gap to aperture ratio = 0.5 and O and I positioned, respectively, 4 and 3 aperture diameters from the midfocal plane of the lens).

accelerating beam, the principle of time reversal of electron trajectories must be exploited in order to construct the zoom curves in the retardation region $[1 < V_3/V_1]$. Remember also that in constructing and interpolating from curves of this kind the V's refer to electron potential relative to zero. Thus they only become the applied voltages when the electron source (filament) is at zero. For the present needs, the filament must be biased negatively in order to achieve the desired beam energy at the crystal, E_0 (Section 4).

Two features of zoom curves of this kind are relevant to the present needs. First, there are, except for limiting retardation or acceleration, two focus potentials, V_2, for any allowed ratio, V_3/V_1. In the Einzel conditions ($V_1 = V_3$), for example, these are found at $V_2/V_1 \sim +8.0$ and -0.45. With the low electron energies used in EEL the high positive potential does not lead to electrical problems and is preferable since, by filling less of the lens, this reduces the spherical aberration of the lens.[24] The second feature to note from the general shape of zoom curves is that for given object and image distances the acceleration (or retardation) ratio is not unlimited. For the example shown in Figure 6 the lens is constrained to operate in the range $1/12 < V_3/V_1 < 12$. Perusal of the nomographs given by Harting and Read[23] shows that in order to increase this range without unduly magnifying the object (in the retardation mode) both the object and image must be moved closer to the midplane of the lens. However, there are practical and theoretical[23] restrictions on the extent to which this may be done and the better compromise is in a tandem pair of somewhat weaker lenses.

As already discussed, equation (3) shows that the resolution of the energy selector depends, in part, on the angular divergence of the beam entering the selector, and some consideration must be given to this parameter to ensure that serious resolution degradation does not occur. With the low electron energies used in EEL spectrometers, divergence limiting pupils are better located on the high potential side of the retarding lens since here there will be the least disruption of the beam from field inhomogeneities. An arrangement of circular window and pupil is shown in Figure 7. Two angles are defined, viz., the pencil angle, α_p, and the beam angle, α_B. As pointed out by Read et al.,[7] it is the pencil angle that appears in the law of Helmholz and Lagrange [equation (9)], while it is the total divergence, $\alpha_p + \alpha_B$, that is appropriate when considering the resolution degrading effect of a dispersive beam entering the energy selector. While it is apparent that the pencil angle must be finite, the beam angle through the lens may be reduced to almost zero by

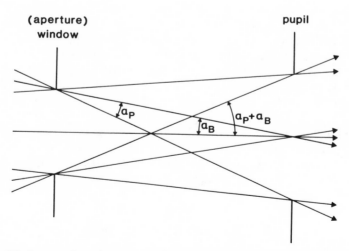

Figure 7. Window–pupil arrangement used to define the divergence of the beam entering the energy selector of the monochromator.

arranging for the pupil to be imaged at infinite—i.e., rays passing through the center of the pupil emerge from the subsequent electron optics as a nondivergent beam. Therefore, as far as possible, the pupil should be positioned on the image (in the retardation mode) focal plane. For the present monochromator the location of this pupil, P2 (Figure 4), was established from the data of Harting and Read[23] for a retardation ratio of ~4.

As discussed earlier the 30-fold retardation of the beam between the anode (A) and the energy selector cannot satisfactorily be achieved with a single three-element zoom lens. Instead, the present arrangement retards the beam in two independent zoom lenses. The first, ZR1, acts as a strong lens retarding the beam by a factor of 8–10 while the second, ZR2, operates more weakly in order to restrict the divergence of the beam defined by the W1–P2, window–pupil, arrangement. For W1 and its image, respectively, working distances of three and four lens aperture diameters were chosen to ensure approximately unit magnification (actually ~1.2 for $V_3/V_1 = 0.25$) of W1 on the equatorial plane of the selector. Under these conditions the window W1 (0.5 mm) is magnified to 0.6 mm at the selector input. With a pupil, P2 (0.7 mm), located to be imaged at infinity, the law of Helmholtz and Lagrange shows that α, the divergence angle appropriate to equation (3), becomes, as required, ~0.05 for a fourfold retardation. A similar divergence limiting pupil, P1, located in ZR1 is not designed to influence directly the divergence, but instead to decrease many of the unwanted electrons with excessive transverse momentum before they enter the retardation lenses, ZR1 and ZR2.

The zoom lens ZA accelerates the monochromatized beam emerging from the energy selector to the desired incident energy at the crystal target, E_0. With its large aperture diameter (11 mm) it is a strong lens with a relatively long working distance and capable of imaging W2 on the crystal surface with a linear magnification of ~2 and acceleration ratios up to 16.

Zoom curves of the type shown in Figure 6 for ZR2 were also constructed for ZR1 and ZA from the nomographs of Harting and Read.[23] These curves were used to set the focus voltages prior to iterating the spectrometer for optimum sensitivity and resolution (Section 5.1).

2.3.2. Current Available from an Electron Monochromator

With the low kinetic energies of the electrons at the input to the monochromator energy selector, excessively high current densities will lead to space-charge Coulomb repulsion and a concomitant degradation of the output current (and to a lesser extent resolution).[7,8] For given physical dimensions of the energy selector, increased resolution can only be achieved by reducing the pass kinetic energy and with it the available current. For CHA based monochromators, Ibach and Mills[8] have estimated that the maximum available current, I, is given by

$$I \propto (\Delta E_M^2)(\Delta E_k)^{-1/2} \qquad (10)$$

where ΔE_k is the thermal energy spread before monochromatization. Equation (10) shows that the compromise between current and resolution is severe and as will be discussed in Section 5 must be carefully tailored to the demands of the experiment.

2.3.3. The Electron Source

In a typical EEL spectrometer space charge effects at the input of the monochromator energy selector (Section 2.32) limit the input electron current to a level well below that available from a typical thermal emitter. As a result of this, there is not the usual demand for precollimation of the electron source, and the conventional (for high kinetic energy sources) immersion arrangement of the emitter located within a negatively biased Wehnelt lens is not usually employed in EELS monochromators. Instead, low kinetic energy sources favor a Pierce[25] arrangement in which the emitter is located in front of a negatively biased repeller, R, as shown in Figure 4. For the present needs, this arrangement has the advantage of increasing the electric field at the emitter surface, thereby minimizing the disturbing effect of the local magnetic field generated by the emitter heating current. Such deflection as does occur is immediately trimmed with the electrostatic deflectors D1. Similarly, direct current must be used to heat the filament, and the electrical connection to the vacuum feedthrough should be noninductively twisted together. According to equation (10), the current available from the monochromator may be increased by reducing the energy spread from the emitter. While low work-function oxide coated and lanthanum hexaboride emitters have been used with this advantage in mind, tungsten hairpin emitters are often preferred for their low cost, convenience, and robustness. Replacement electron microscope emitters of both the tungsten hairpin and lanthanum hexaboride types are ideally suited to the present needs.

2.4. The Energy Analyzer

The energy analyzer used in the present spectrometer is based, like
the monochromator, on a 180° concentric hemispherical energy selector.
In its original form, predating its application to the present EELS work,
the input two-element concentric cylinder lens was designed according to
Bassett *et al.*[26] for fixed tenfold retardation, i.e., for operation in a fixed
resolving power, $\Delta E_H/E_p$ mode. For EELS work, the two lens elements
were each split into two for greater control over the beam and a double
input window (W3 and W4) incorporated (Figure 4). The first element of
the new four concentric cylinder lens stack was made a composite
focus-deflection lens (FA-D) to achieve the necessary axial control over
the beam. Input and output windows, W5 and W6, to the energy selector
were incorporated, thereby necessitating a pair of deflectors mounted
within the hemispheres (HD) to provide trim deflection normal to the
plane of mean deflection. These deflectors were positioned at the mean
radius of the transmitted beam ($R_0 = 35$ mm) and about 20 mm either
side of it.

Unlike the monochromator electron optics, an appraisal of the
analyzer optics cannot be made from data of the kind provided by
Harting and Read.[23] Instead, the operating voltages that are included in
Table 2 were found by iterative procedures of the type described in
Section 5.

2.4.1. The Suppressor

It has long been recognized in electron energy analyzers that
electrons at other than the pass energy of the selector may be scattered
from the internal surfaces of the analyzer and adventitiously reach the
detector. This has two undesirable consequences. First, ghost peaks may
be seen in the region near the elastic peak. These peaks arise from
electrons being scattered from the hemispheres of the energy selector to
the detector. Ghost peaks of this kind are sensitive to the details of the
tuning operation (Section 5.12) which result in subtly different trajectories through the selector. As a result of this, loss bands of this kind in
the EELS spectrum may be difficult to distinguish from genuine bands
arising from inelastic scattering at the crystal surface. Second, there is a
general increase in the background count rate throughout the entire
spectral region arising from the general level of directionally nonspecific
electrons scattered within the electron optical components.

An increase in the background count rate will have the effect of
decreasing the signal-to-noise ratio of the spectral bands (S/N). For a

band with a maximum count rate of S counts s^{-1} superimposed on a background of B counts s^{-1}, S/N becomes $S/(B + S)^{0.5}$. In this way, weak loss bands become difficult to detect on excessively large backgrounds.

Many contrivances have been proposed to counter this problem. These include (i) a mesh covered hole in the outer element of the energy selector (OH in the present instrument) in line with the input beam into which those electrons with energies well outside the selected pass energy may be dumped; (ii) the energy selector field defined by fine mesh rather than solid components so that unwanted electrons may be transmitted rather than scattered; and (iii) "Sawtooth" serations milled into the selector elements in a manner designed to ensure that electrons scattering from these elements will mostly be directed away from the detector. This solution has been employed with success to the 127° CDA based analyzers[4] but is not practical in the 180° CHA based analyzers of present interest.

A better solution to this problem has been found[27] by recognizing that electrons adventitiously scattered from the field defining components of the energy selector will not appear at the output window (W6) with the same near normal incidence as the electrons with the desired pass energy. Instead, by being scattered from these components (principally the inner and outer hemispheres, in the present case), they necessarily appear with somewhat greater divergence and may efficiently be filtered with a second suppressor window W7 positioned in the manner shown in Figure 4. An intermediate composite focus–deflector lens, FS-D, provides additional control over the beam in this region during optimization tuning of the spectrometer (Section 5.12). In the absence of the suppressor the present spectrometer produced tunable "ghosts" in the <400 cm^{-1} loss region and a broad background of ~2% the elastic peak at 1500 cm^{-1} energy loss. With the incorporation of the suppressor no further problems with low frequency "ghosts" were experienced and the background, after scattering from a Pt(111) surface, was reduced to ~0.2% of the elastic peak at 1500 cm^{-1} energy loss.

2.4.2. Electron Detectors

In view of the small currents (~10^{-10} A) available from a monochromator operating at the required resolution and the further attenuation of the beam that occurs on reflection from the crystal surface, the use of electron multiplier detection at the spectrometer output is essential. For their compactness, air-insensitivity, and high gain, continuous dynode multipliers have usually been preferred over the very much bulkier discrete dynode multipliers of either the venetian blind or box and grid types. When operated in a particle-counting mode the maximum current that may be accepted by continuous dynode multipliers is typically $\sim 3 \times 10^{-15}$ A or 2×10^5 counts s^{-1}. The saturation that occurs at higher count rates arises from the finite time required to recharge the internal emitting surface of the multiplier after the passage of an electron pulse. Although this is almost an order of magnitude less than that of typical discrete dynode multiplier, the air sensitivity and excessive bulk of the latter has been responsible for the preference for the continuous dynode type in EEL spectroscopy. However, a new discrete dynode multiplier, manufactured commercially by ETP under license to CSIRO, is both compact and air insensitive and with the additional feature of replaceable dynodes should prove to be an economical and versatile alternative to the continuous dynode multipliers.

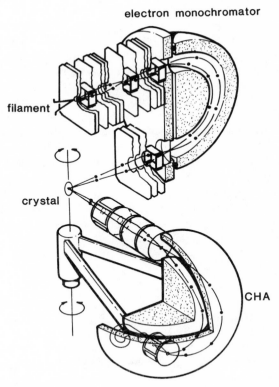

Figure 8. Cut-away isometric view of the assembled EEL spectrometer.

A cut-away isometric diagram of the assembled EEL spectrometer is shown in Figure 8.

3. Construction Material and Methods

3.1. Materials

With electron kinetic energies as low as $\sim300\,meV$ in the energy selector, magnetic fields, both static and alternating, are potentially detrimental to the resolution and sensitivity of an EEL spectrometer. For example, it may be estimated that for a path length of 10 cm, a 300-meV electron beam will be deflected $\sim1.7\,mm$ per mG of magnetic flux density. Methods of reducing the ambient magnetic field will be discussed later, while for the moment the need for nonmagnetic structural materials that are also ultrahigh vacuum compatible will be addressed.

Traditional ultrahigh vacuum materials include the 18Cr, 8Ni austenitic type stainless steels. Annealed, they are nonmagnetic but the cold working procedures that are used to produce bar and plate stock induces a partial recrystallization to the magnetic ferritic phase. As a result of this, if 18:8 stainless steel is to be used it should be annealed at $\sim1050°C$ and water quenched before machining. Final degaussing and checking for residual magnetization by passing the fabricated components past a stationary Hall probe is a sound precaution. Similarly, commercial stainless steel nuts and bolts often are cold formed and highly magnetic but again may be annealed, preferably in a vacuum or hydrogen furnace to prevent undesirable surface oxidation. Alternatively, many UHV-compatible nonmagnetic metals are available and should be considered in the fabrication of spectrometer components. I have found copper and aluminum alloys to be satisfactory, while for higher tensile components, like nuts and bolts, fabrication from copper–beryllium alloy is ideal.

In the construction of an EEL spectrometer there is also a need for insulating components. Here, commercially available ceramic components are sometimes usable while the more specialized, custom components may be fabricated from the machinable glass ceramic, Macor (registered trademark of Corning glass works). For precision spacing of components, ruby balls captured in holes or grooves are particularly useful.

Electrical connections of the electron optical components to the vacuum feedthrough may be made with almost any soft wire, such as, for example, copper (bare or tinned) or tantalum. Insulation of these connecting wires is commonly achieved with flexible glass spaghetti tubing. However, this material has the undesirable tendency to fray at the ends. While this may be reduced by flame-polishing the ends, there is always the risk of insulating fragments lodging in crucial regions of the electron optics and, on charging-up, distort the electron trajectories in an erratic manner. More recently, Teflon tubing has been used, and with bake-out temperatures in excess of ~120°C being unnecessary for the generation of ultrahigh vacuum, it is proving to be a superior alternative to glass spaghetti.

3.2. Fabrication Methods

The principal demands of the electrostatic lens stacks are that they be both rigid and accurately aligned. The methods that have been used to do this, along with the fabrication of the spectrometer in general, are as many and varied as the imaginations of the design engineers involved. Consequently, it is impracticable to give a comprehensive treatment of these methods. Instead, some examples only of established methods will be given.

The aperture lenses used in the present monochromator were made from 37 × 37-mm guillotined squares of annealed and polished 0.5-mm copper sheet. An approximately central 0.5-mm pilot hole was located in a suitably constructed die which allowed four 6-mm-diam holes at the corners of a 20 × 20-mm square centered on the pilot-hole to be punched reproducibly. The pilot hole could be enlarged to produce the required components of the lens stack. The lens stack was assembled in the manner shown in Figure 9 by threading the lens components onto precision ground ceramic tubes. Insulating spacers are conveniently made from machinable glass ceramic and should be coated with graphite to prevent charging. After coating, breaking the edge on one side with a file

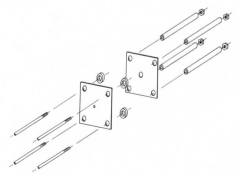

Figure 9. Exploded view of two elements of a typical aperture lens stack.

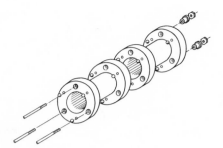

Figure 10. Exploded view of two elements of
a typical ruby ball spaced cylinder lens stack.

ensures electrical insulating of adjacent components. The required
alignment of the stack is achieved with precision ceramic tubes (Figure
9), which may be ground from commercially available high-density
alumina thermocouple sheaths. Tensioning a suitable terminated
through-bolt (copper–beryllium) within the alumina tubes ensures rigidity of the lens stack. This method of alignment is particularly suitable
for the many-element and relatively lightweight lens stacks of the present
monochromator.

For cylinder lenses the greater weight of the components may
fracture the alumina rods and the ruby ball spacer or optical bench
methods are usually superior alternatives (Figures 10 and 11, respectively). For the former, cylinder lens of the desired internal diameter and
length are spaced and insulated with ruby balls located by precision
drilled holes. On compression, again with a suitably terminated bolt, a
rigid and accurately aligned lens stack is achieved. This is the construction method used for the lens on the energy analyzer shown in Figures 4
and 8. For the optical bench method (Figure 11) cylinder lenses are
positioned on a pair of precision ground ceramic rods (high-density
alumina is ideal) located in grooves milled in the optical bench. The lens
elements are located as required on the bench and fastened from the
underside with ceramic insulated bolts. Apertures in cylinder lens systems
may be fastened directly to the cylinder.

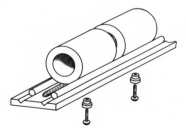

Figure 11. Electron optical bench method of locating cylinder lens stack.

Figure 12. Trim deflector assemblies suitable for aperture (left) and cylinder lens (right) stacks.

In view of the importance attached to electrostatic deflectors to correct for inevitable mechanical and field imperfections, some consideration will be given to their design and fabrication. Two deflector systems suitable for aperture and cylinder lens systems are shown in Figure 12. Both are based on machined glass ceramic blocks with the X and Y deflector pairs fitted in the manner shown. Electrical connection may be made with the securing nut. For incorporation in an aperture lens stack, the ceramic block is jig drilled with the same 20×20-mm square array of 6-mm holes used for the other electron optical components and takes its place as required in the lens stack (Figure 9). The X and Y deflector pairs for the cylinder lens are similarly fastened in the ceramic block, which in this case may be secured directly to the adjacent lens component. In the arrangement shown in Figure 12 the inside profiles of the deflectors form a cylinder, which may then also act as a lens in the manner required of FA-D and FS-D in the present analyzer. Similarly, the rectangular aperture lenses used in 127° CDA based spectrometers may be split about the long axis to give transverse deflection of beam. In these spectrometers deflection about the short axis is not usually necessary.

For lenses with cylindrical symmetry like the aperture and cylindrical types just described, the manufacture of small apertures for windows and pupils (down to 0.5 mm) is a simple machine-shop operation. However, the CDA based EEL spectrometers calls for fine input and output slits, which require nonstandard methods of construction. Here a suitable procedure has been found whereby guillotined and polished metal strips are spotwelded together in the manner shown in Figure 13. A temporary wire of the appropriate diameter separates two of the strips during this operation so that when it is removed a fine parallel-sided slit is formed. Needless to say, copper with its poor spot weldability is an unsuitable

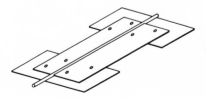

Figure 13. Typical method of manufacturing metal slits for CDA-type energy selectors. Here four metal strips are spot-welded in the manner shown with a temporary wire of the required diameter defining the slit width.

material for this purpose. Tantalum or copper–beryllium would be ideal. During any spotwelding operation, the quality of the weld will usually be improved if a drop of a suitable flux, like ethanol, is allowed to cover the areas to be welded. Similarly, when highly refractory metals are to be used, the weld may be improved if an intermediate sliver of lower melting metal, like nickel or gold, is incorporated in the weld.

In order to ensure that the electrostatic field gradients experienced by the electron beam in the EEL spectrometer do not contain resolution degrading discontinuities it is necessary to first eliminate as far as possible, all edges of small radius where the field gradient will be concentrated. This is particularly important at windows and pupils that are not located in field free space. The edges of all electron optical components should be broken and at least mechanically polished. Final electrolytic polishing is desirable since it is in the nature of this procedure to preferentially remove material from these small radii edges. Second, inhomogeneuous fields arising from patchy work function distributions on real, polycrystalline metal surfaces is potentially serious, particularly in regions when the electron beam closely approaches the surface or is at a low kinetic energy. This problem is best overcome with a thin layer of graphite, which may be applied as an isopropyl alcohol dispersed aerosol or directly by brush. Graphite has the effect of homogenizing the work function by virtue of its laminae habit, which preferentially exposes the basal plane. Additionally, it has the further advantages of minimizing secondary electron emission, relative chemical inertness, and good electrical conductivity. Indeed, the graphite coating of all conducting electron optical components that can see the electron beam should be regarded as part of the dogma of EEL spectroscopy. The earlier practice of electroplating a porous gold film onto the field-defining surfaces is less convenient and certainly no better than a graphite coating.

3.3. Shielding

In a well-designed EEL spectrometer it is essential to reduce, as far as possible, the undesirable effects resulting from adventitious magnetic and electrostatic fields as well as soft x-radiation in ion-pumped systems.

As discussed in Section 3.1, the appropriate choice and treatment of construction materials may, with the exception of directly heated thermal emitters, (Section 2.33), eliminate the intrinsic magnetic fields within the spectrometer and scattering region near the crystal target. Extrinsic fields arising from outside the apparatus may be either direct or alternating. Direct fields are usually dominated by the Earth's magnetic field, which is of the order 500 mG, depending on location. While fields of this kind may satisfactorily be nulled with Helmholtz coils (which should have a diameter-to-separation ratio of 2 for optimum homogeneity), this is usually not necessary, because, in shielding the more troublesome alternating fields, the residual direct field is reduced to a level that is easily compensated with the electrostatic trim deflectors within the EEL spectrometer. The major source of alternating magnetic field is usually at mains frequency (50 or 60 Hz) and arises from electronic equipment in the laboratory. It is difficult to null this component by superimposing an alternating field, with appropriate magnitude and phase control, on the direct field of a set of Helmholtz coils. Instead, the most satisfactory solution is to enclose the spectrometer and scattering region in a high-permeability magnetic shield (e.g., mu-metal annealed after fabrication in hydrogen or cracked ammonia at 1060°C). It is usual to fit a shield inside the vacuum chamber, in which case it must be constructed in two sections, which may slide together with good overlap in order to form a low reluctance magnetic pathway around the spectrometer and scattering region. Inevitably, openings must be made in the shield, but by keeping these to a minimum and ensuring that particularly magnetically leaky items of electronic equipment are well removed from the spectrometer, a well-designed mu-metal shield reduces both the direct and alternating magnetic fields to a level that is not noticeably resolution-degrading.

For electrostatic shielding it is usually only necessary to ensure that the fields generated by the electron optical components and multiplier are screened from the field free region near the crystal. A negatively biased shield about the multiplier should lower the dark count rate by repelling electrons other than those that have passed through the analyzer energy selector and suppressor.

Finally, when ion-pumped systems are used it is necessary to provide sufficient line-of-sight baffles to prevent soft x-rays generated in the pump from stimulating photoelectron emission in the region of the spectrometer. Electrons from this source may add greatly to the dark count rate, thereby reducing the signal-to-noise sensitivity of the spectroscopy, as discussed in Section 2.42.

4. Monochromator and Analyzer Power Supplies

To scan an electron energy loss spectrum either the incident beam energy, E_0, or the scattered electron energy, $E_0 - \Delta E$, may be swept while the other is held constant. Since the reflectivity of the crystal can be strongly dependent on E_0, the latter method is usually chosen. Additionally, the analyzer may be operated in either a constant resolving power,

$\Delta E_H/E_p$, or constant resolution, ΔE_H, mode. The former is achieved by proportionally sweeping the potential on all electron optical elements[26] but leads to the undesirable feature of variable resolution throughout the spectral range. Instead, it is more common to operate in the constant resolution mode by maintaining the same relative potentials on all analyzer elements after and including FA-D and adding to each the sweep voltage. In this way, during the sweep the electron trajectories are disturbed only in the first, or adder lens, FA.

In the design of the spectrometer it was emphasized that the various stages in the electron optical pathway should as far as possible be made independent of each other. This same design philosophy has been extended to the power supplies that set the voltages on the electron optical elements. For example, deflector power supplies referenced to ground will lead to electron deflection amplitudes that change with the operating lens potentials and should be avoided. Similarly, it is highly desirable to be able to change the beam energy at the crystal, E_0, without greatly affecting the resolution, ΔE_H, and vice versa. Operational schematic diagrams for the monochromator and analyzer power supplies that fullfill these objectives are shown in Figures 14 and 15, respectively.

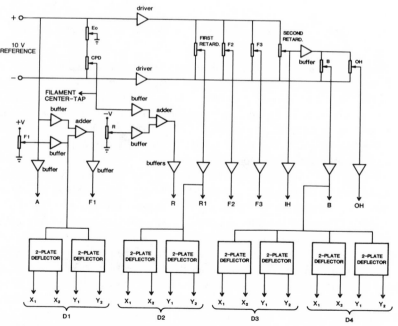

Figure 14. Operational schematic diagram for the electron monochromator power supply.

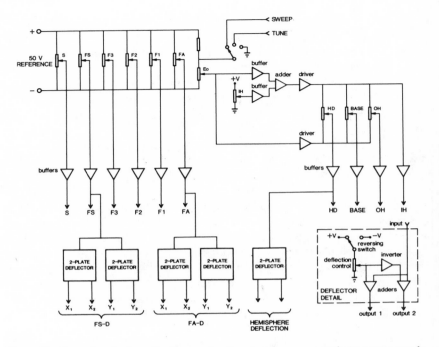

Figure 15. Operational schematic diagram for the electron energy analyzer power supply.

For the monochromator, all voltages are referenced to the positive and negative rails of a precision 10-V reference with only F1 and R requiring additional voltage. The emitter filament is heated from a separate dc power supply, the center tap of which, along with the repeller, R, is biased (typically ~0.5 V) from the negative rail by the CPD potentiometer. The CPD control nominally corrects for the contact potential difference between the tungsten emitter and the graphite coating on the monochromator elements but additionally also accommodates any difference in the potential drops along the two electrical connections between the filament power supply and the emitter. The role of the CPD control is that, by referencing all monochromator element voltages to the negative rail, the electron potential in any region of the monochromator may be measured directly as required by, for example, the zoom curves of the type described in Section 2.3.1. The incident beam energy, E_0, is established by the E_0 tap to ground and of course, by measuring from the negative rail, will be relative to graphite. As discussed by Katz et al.,[28] electrical noise on the electron optical elements can seriously degrade the resolution. In the present system, noise was minimized by choosing low potentiometer resistors (typically ~200 Ω), necessitating low impedance buffers and current drivers in the voltage rails. The buffers also serve to electrically isolate otherwise interacting parts of the circuit. With this scheme, the beam energy is established explicitly by E_0 relative to graphite, while the resolution is set almost independently, as required, by the second retardation control. Similarly, deflector power supplies (detail in Figure 15) are referenced to the potential of the electostatic element in which they are located.

The equivalent functional diagram for constant resolution operation of the analyzer is shown in Figure 16. Again, low resistance potentiometers and buffers have been used to suppress electrical noise. As discussed in Section 3.2 the adder and suppressor lenses are split to act also as deflectors. While, for example, FA gives the mean focus voltage in the adder lens the actual voltages are established by the deflector control FA-D. The hemisphere deflector, HD, was required because of the use of input and output apertures on the energy selector. While the deflector voltage maintains alignment of the beam, a small gain in spectrometer performance was found by driving the mean potential more

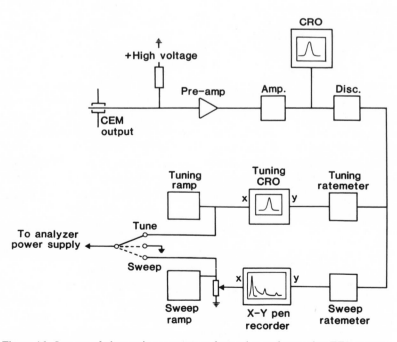

Figure 16. Layout of electronic components for tuning and sweeping EEL spectra.

negative that B with HD. This apparently compresses the beam somewhat towards the plane of mean deflection, thereby reducing the overall
divergence of the beam. Analyzer pass energy and therefore resolution is
adjusted with the IH control, while the EEL spectrum is swept by setting
E_0 and adding the appropriate scan voltage at the sweep–tune–ground
mode selector.

On both the monochromator and analyzer control front panels the
deflector reversing switches are oriented up–down and right–left with the
same deflection sense experienced by the electron beam. In this way, a
glance at the switches and magnitude potentiometers gives the operator a
useful feel for deflection being imposed on the beam at any point on the
beam pathway. For example, oscillation of the beam in the input
retardation optics of the monochromator could be detected by, say, large
magnitude up–down–up deflection in D1, D2, and D3. In this case it may
be necessary to initiate a new iterative procedure for optimum tuning
(Section 5.1).

While low noise on all electron optical components is clearly
desirable, the requirement for this will be most stringent in the energy
selectors of both the monochromator and analyzer. An even more
stringent demand is put on the emitter potential, for any electrical noise
or fluctuations here will be transferred directly to the beam energy, E_0. It
therefore is desirable to capacitively couple these potentials to ground,
although for the analyzer, which must be swept, there is a limit to the
extent to which this may be done. Alternatively, E_0 drift may be
compensated by referencing the sweep voltage to the emitter center tap,
although this has not been found to be necessary for the arrangement used
in this laboratory. An objective assessment of the worth of capacitive
filtering may be obtained by observing the noise level superimposed on
the elastic peak during tuning (Section 5.1). For example, it may be
found that capacitors are better located on the vacuum feed-through
rather than in the control box. Similarly, observing the elastic peak in this
way is useful for identifing sources of noise, which may then be
eliminated rather than filtered.

5. Operating an EEL Spectrometer

After assembling the EEL spectrometer and associated shields in the
vacuum chamber it is necessary to position the crystal as near as possible
to the point of reflection defined by the spectrometer scattering geometry. Extreme accuracy at this stage is not usually necessary, and light
metal or cardboard jigs will usually suffice. Crystal rotation to establish

$\theta_i = \theta_r$ can be set in similar fashion. If the crystal manipulation provides tilt, the crystal surface normal can be set in the scattering plane by viewing the crystal in this plane and adjusting until an image of the observers eye is seen in the crystal. In the absence of manipulator tilt the crystal support leads must be distorted until this condition is attained.

Significant features in an EEL spectrum appear at the electron multiplier with currents as low as $\sim 10^{19}$ A. Detecting currents of this order is best done with conventional pulse-counting techniques of the general configuration shown in Figure 16. Here the amplified and discriminated pulses are converted to an analog signal proportional to the count-rate in a ratemeter. Monitoring the amplified pulses with a cathode ray oscilloscope (CRO) not only assists in setting the discriminator levels but also gives a continuous indication of intermittent noise which may interfere with the EEL spectrum. Two modes of operation are selectable with the tune-sweep mode switch. For tuning (Section 5.1) a 0.03-V sweep at ~ 3 Hz is used to provide a more or less continuous display of the elastic peak on the tuning CRO. Once suitably tuned for optimum peak shape, resolution, and sensitivity the sweep ramp is selected to record the EEL spectrum. This -0.02 to $+0.5$ V ramp is continuously adjustable from ~ 0.06 to 0.18 mV s^{-1} (0.5 to $15 \text{ cm}^{-1} \text{ s}^{-1}$) with additional fast scan, reverse, and reset capability.

Although digital data acquisition methods, e.g., multichannel analyzers and computers, may also be used for EELS, analog methods of the type described here are more widely exploited by practicing EELS experimentalists.

5.1. Tuning

In spite of the modern indulgence in computerization and automation of scientific experiments, tuning an EEL spectrometer remains very much a manual and indeed often subjective operation. This section will attempt to give hints and advice I have found useful for first aligning the beam and subsequent tuning for optimum sensitivity, resolution, and elastic peak cutoff.

5.1.1. Aligning the Beam

For the desired V_1/V_3 ratios, the required focus voltages on the monochromator zoom lenses ZR1, ZR2, and ZA may be estimated from zoom curves of the type shown in Figure 7 for ZR2. Similarly, the relative potentials on B, IH, and OH may be estimated from equations

(4)–(6). Preliminary optimization of the electron source, ZR1, and ZR2 is best done by monitoring the current directly to OH. This is typically of the order of 10^{-7} A and easily detected by analog means. With OH reconnected to its operating voltage, the current to the crystal may be monitored with a picoammeter. If difficulty is experienced here, the beam energy, E_0, and the selector pass energy, should be increased and D3 adjusted until a current at the crystal is detected. In view of the noise usually encountered in detecting currents that initially may be as low as 10^{-12} A, it is helpful to flick the deflection-reversing switch in D1, thereby interrupting the beam and watching for the corresponding change in current to the crystal. Once detected the crystal current may be optimized, for now.

Since only a few percent of beam incident on the crystal may be elastically reflected and not necessarily all this intercepted by the analyzer input, electron currents in the energy selector rarely exceed 10^{-13} A, even when well tuned. As a result of this I have never had success in optimizing the scattering geometry and analyzer voltages by attempting to detect a current to OH of the analyzer. Instead, the preferred procedure is to set the analyzer selector for high pass energy (low resolution) and to attempt to detect transmitted electrons digitally with the electron multiplier. Here, as little as 10 counts s^{-1} or $\sim 10^{-19}$ A can be detected. However, in this phase of preliminary tuning, the spectrometer can be frustratingly opaque, particularly if the incident beam energy and selector pass energies are initially set too low. If the operating voltages of the input retardation and suppressor optics are known or may be estimated, then these voltages should be applied accordingly; otherwise somewhat positive potentials will be best at this preliminary stage of tuning. Since it is usually not practical to attempt to predict the selector pass energy of the analyzer (or monochromator) precisely, it is advisable to scan the analyzer pass energy with the tune sweep. When this is done, better detection sensitivity may be obtained by viewing the CRO display of the amplifier output (Figure 16) rather than observing the analog output of the rate meter. Eventually, after an often protracted iteration of scattering geometry defined by crystal translation, rotation, and beam deflection as well as analyzer lens voltages, a miniscule portion of the elastic beam will be detected at the multiplier. Once found, and confirmed by again interrupting the beam by flicking a D1 reversing switch, iteration should proceed cautiously lest the beam be lost again. Usually small changes in the scattering geometry lead to rapid improvement in the elastic beam. At this stage, it is not necessary to spend too much time iterating lens voltages. Instead, the better procedure is to incrementally reduce the incident beam energy and selector pass energies to the desired level while at the same time ensuring that the elastic peak is always detectable by continual iteration of the scattering geometry and focus voltages.

Of course, if the energy analyzer is rotatable to the extent that the crystal may be translated out of the way and the monochromator beam directed straight into the analyzer, higher sensitivity will be achieved, greatly facilitating the iteration of operating parameters necessary to get the beam to the detector. Once thus established, the analyzer may be rotated to the desired angle of incidence, θ_i when only the scattering geometry need be iterated.

5.1.2. Optimizing of the Beam

Optimum intensity and resolution are dictated by the usual spectroscopic compromise between sensitivity and resolution. With the commonly used continuous dynode electron multipliers (Section 2.43) operating in a particle counting mode the maximum usable elastic count rate is limited to $\sim 2 \times 10^5$ counts s^{-1} (Section 2.43). Count rates greater than this may be reduced by limiting the current into the monochromator energy selector with D1 or R, which keeps in reserve some sensitivity if it is believed that it may be required at some later stage in the experiment. Alternatively, often the better procedure is to reduce the pass energies in the energy selectors with the concomitant improvement in resolution as expressed by equation (10).

The elastic beam intensity which is passed by the analyzer will necessarily depend on the crystal reflectivity, which in turn depends on the level of surface order and work function. As these parameters may change with the adsorption and reaction of molecules on the surface, routine optimization tuning is necessary and must be made as convenient as possible. By displaying the elastic peak on the tuning CRO at ~ 0.3-intervals, more or less continuous iteration of the scattering geometry and operation voltages may be achieved by a simple hand–eye operation. With experience, routine retuning of this kind can usually be achieved within a minute. When changes in crystal work function occur, some EELS experimentalists advocate compensation by applying a counterpotential to the crystal. However, I have never found this method to be as satisfactory as retuning, probably because it is never possible to apply a counterpotential to the crystal alone. Instead, the potential inevitably is also applied to all conducting surfaces in electrical contact with the crystal (heater leads, supports, etc.), which, however, have not experienced a work function change but nevertheless, by their proximity, influence the electron trajectory. If a work function change occurs to the

extent that no elastic beam is detectable, I found that the best first move
is to translate the crystal along the surface normal. Once the beam is
reestablished, tuning the elastic peak usually proceeds rapidly with
iteration of the scattering geometry.

As pointed out by Ibach and Mills,[8] the detection of weak bands
with low energy loss ($<300\ cm^{-1}$), as often encountered in surface
phonon spectra, requires good cutoff of the elastic peak, and for
experiments of this kind is a criterion of comparable importance to good
resolution. Similarly, the suppression of adventitiously scattered electrons
throughout the entire loss range is also essential (Section 2.42). Thus
during optimization tuning, the EELS operator should be mindful of
these advantages and watch not only for improvement to resolution but
also the shape of the elastic peak, particularly on the energy loss side.
Comparison of the present spectrometer with that at Caltech[13] suggest
that W3 and W4 windows smaller than those used in the present analyzer
favor good elastic peak cutoff, although the extent to which this may be
exploited is limited by demands on sensitivity when less highly reflecting
crystal surfaces are encountered.

An example of an EEL spectrum, of nondissociatively adsorbed
cis-2-butene on Pt(111) at 170 K, carefully optimized for resolution and
elastic peak cutoff is shown in Figure 17.

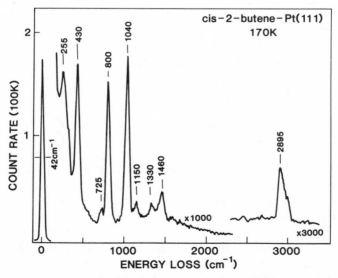

Figure 17. A typical high-resolution EEL spectrum for nondissociately adsorbed cis-2-
butene on Pt(111) at 170 K.

While more intense EEL spectra are usually obtained with low incident beam energies, there is a limit to the extent to which this may be exploited. If the beam energy is set too low, a high work function crystal surface will reflect the electron beam without inelastic surface interaction. This condition is called beam lift-off and may be recognized by (i) abnormally high elastic count rate, (ii) improved resolution, (iii) high susceptibility to vibration of the crystal, and (iv) excellent cutoff of the elastic peak. Needless to say, no EEL spectrum will be obtained when this condition occurs. With higher incident beam energies lift-off may be induced by applying a negative bias potential to the crystal. While this greatly enhances the intensity of the elastically reflected beam, it is of little usefulness in the early stages of aligning the beam owing to the strong dependence on crystal position and bias potential. By inducing beam lift-off in this way the spectrometer can usually be tuned easily to an elastic peak FWHM as low as 25 cm^{-1} with excellent cutoff. However, this performance can never be repeated under the more normal operation involving surface interaction, presumably because of the degrading effect of incoherent scattering from real surfaces. As an aside to this effect, with a FWHM of 3.5–5 meV the electron beam from the monochromator is ideally suited for work function measurements by the retarding potential method.

5.2. Calibration

As discussed in Section 5, the electron energy loss axis of an EEL spectrum often is taken directly from the sweep ramp voltage (Figure 16), which presumes a direct correlation between the applied potential and the energy added to the electron beam in the analyzer. While this chapter follows the usual practice of referring to spectroscopic *energy* analysis, devices of this kind strictly are *momentum* analyzers. As a result of this, only the energy associated with the axial component of the momentum will be affected by the accelerating potential in the adder lens. Details of the actual electron trajectories in the retardation region between the adder lens and the energy selector are unimportant since all electrons have the same kinetic energies and will follow virtually the same trajectory in this region. A loss electron with energy $E_0 - \Delta E$ inclined at an angle of ϕ to the adder field will be accelerated to the analyzer pass energy with an adder voltage of $\Delta E(1 + \sin^2 \phi)$ i.e., the applied adder voltage overestimates the loss electron kinetic energy by $\sin^2 \phi$. At $\phi \sim 5°$ an error of ~1% is expected, which at the high loss energy end of the spectrum, say 3000 cm^{-1}, is comparable to the accuracy of measuring EELS bands. With the spherical windows, in the present instrument ϕ is not expected to exceed ~2°, in which case the error in the energy loss scale is trivial (~0.1%). However, with the elongated slits used in the CDA based instruments the scope for error must be somewhat greater. In any event, calibration of the instrument is a sound practice.

Two methods are available for calibrating an EEL spectrometer. The elastic peak shift method exploits the fact that changes in the incident beam energy are given explicitly by the E_0 potentiometer on the monochromator power supply (Figure 14). The method consists of measuring the beam energy with a digital voltmeter sensitive to better than 1 meV and recording the elastic peak. By reducing the beam energy in accurately measured increments of 50–100 meV (leaving all other voltages fixed) and recording the new elastic peak on the same chart, the shift may be measured and compared with the change in E_0 as measured with the digital voltmeter. Additionally, the intensity attenuation of these elastic peaks is a measure of the decreasing spectrometer transmission

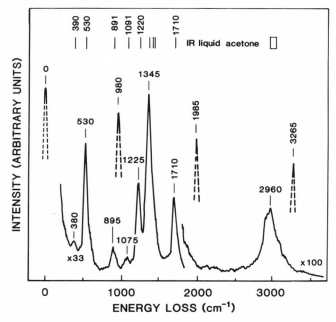

Figure 18. EEL spectrum (solid curve) of acetone condensed on Pt(111) at 130 K correlated with the infrared spectrum of liquid acetone. The dashed curves are the tips of elastic peaks obtained by decreasing the incident beam energy by the amount indicated. Both procedures indicate that the energy loss scale taken directly from the sweep ramp is accurate to $<10\,\mathrm{cm}^{-1}$.

with electron energy loss, provided of course the crystal reflectivity is not a strong function of E_0 in the region of interest. A second method of spectrometer calibration involves condensation of 5–10 monolayers of a suitable organic molecule on the surface at low temperature ($\sim$100 K) and comparing the EEL spectrum with a more accurately measured infrared spectrum of the same molecule in the same state, usually liquid or solid. A suitable molecule is one that has a good distribution of sharp bands throughout the required spectral range.

Examples of these methods are shown in Figure 18. The solid curve shows the spectrum of $\sim$5 monolayers of acetone [$(CH_3)_2CO$] condensed on a Pt(111) surface at 130 K. The positions of prominent bands are measured directly by their displacement from the elastic peak, $\Delta E = 0$, and are correlated with the infrared spectrum of liquid acetone. For acetone, the nondegenerate methyl stretches ($\sim$2960 cm^{-1}) and deformations ($\sim$1345 cm^{-1}) are, for EELS, too broad for this purpose. The dashed curves show the tips of the elastic peaks obtained by reducing the incident beam energy by accurately known amounts. In these ways it can be shown that, as expected for the relatively small windows in the adder lens, the sweep voltage gives an accurate measure of the electron loss energy.

6. Application of EELS to Surface Chemistry

As pointed out in Section 1, high-resolution EELS in the 0–500-meV energy loss region is a vibrational spectroscopy. While this will inevitably attract the interest of many scientific disciplines, it offers the surface chemist in particular a powerful means for studying the molecular geometry and bonding configuration of relatively large adsorbed molecules.

6.1. Experimental

In this section, I address some of the more important experimental techniques that have been devised for handling these often "sticky" molecules in ultrahigh vacuum system. Compared with the simpler, noncondensible molecules, procedures should be adopted that allow the crystal surface to be first exposed to these vapors without seriously fouling the vacuum and later to obtain EEL spectra after thermal reaction that are not contaminated by volatilization from cold surfaces within the vacuum chamber. Additionally there is a need in all surface spectroscopy work to avoid contamination of the crystal by the residual gases during excessively long cool-down times. Even in the most recent literature it is apparent that all too often little effort is made to circumvent problems of this kind.

An arrangement for mounting the crystal in a manner conducive to rapid heating and cooling is shown in Figure 19. Here, a hollow copper tank is filled with liquid nitrogen by pumping through ~1.5-mm-diam flexible stainless steel tubes from a Dewar outside the vacuum system. With two turns of these tubes there is sufficient flexibility to obtain the necessary X, Y, and Z translation and rotation with a conventional manipulation. With appropriately designed manipulations some tilt is also

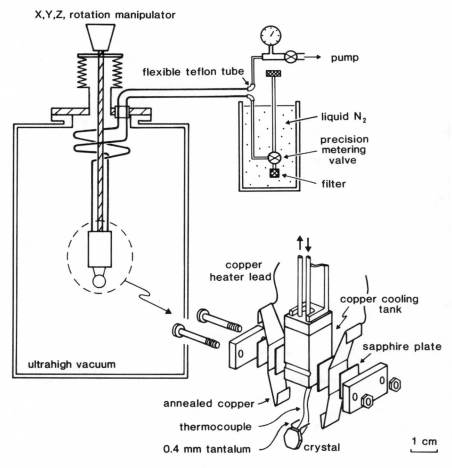

Figure 19. A method for crystal mounting which permits rapid cooling and heating.

possible. Electrical insulation of the annealed copper support strips from the liquid nitrogen tank is achieved with $10 \times 10 \times 0.5$ mm sapphire plates, which at low temperatures give excellent thermal conductivity, indeed better than copper itself. Good thermal contact of these components is achieved by clamping in the manner shown in Figure 19. The crystal is mounted by spot welding two 0.4-mm-diam tantalum wires to the back side of the crystal. The tantalum is then shaped and crimped into the copper support strips so that about 2 mm of tantalum is exposed. This method of supporting the crystal has several desirable features including (i) Damage to the crystal during spot welding is not only minimized but confined to the edge. (ii) Stress on the crystal is minimized. (iii) Nondestructive demounting of the crystal is obtained by prizing apart the crimped copper support strips. A thermocouple may be spot welded to the back of the crystal directly with a conventional spotwelder. For highly conducting crystals, like copper, spot welding is not possible and both the thermocouple and tantalum heater leads must be mechanically attached in holes or grooves cut into the crystal, preferably near the edge.

Liquid nitrogen is pumped into the cooling tank through a precision metering valve fitted with a $60\,\mu\text{m}$ sintered metal filter to prevent the valve blocking with ice condensed in the liquid nitrogen. The metering valve was modified so that the control knob extends above the liquid nitrogen Dewar. With the metering valve fully open, excess liquid nitrogen was purged through the cooling tank and a $7\,\text{mm} \times 0.7\,\text{mm}$ Pt(111) crystal, for example, could be cooled from 600 to 100 K in three minutes. Under these conditions a base temperature of 88 K could be achieved in an additional minute. Alternatively, by restricting the supply of liquid nitrogen with the metering valve, the pressure in the cooling tank could be reduced and a base temperature of 77 K was attainable. A limit to the extent to which this may be done is imposed by the requirement that sufficient liquid nitrogen is passed through the system to maintain the cooling tank at the base temperature. This condition is easily established by erratic pressure fluctuations on the pump vacuum gauge. Similarly, ice blocking the filter could be detected by a pressure decrease. Although this has not proven to be common occurrence, it could usually be cleared by closing the pump valve and allowing the liquid nitrogen in the cooling tank to expand violently back into the Dewar and in so doing clear the filter.

In order to expose the crystal surface to the desired vapor without fouling the vacuum or initiating excessive hydrogen and carbon monoxide build-up in ion pumped systems, a pair of movable microcapillary array dosers were used (Figure 20). The microcapillary array consists of bundles of aligned glass capillaries, which at short range produce a nearly

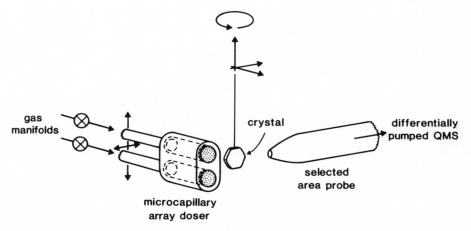

Figure 20. Diagrammatic illustration of the movable dual microcapillary array doses for exposing the crystal surface with minimum fouling of the ultrahigh vacuum and the differentially pumped selected area probe used for obtaining thermal desorption spectra.

collimated vapor flux. When moved with an externally controlled bellows to within a few millimeters of the crystal surface, high local fluxes are produced without an undesirable massive increase in background pressure. With collimated fluxes of this kind the copper support strips for the crystal (Figure 19) are oriented to reduce their cross section to the flux and in so doing minimize the amount of vapor condensed in the vicinity of the tantalum heaters. Each of the two dosers is connected by glass-lined stainless steel tubing to independent leak valves and gas handling manifolds. With this arrangement a Pt(111) surface, for example, could be given a monolayer exposure of vapor without the appearance of *any* adsorbed carbon monoxide in the resulting EEL spectrum as may be seen from the example in Figure 17. Similarly, it has not been found necessary to internally heat the spectrometer to maintain either ultrahigh vacuum or optimum performance of the spectrometer. However, at this point it may be noted that in normal operation the spectrometer is quite warm owing to heat derived from its own and other hot filaments in the vacuum chamber.

As suggested by Feulner and Menzel,[27] a selected area probe for thermal desorption spectroscopy, TDS, has several advantages over the earlier method of desorption into the large volume of the vacuum system. Briefly, the method uses a nozzle, or probe, which is positioned near, and in front of, the crystal so that desorbing molecules are directed into a small differentially pumped chamber containing the quadrupole mass spectrometer. Differential pumping may simply be back into the main chamber. If the probe hole is significantly smaller than the crystal a selected area is probed and heater wire and crystal edge and back effects are minimized. Additionally, the smaller volume of the mass spectrometer chamber results in greater sensitivity and low heating rates may be used $(1-3 \, \mathrm{K \, s^{-1}}$ are now common). With heating rates of this order, complementary EEL spectra may be taken that correlate directly with significant features in the TD spectra (Section 6.2).

With the differentially pumped selected area probe, care must be taken to ensure that unduly low pumping speeds do not distort the thermal desorption spectra. To check for this, the first-order pumping time constant may be measured by tuning the quadrupole mass spectrometer to either hydrogen or carbon monoxide and switching the emitter off for a few minutes to allow adsorption of these gases. On reactivation of the emitter the desorption burst and subsequent decay of the monitored gas could be followed with a time-base pen-recorder. From this, the first-order time constant, k $(\mathrm{s^{-1}})$, characteristic time, $\tau \equiv 1/K$ (s), or half time, $t_{1/2} \equiv 0.693\tau$ (s), could be estimated. In the present system $t_{1/2}$ for carbon monoxide was estimated to be 0.2 s, which will not noticably damp the thermal desorption peaks with the heating rates used here $(\beta = 1-3 \, \mathrm{K \, s^{-1}})$.

In summary, the experimental procedures described in this section have been developed in order to minimize undesirable artifacts in taking EEL and associated TD spectra. Also, it is becoming increasingly apparent that there is considerable advantage in complementing EELS studies of surface chemical reaction with good TDS data in order to better define the phenomenological, thermodynamic, and kinetic aspects of the interaction. Indeed, the need for complementary data of this kind will become more acute as increasingly more complex molecular-surface interactions are studied.

6.2. Surface Reaction Intermediate and Kinetics; Time-Resolved EELS

With TDS experiments of the type described in Section 6.1, the kinetics of the desorbing products will only correlate with the kinetics of a surface reaction if the reaction products are not adsorbed at the temperature of the reaction. Under these reaction-limited conditions, TDS is a powerful means of following surface kinetics. More commonly, however, this is not the case and reaction products appear in the TD spectrum with desorption-limited kinetics. EELS, on the other hand, observed directly the adsorbed species and is ideally suited to surface kinetic studies.

The simplest method of obtaining qualitative surface kinetics is by the anneal and quench method. Here the crystal surface is dosed with the desired reactant(s) at low temperature, annealed to the required temperature for a specified time (typically a few seconds), and quenched again to low temperatures to both suppress further reaction and record the EEL spectrum. By annealing to progressively higher temperatures, changes in the surface concentration of reactants, intermediates, and products may be followed. Additionally, since the spectra are recorded with the reaction quenched there is ample time to retune the spectrometer for optimum performance and the best possible opportunity to identify and characterize the structure and reactivity of surface species. For example, anneal and quench EEL spectra have been correlated with TDS to characterize the two forms of nondissociatively adsorbed oxygen on Pt(111).[30] In particular, it was shown that the low-frequency form $[\bar{v}(00) = 690\,\text{cm}^{-1}]$ was more reactive and the precursor to dissociation, whereas the more stable, higher-frequency form $[\bar{v}(00) = 860\,\text{cm}^{-1}]$ desorbed reversibly. Similarly, EELS has been able to identify an intermediate $\eta^2\text{-}c\text{-}C_5H_8$ species in the dehydrogenation of $c\text{-}C_5H_{10}$ to $\eta^5\text{-}c\text{-}C_5H_5$ on Pt(111).[31] Although anneal and quench procedures of this kind yield only qualitative kinetics, the latter in particular is a good example of the need for good quality spectra to characterize relatively complex adsorbed intermediates.

In order to follow the kinetics of surface reactions more quantitatively it is preferable to sample EEL spectra at the reaction temperature and at frequent intervals. Direct ohmic heating of the crystal (Section 6.1) will catastrophically distort the electron trajectories near the crystal surface, necessitating time-sharing (between data acquisition and heating) techniques. Alternatively, crystal-heating techniques like shielded electron beam, hot filament, or focused light illumination should eliminate the need for time sharing. Depending on the temperature and with it the rate of reaction, the spectra sampling frequency may be of the order of seconds or, as in recent spectrometers built for the purpose, as high as $10^3\,\text{s}^{-1}$. In order to achieve the high spectroscopic sensitivity required for the shorter data acquisition times, two approaches have been adopted. Both exploit the linear energy dispersion of a spherical sector (90° and 180°) energy selector (Section 2.2) by removing a resolution-determining window to transmit more current and recovering the resolution by either position sensitive detection or dispersion compensation.

In the former approach, a fairly conventional position sensitive detector is located on the output plane of the 180° analyzer.[18] The detector consists of screening grids, tandem channel plate electron multipliers, and a linear array of 96 anode collectors, each of which is connected to independent pulse-counting electronics. Thus for fixed voltages on the analyzer 96 adjacent energy intervals are simultaneously and independently sampled. If required, a larger energy range than the 96 channels provide may be obtained by stepping the analyzer voltages. Needless to say, a microcomputer is required to collate the data acquisition of the 96 channels with the analyzer voltages, but it is a relatively standard procedure in many electron spectroscopies. With this arrangement a CO on Ni(110) EEL spectrum has been recorded with a 5-ms data collection time giving a $S/N \sim 5$ and FWHM resolution of $120\text{--}180\text{ cm}^{-1}$.[18]

In the dispersion compensation approach,[16] 90° spherical energy selectors were chosen so that monochromatized electrons are focused along the extension of a line defined by the electron source (actually a window) and the center of curvature of the spherical sector. By arranging for the crystal surface rather than a resolution-determining window to intersect this line, monochromatized electrons will be focused and energy dispersed across the crystal surface. In the preferred out-of-plane dispersion geometry the crystal is oriented so that energy dispersion occurs transverse to the plane of mean specular reflection. The trajectory of electrons reflected from the crystal is a mirror image of their individual incident paths (essentially time reversal) and will fully compensate for the energy dispersion at the crystal if the analyzer is of identical design to the monochromator. A prototype of such an arrangement has been shown to perform to expectation but with the requirement that the selector pass energy be approximately the same as the incident beam energy ($E_p \sim E_0$) the spectrometer resolution was limited to $\sim 180\text{ cm}^{-1}$.

While EEL spectrometers of either the position sensitive detection or dispersion compensation types should always be capable of much higher sensitivity than the more conventional designs, upgrading to comparable resolution may prove difficult owing to the stringent demand on maintaining the required field over a larger volume of the spectrometer. Indeed, it is possible that the somewhat subjective tuning procedures that have proved suitable for conventional spectrometers (Section 5.1) may not be usable. With their limited resolution and short data acquisition times it appears that spectrometers of this type will not be suitable for identifying and characterizing new surface intermediates. Instead, they are more likely to find a role in studying the kinetics of phenomenologically known surface reactions. However, many problems may be anticipated, not the least of which is the need to establish the coverage dependence of the loss intensity and the effect on this of other adsorbates or intermediates which may be incapable of reproduction in calibration experiments. At the experimental level, a major difficulty will occur in cases where reaction proceeds with a concomitant work function change, thereby detuning the spectrometer–crystal geometry. Since it is not feasible to retune on the time scale of these measurements, the loss intensity must be normalized to the elastic count rate at each sample interval. While this is a relatively straightforward computer-controlled operation, the intensity loss suffered in unfavorable cases of large work function change may negate the initial sensitivity advantage gained by these spectrometers. While the development of these spectrometers is striving to achieve shorter data acquisition time intervals, it is of interest to note that no effort has yet been expended on determining surface kinetics on the technically simpler time scale of a second in the lower-temperature regime of the reaction.

References

1. L. H. Little, *Infrared Spectra of Adsorbed Species,* Academic, New York (1966).
2. F. M. Probst and T. C. Piper, Detection of vibrational states of gases adsorbed on tungsten by low-energy electron scattering, *J. Vac. Sci. Techol.* **4,** 53–56 (1967).
3. H. Froitzheim and H. Ibach, Interband transitions in zinc oxide observed by low-energy electron spectroscopy, *Z. Phys.* **269,** 17–22 (1974).
4. H. Froitzheim, H. Ibach, and S. Lehwald, Reduction of spurious background peaks in electron spectrometers, *Rev. Sci. Instrum.* **46,** 1325–1328 (1975).
5. C. E. Kuyatt, and J. A. Simpson, Electron monochromator design, *Rev. Sci. Instrum.* **38,** 103–111 (1967).
6. W. Steckelmacher, Energy analysers for charged particles, *J. Phys. E* **6,** 1061–1071 (1973).
7. F. H. Read, J. Comer, R. E. Imhof, J. N. H. Brunt, and E. Harting, The optimization of electrostatic energy selection systems for low energy electrons, *J. Electron. Spectrosc. Relat. Phenom.* **4,** 293–312 (1974).
8. H. Ibach and D. L. Mills, *Electron Energy Loss Spectroscopy and Surface Vibrations,* Academic, New York (1982).
9. D. Roy and J. D. Carette, Design of electron spectrometers for surface analysis, in: *Electron Spectroscopy for Surface Analysis* (H. Ibach, ed.), Chap. 2, pp. 13–58, Springer-Verlag, Berlin (1977).
10. S. Andersson, Surface vibrations of oxygen and sulphur on Ni, *Surf. Sci.* **79,** 385–393 (1979).
11. B. A. Sexton, High resolution electron energy loss spectrometer for vibrational surface studies, *J. Vac. Sci. Technol.* **16,** 1033–1036 (1979).
12. L. L. Kesmodel, New high resolution electron spectrometer for surface vibrational analysis, *J. Vac. Sci. Technol. A* **1,** 1456–1460 (1983).
13. G. E. Thomas and W. H. Weinberg, Versatile electron spectrometer for surface studies, *Rev. Sci. Instrum.* **50,** 497–501 (1979).
14. N. R. Avery, Vibrational spectroscopy of CO adsorbed on a Pt(111) surface, *Appl. Surf. Sci.* **13,** 171–179 (1982).
15. P. Thiry, J. J. Pireaux, and R. Caudona, A. versatile electron spectrometer for the study of solid surfaces, *Phys. Mag.* **4,** 35–47 (1981).

16. S. D. Kevan and L. H. Dubois, Development of dispersion compensation for use in high-resolution electron-energy-loss spectroscopy, *Rev. Sci. Instrum.* **55**, 1604–1612 (1984).

17. S. Lewhald, H. Ibach, and J. E. Demuth, Vibration spectroscopy of benzene adsorbed on Pt(111) and Ni(111), *Surf. Sci.* **78**, 577–590 (1978).

18. W. Ho, Time resolved electron energy loss spectroscopy of surface kinetics, *J. Vac. Sci. Technol. A* **3**, 1432–1438 (1985).

19. D. W. Turner, High resolution molecular photoelectron spectroscopy, *Proc. R. Soc. London* **A307**, 15–26 (1968).

20. M. E. Rudd, in *Low Energy Electron Spectrometry* (K. D. Sevier, ed.), Wiley–Interscience, New York, pp. 17–32 1972.

21. J. N. H. Brunt, F. H. Read, and G. C. King, The realization of high energy resolution using the hemispherical electrostatic energy selector in electron impact spectrometry, *J. Phys. E* **10**, 134–139 (1977).

22. H. Wollnik and H. Ewald, The influence of magnetic and electric fringe fields on the trajectories of charged particles, *Nucl. Instrum. Methods* **36**, 93–104 (1965).

23. E. Harting and F. H. Read, *Electrostatic Lenses,* Elsevier, New York (1967).

24. A. Adams and F. H. Read, Electrostatic cylinder lenses III: Three element asymmetric voltage lenses, *J. Phys. E* **5**, 1500–155 (1972).

25. J. R. Pierce, *Theory and Design of Electron Beams,* Van Nostrand, New York (1954).

26. P. J. Bassett, T. E. Gallon, and M. Prutton, A high energy resolution Auger electron spectrometer using concentric hemispheres, *J. Phys. E* **5**, 1008–1013 (1972).

27. A. Lahman-Bennani and A. Dugult, Reduction of energy-loss "Ghost structures" observed in electrostatic deflection type electron analysers, *J. Electron Spectrosc. Relat. Phenom.* **18**, 145–152 (1980).

28. J. E. Katz, P. W. Davies, J. E. Crowell, and G. A. Somorjai, Design and construction of a high-stability, low-noise power supply for use with high-resolution electron loss spectrometers, *Rev. Sci. Instrum.* **53**, 785–789 (1982).

29. P. Feulner and D. Menzel, Simple ways to improve "flash desorption" measurements from single crystal surfaces, *J. Vac. Sci. Technol.* **17**, 662–663 (1980).

30. N. R. Avery, A EELS and TDS study of molecular oxygen desorption and decomposition on Pt(111), *Chem. Phys. Lett.* **96**, 371–373 (1983).

31. N. R. Avery, Adsorption and reactivity of cyclopentane on Pt(111), *Surf. Sci.* **163**, 357–368 (1985).

7

Reflection Absorption Infrared Spectroscopy

Brian E. Hayden

1. Introduction

1.1. Historical Development

Much of the information concerning the bonding and symmetry of molecules in the gaseous, liquid, and solid states has come from the study of molecular vibrations using infrared spectroscopy. It seemed, therefore, only natural that infrared spectroscopy should be extended to provide a vibrational spectroscopic technique for the study of the adsorbed phase. The first step in this direction appears to have been taken by Buswell *et al.* in 1938,[1] extending the traditional transmission ir technique to the study of adsorbed water on montmorillonite clay surfaces. It was not until the 1950s that the ir transmission method was again taken up by several groups to study adsorbed species on high surface area materials.[2–4] Perhaps the most remarkable of these experiments was the work of Eischens and co-workers in 1956,[4] who studied the adsorption of CO on supported nickel, palladium, and platinum catalysts. A comparison of the spectra obtained on the catalysts with those of various metal carbonyl complexes enabled then to assign bands to various adsorption configurations (i.e., linear and bridging), and to recognize the influence of

Brian E. Hayden • Department of Chemistry, University of Bath, Claverton Down, Bath BA2 7AY, England, and Fritz Haber Institute of the Max Planck Society, D-100 Berlin 33, West Germany.

vibrational coupling on the C–O stretch vibration of adsorbed CO. Indeed, their results and interpretations have been largely substantiated in work on single-crystal surfaces, as will be seen in Section 3 of this chapter. A considerable number of transmission ir experiments ensued, and several reviews have appeared on this subject,[5,6] the reader also being referred to Chapter 3, by A. Bell in this volume.

Since the number of molecules with which radiation can interact in the "two-dimensional" adsorbed phase is orders of magnitude lower than typical bulk samples, it was important that high surface area systems were used in the early ir studies of adsorbates to ensure a considerable contribution from the surface as distinct from the bulk of the sample. This was indeed the rationale behind the use of thin $(100-400 \text{ Å})$ porous evaporated metal films for the study of adsorbed molecules on metals by ir transmission.[7,8] These samples transmitted a reasonable proportion of the incident radiation (absorption due to the bulk metal) and exhibited a large effective surface area so as to increase the proportion of surface absorbers. The high-area porous film and supported metal catalyst, however, exhibit a rather heterogeneous and difficult to define surface, and a growing awareness of the structural influence of the surface on chemisorption and catalysis led to the study of adsorption on clean, annealed polycrystalline, and ultimately well-defined single-crystal, metal surfaces prepared under ultrahigh vacuum (UHV) conditions. Could ir spectroscopy be applied to study vibrations of adsorbates on such surfaces? The two main experimental considerations arising from such a proposal were the following:

1. Since the metal is nontransparent to ir radiation, the experiment would have to be carried out in reflection.
2. The absorption experiment would have to be capable of detecting very low absorption intensities: the transmission experiment on high surface area supported catalysts and porous metal films, although having fewer surface absorbers in a typical sample $(\sim 10^{16})$ than the bulk phase $(\sim 10^{18})$, has considerably more than in a few monolayer on a well-annealed polycrystalline surface or the surface of the 10×10-mm single crystal $(\sim 10^{14})$.

Initially it was believed that the only way to overcome the sensitivity problem was to reflect the radiation several times from the surface, increasing the number of adsorbate molecules with which the radiation could interact. The first use of multiple reflection for the study of adsorbed monolayers on annealed metal films, and therefore (strictly speaking) the first use of reflection absorption infrared spectroscopy or RAIRS, was by Pickering and Eckstrom[9] and Francis and Ellison,[10] both in 1959. Multiple reflection RAIRS was subsequently exploited by a number of groups on films prepared under increasingly better conditions.[11–15] It was Greenler who investigated not only experimentally[12] but theoretically[16] the conditions necessary to produce the highest absorption in the multiple reflection technique. He also considered[16] the effect of the molecular adsorbates orientation with respect to the metal surface on the vibrational excitation, an effect alluded to by previous workers.[10] It later become evident, however, that a single reflection was sufficient (and in most cases near optimum) for obtaining good ir reflection spectra,[17,18] and it is indeed the single reflection technique that one associates with today's RAIRS experiments on single-crystal metal surface. Since 1970, when it was shown experimentally that good spectra could be obtained of adsorbed CO on Cu using a single reflection,[19] the technique has been used extensively in the last 15 years in studies of a variety of adsorption systems.

1.2. General Principles

It is the interaction of the electromagnetic field of infrared radiation with the oscillation dipole associated with a particular normal vibrational mode that allows the vibrational excitation of molecules in both the bulk and adsorbed phases. This excitation manifests itself in the absorption of a proportion of the transmitted or reflected radiation in the infrared experiment. The dipole selection rules, and factors influencing absorption intensities and frequencies of the infrared absorption in the gas, liquid, and solid state, therefore form the basis for the interpretation of RAIRS spectra. The electromagnetic field experienced by the adsorbed molecule, however, is dominated by the dielectric response of the substrate at infrared frequencies, with which it is in such close spacial proximity relative to infrared wavelengths. Not only is the first monolayer under the substrates influence, therefore, but so too are molecules adsorbed in the multilayers of thicker films. A most convincing experimental demonstration of this was the continuing operation of the so-called "surface selection rule" (a consequence of the dielectric properties of a metal) in the RAIRS spectra of copper oxalate films on copper[20] which were up to 25 nm thick.

The dielectric properties of the substrate will also determine the condition under which the RAIRS experiment should best be carried out. The influence of a *metallic substrate* on the absorption of infrared radiation by adsorbed molecules was originally treated theoretically by Francis and Ellison,[10] who also measured ir spectra of Blodgett films adsorbed on metal mirrors. They employed a classical electrodynamic, and therefore macroscopic model, which was subsequently pursued by Greenler,[16] and the results of this relatively simple model provided the foundation of the RAIRS experiment. Several other theoretical treatments of the RAIRS experiment have since appeared in the literature.[12,17,18,21,22]

Consider the reflection of infrared radiation from a clean and highly reflecting metal surface. The incident beam impinges at an angle ϕ relative to the surface normal: the incident and reflected beam and the surface normal lie in the incident plane (Figure 1). The interaction of the light with the surface is described by the Fresnel equations,[23] which incorporate the appropriate boundary conditions in the electromagnetic wave equations of the incident, reflected, and refracted wave fronts, providing the amplitude r and phase δ of the reflected wave with respect to the incident in terms of the complex index of refraction $\bar{n} = n + ik$ of the phases making up the interface. The amplitude and phase changes experienced on reflection depend upon the direction of the electric field vector of the wave fronts, and it is convenient to resolve the electric field vector into components in the incident plane (P polarized) and normal to the incident plane (S polarized) (Figure 1), likewise resolving the reflection coefficients (r_s and r_p) and phase changes (δ_s and δ_p) yielded in Fresnel's equations. If $n^2 + k^2 \gg 1$, which is true for metals in the infrared wavelength region, the following formula can be derived:[20]

$$R_p = r_p^2 = \frac{(n - \sec \phi)^2 + k^2}{(n + \sec \phi)^2 + k^2} \tag{1}$$

$$R_s = r_s^2 = \frac{(n - \cos \phi) + k^2}{(n + \cos \phi) + k^2} \tag{2}$$

$$\Delta = \delta_p - \delta_s = \arctan\left(\frac{2k \tan \phi \sin \phi}{\tan^2 \phi - (n^2 + k^2)}\right) \tag{3}$$

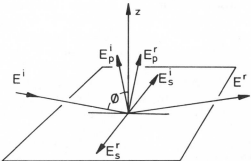

Figure 1. The reflection geometry showing the s and p components of the electric fields of incident (E^i) and reflected (E^r) radiation.

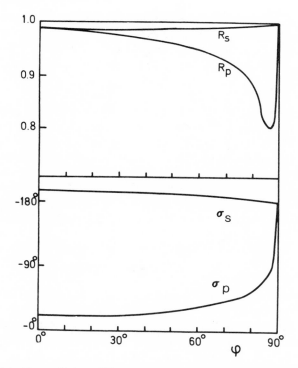

Figure 2. The intensity coefficients (R) and phase shifts (σ) of the s and p components of the infrared radiation on reflection from a metal ($n = 3$, $k = 30$) surface.

where R_p and R_s are the intensity coefficients, σ_p and σ_s the phase shifts on reflection, and ϕ is the angle of incidence (Figure 1).

Figure 2 shows a plot of R_p and R_s, σ_p and σ_s as a function of ϕ under the appropriate conditions of a highly reflecting metal in the infrared ($n = 3$, $k = 30$). The electric field at the surface is the vector sum of the electric field components due to the incident, reflected, and refracted waves. The optical properties of the metal ($n^2 + k^2 \gg 1$) lead to most of the incident intensity being reflected, and the refracted wave contribution to the surface electric field is negligible. If the amplitude of the incident electric field is $E^i \sin \theta$ (where θ is an arbitrary phase), the field due to the reflected wave is $E^i r \sin(\theta + \delta)$. The resulting field at the surface is therefore given by

$$E = E^i[\sin \theta + r \sin(\theta + \delta)] \tag{4}$$

The P and S polarized electric field components can be considered separately as before, and we are interested in the direction of the resulting surface electric field E_p or E_s with respect to the metal surface. Note that for all incident angles ϕ, E_s^i and E_s^r remains parallel to the surface (Figure 1), i.e., the resulting electric field which is parallel to the metal surface is given by $E_s = E_s^i + E_s^r$, and from equation (4) we have

$$E_s = E_s^i[\sin \theta + r_s \sin(\theta + \delta_s)] \tag{5}$$

Since δ_s is close to 180° and $r_s \approx 1$ for all ϕ (Figure 2) it can be seen from equation (5) that the 180° phase change leads to destructive interference and a vanishingly small electric field at the surface. No interaction of the surface electric field due to E_s with surface dipoles is possible.

P polarized radiation, however, behaves quite differently since the incident and reflected electric wave fields have components both parallel and normal to the surface (inset in Figure 3) and sum to yield parallel $E_p^{\rightrightarrows}$ and normal $E_p^{\perp}$ components of the surface electric field given by

$$E_p^{\rightrightarrows} = E_p^i \cos \phi \, [\sin \phi - r_p \sin(\theta + \delta_p)] \tag{6}$$

$$E_p^{\perp} = E_p^i \sin \phi \, [\sin \theta + r_p \sin(\theta + \delta_p)] \tag{7}$$

Note that for a wide range of angles ϕ, δ_p remains small (Figure 2) and only increases to $-180°$ near grazing incidence. The *parallel components* of E_p^i and E_p^r combine to give a very small resultant field $E_p^{\rightrightarrows}$ at low angles, even though they are in phase, similar in magnitude ($r_p \sim 1$), and their contribution to E_p is largest [$\cos \phi$ term in equation (6)] simply because they are in the opposite direction. The normal components combine constructively [equation (7)] but at low angles $E_p^{\perp}$ remains small because only a small proportion is resolved in the surface normal direction [$\sin \phi$ term in equation (7)]. As ϕ increases, so does this normal component, with a concomitant decrease in the parallel component. Constructive interference yields a normal component of $\sim 2E_p^i$ before the sharp change in phase towards $-180°$ (Figure 2) causes mutual cancellation [equation (7)]. The effect of ϕ on the component $E_p^{\perp}$ is shown in Figure 3 again for reflection from a metal with $n = 3$, $k = 30$. One must conclude, therefore, that incident P polarized radiation can give rise to significant electric fields at the metal surface, but only in a direction normal to the surface, and only at grazing angles of incidence (high ϕ).

Once the enhanced electric field $E_p^{\perp}/E_p^i$ has been calculated, one notes that the number of molecules with which the incoming incident ray can interact is proportional to $\sec \phi$, and the absorption intensity is proportional to $(E_p^{\perp}/E_p^i)^2$.[24] Therefore the total absorption intensity in the RAIRS experiment is given by $\Delta R = (E_p^{\perp}/E_p^i)^2 \sec \theta$, the function plotted in Figure 3. This function has the same shape for all metals reflecting in the infrared, but ΔR is largest for the most highly reflecting

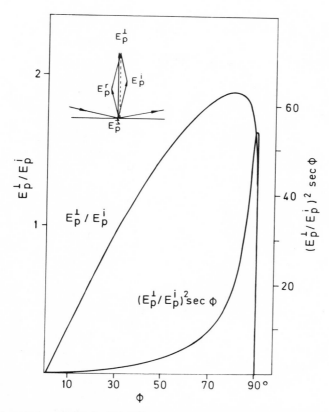

Figure 3. The relative amplitude $(E_p^\perp/E_p^i)$ of the electric field perpendicular to the surface as a function of incident angle ϕ, together with the quantity $(E_p^\perp/E_p^i)\sec\phi$. The inset shows the dominance of the normal component of the field of the surface arising from the p component. (After Ref. 24.)

metals. It can be seen immediately from Figure 3 why it is important to carry out the RAIRS experiment at high angles of grazing incidence.

A more sophisticated classical electrodynamical modeling of the RAIRS experiment necessitates the introduction of a third phase in the analytical problem, i.e., that of the optical constants of the adsorbate phase itself, in determining the induced electric field at the surface. Greenler[16] extended the two-phase (vacuum/solid) macroscopic models already outlined by solving the wave equations with the appropriate boundary conditions for a three-phase (vacuum/adsorbate/solid) system shown schematically in Figure 4a. The adsorbate layer was described by an isotropic dielectric constant $\tilde{\varepsilon}_2 = (n_2 + ik_2)^2$ and thickness d. This

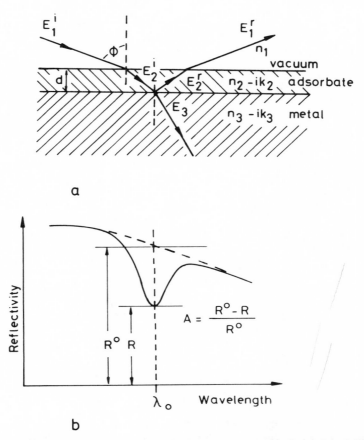

Figure 4. The classical three-phase model (a) for an adsorbate layer ($h_2 - ik_2$) of thickness d on a metal ($n_3 - ik_3$) developed by Greenler[16] to calculate the relative absorption A defined in (b). (After Ref. 16.)

provided a means of calculating an absorption function $A = \Delta R/R^0$, where $\Delta R = R^0 - R$; R^0 is the reflectance in the three-phase systems with a nonabsorbing adsorbate, and R is the reflectance of the modeled adsorbate (Figure 4b). These results demonstrated the advantage of RAIRS carried out using high incident angles, and it was shown that under optimum conditions one could expect a gain in sensitivity over a transmission infrared experiment on an unsupported metal film of up to a factor of 25. This advantage reduces to a factor of ~17 for a more realistic experimental situation where a finite divergence in the spectrometer optics gives rise to a spread of incident angles of ±5° about

$\phi = 85°$ (Section 2.1). The three-phase model of Greenler[16] also allowed an estimate to be made of the effect the optical response of the adsorbate medium $\bar{\varepsilon}_a$ may have on the position, shape, and intensity of the absorption band due to changes in *both* n_a and k_a in the region of an absorption feature. This problem was pursued both theoretically and experimentally by Greenler *et al.*,[21] who concluded that the typical RAIRS spectra of a moderate absorber should be directly comparable with a transmission spectrum, in agreement with the calculations of Ito and Suetaka[25] on the basis of a different model.

The series of curves A vs. ϕ for different absorbate film thicknesses d on a reflecting metal surface from the three-phase model of Greenler[16] have the same form as those for the two-phase model in Figure 3, although the function A is now a relative reflectivity change rather than simply ΔR. The three-phase model of RAIRS has also been treated by McIntyre and Aspnes[22] and a more detailed review of their approach, and its relationship to the previous models, can be found elsewhere.[26] The essence of the theoretical modeling of the RAIRS experiment on metals is contained, however, in the simplest two-phase models insofar as it shows the following:

1. Only the p component of the incident light can interact with the adsorbate.
2. Only molecular vibrations with a finite component of their dynamic dipole perpendicular to the surface are observable.
3. The experiment will be most efficient at high angles of incidence ϕ, i.e., at near grazing incident angles.
4. The most highly reflecting metal surface will yield the highest absorbance.

Clearly, 1, 3, and 4 above will form the experimental ground rules of RAIRS, and the optimum experimental conditions were indeed analyzed explicitly using the three-phase model by Greenler.[17,18] Since the sizes of the absorption bands were expected to be very small, multiple reflection experiments on polycrystalline surfaces were carried out in an attempt to increase the number of absorbers with which the beam would interact.[11–15] Greenler considered the theoretical advantage of multiple reflection in the detector noise (P_{DN}) sensitivity limit where it is simply a matter of optimizing the size of ΔR (Figure 4b). Because of the increasing loss in total intensity with an increasing number of reflections, there is consequently an optimum number of reflections that maximizes ΔR, and this occurs at the point at which the background intensity reduces to $1/e$ of its original value.[12] It was concluded[17,18] that the advantage of multiple reflection to optimize the absorbance ΔR was small for any but the most reflecting metals (Al, Cu, Au, Ag). A calculation of the optimum number of reflections N, and the reflectivity change in the absorption band following the optimum number of reflections ΔR (OPT), and a single reflection ΔR as a function of incident angle ϕ is presented in Ref. 18 for a large number of metals at 2100 and 500 cm^{-1}. These curves provide a means of designing the optimal reflection experiment in the detector noise limit for any spectrometer of zero or finite divergence. Table 1 contains data calculated using the curves of Ref. 18 for realistic experimental conditions of reflection, i.e., $f/4$ spectrometer optics with incident angle $\phi = 82°$. Included is the optimum number of reflections N and the extent of signal loss if one carries out a single rather than a multiple reflection experiment $\Delta R / \Delta R$(OPT). A quantity ΔR is also included, which is the integral of ΔR over the angles defined by the conditions above, and when compared between metals allows a comparison of the relative efficiency of the RAIRS experiment. It is interesting to note that to date RAIRS experiments have been carried out at ~2100 cm^{-1} on Cu, Ni, Pd, and Pt surfaces, all of which have relatively high values of ΔR^*. It can also be seen that ΔR^* at 500 cm^{-1} is significantly less than for the same metal at 2100 cm^{-1}, and this provides

Table 1. The Optimum Number of Reflections in RAIRS, and the Efficiency of a Single Reflection[a]

Metal	2100 cm^{-1}			500 cm^{-1}		
	N(OPT)	$\Delta R / \Delta R$(OPT)	ΔR^*	N(OPT)	$\Delta R / \Delta R$(OPT)	ΔR^*
Cr	2–4	69	770	1–7	61	270
CO	2–5	66	717	3–22	28	426
Cu	4–17	26	705	1–22	27	400
Au	3–12	38	657	1–16	36	415
Fe	1–3	78	712	1–10	48	320
Mo	2–5	60	925	1–8	60	290
Ni	2–5	70	895	1–15	37	410
Pd	2–6	62	900	1–12	45	358
Ag	6–26	22	402	2–30	22	240
Ta	1–2	92	510	1–6	65	252
Ti	1–2	97	435	1–3	85	192
W	2–4	69	828	1–12	47	355

[a] Extracted from data in Refs. 11 and 26. Calculations are made assuming an incident angle $\phi = 82$, spectrometer optics $f/4$. N(OPT) is the theoretically calculated optimum number of reflections and $\Delta R / \Delta R$(OPT) is the relative efficiency of a single reflection RAIRS experiment to that carried out using the optimum number of reflections (expressed as a percentage). R^* is the integral of ΔR over the spread of incidence angles (expressed as arbitrary units) and is a measure of the relative ease of a single reflection RAIRS experiment.

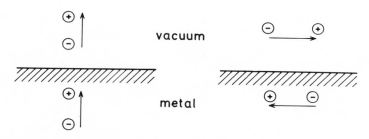

Figure 5. The image dipole picture of the metals' screening of a dipole orientated parallel to the surface, and the enhancement of a perpendicular dipole.

another reason (Section 2) why RAIRS is more difficult experimentally at long wavelengths. The results of the classical optical reflectivity treatment in restricting absorption to only those vibrations with a dipole moment perpendicular to the metal surface forms the basis of the so-called "surface selection rule" and is entirely equivalent to the result obtained using image dipole theory, which is an alternate statement of the electromagnetic boundary conditions.[28] The image dipole theory of the surface selection rule is expounded in Figure 5. The long-range electromagnetic field of the infrared radiation cannot distinguish the dipole and its image, and interacts with the sum of their dipole fields. In the case of a perpendicular dipole (left-hand side of Figure 5) this leads to an increased response (and hence absorption). In the case of a dipole parallel to the surface, the net summation yields only a quadrupole component, and no dipolar field remains for interaction (right-hand side of Figure 5). The ramifications of this strict dipole selection rule for metal surfaces in RAIRS are pursued in Section 3.1.

1.3. A Comparison with Other Techniques

In a comparison of any technique with another, it is instructive to review briefly the type of measurements one would ideally like to perform, and the inherent practical strengths and weaknesses of the method being applied. Clearly the vibrational spectroscopy of adsorbates will have important applications in obtaining a wide variety of both physical and chemical information concerning the adsorbed phase. If one wishes to study well-characterized adsorbate systems, a technique sensitive enough to measure monolayers (and submonolayers) on single-crystal surfaces is required, and some of the important applications of vibrational spectroscopy to adsorbate studies on such surfaces are summarized below:

1. A *structural analysis* of the adsorbate layer can be achieved in which *dipole selection rules* can yield information concerning site specificity, molecular conformation, and orientation.
2. Characteristic vibrational frequencies of common bulk phase or adsorbed moieties allows the *recognition of adsorbate species* through the application of *vibrational fingerprinting,* the frequency of such bands also yielding considerable qualitative information (as is the case in inorganic and organic chemistry) on bonding and reactivity.
3. The precise measurement of frequency, intensity, half-width and line shape can yield important information concerning multiplicity of adsorption sites, the influence of vibrational coupling, local chemical environment and bonding site, adlayer growth, order-disorder phenomena, lateral interactions, and dynamical effects.

The application of RAIRS to all of these areas of interest is included in Section 3, and will be more or less effective than other surface sensitive vibrational spectroscopic techniques depending on how they compare in their main factors of merit, which are summarized below for RAIRS:

a. Spectral range $\sim$800–4000 cm^{-1}
b. Resolution   1–5 cm^{-1}
c. Sensitivity $\sim$10^{-3} monolayer (strong dipole absorber)
d. Selection rules   Strict dipole
e. Versatility   Sample temperature and adsorbate coverage easily variable; conducive to simultaneous measurements with other techniques; insensitive to ambient pressure.

The restricted spectral range of RAIRS provides perhaps the most serious drawback since nearly all adsorbate–substrate modes are inaccessible: Only internal molecular modes of adsorbates have been studied to date using RAIRS. Both EELS and Raman spectroscopy, on the other hand, can access the spectral region below 800 cm^{-1}. Those vibrations accessible using RAIRS, however, can be studied with a high spectral resolution (significantly greater than EELS) and high sensitivity (significantly greater than Raman). The high resolution and sensitivity of RAIRS make it particularly suited to studies requiring precise measurements of frequency, intensity, half-width and line shape. Further, a prerequisite of such measurements is often the ability to vary surface conditions (e.g., coverage, temperature) or to undertake simultaneous measurement using another technique (e.g., LEED, $\Delta\phi$), a versatility that it also provides. Perhaps one of the main advantages of RAIRS yet to be fully exploited is its insensitivity to relatively high ambient pressure above the sample surface since it is an optical technique and does not suffer the same vacuum constraints as electron spectroscopies.

2. Experimental Considerations

The RAIRS experiment is ideally designed to obtain the vibrational absorption spectrum from the very small number of adsorbed molecules on a single-crystal metal surface in a single reflection. Typically a crystal can be prepared to provide a 10×10 mm surface for reflection on which $10^{14}-10^{15}$ molecules may be adsorbed in a full monolayer, but, as already pointed out, only a small proportion of the reflected infrared intensity can be absorbed in a particular vibrational mode. For example, $\nu(C-O)$ of chemisorbed CO (a relatively strong dipole absorber) provides an absorption band of peak height of between 1% and 2% of the total reflected ir intensity (Section 3.5). Since one is aiming to measure absorption bands of submonolayer coverages of adsorbate molecules with smaller dynamic dipole moments than that of $\nu(C-O)$ of chemisorbed CO, a sensitivity of instrumentation is required allowing detection of absorptions of <0.1% with satisfactory signal-to-noise ratios. It is the achievement of this order to sensitivity (i.e., $S/N \geq 10^3$) with which the experimentalist carrying out RAIRS is concerned[29] and, as will be seen, many contributing factors have enabled the improvements in signal-to-noise ratio to the point at which down to 10^{-3} of a monolayer of CO adsorbed on a metal single crystal can be detected ($S/N = 10^5$). Within the restraints of this sensitivity requirement, experimentalists continually strive to extend the spectral range over which successful RAIRS can be carried out. The majority of RAIRS experiments on single crystals have been exclusively in the spectral range 2000–3000 cm^{-1}, and invariably concern the measurement of $\nu(C-O)$ in the linear chemisorbed CO (~2100 cm^{-1}). The main reason for this is the relative ease of measurement in this spectral region, and it was perhaps fortunate that the molecular stretch vibration of one of the best studied chemisorbed species on metals had such a convenient vibrational energy. The reason for this spectral region being easily accessible with high sensitivity will become evident, but the most important factors are the lack of atmospheric absorption bands, high detector sensitivity, and reasonable ir source intensity. Improvements in technique and instrumentation have now increased the useful spectral range of RAIRS to 1000–4000 cm^{-1}, and a variety of other adsorbate molecules and reaction intermediates have been studied. Indeed, recently the lower spectral limit was extended to ~800 cm^{-1} in an investigation of molecular oxygen on a recrystallized platinum foil.[30] It is unlikely that the steady improvement in sensitivity and spectral range will not continue, and some of the main problems with which one is concerned to achieve this are set out below.

The sensitivity of the RAIRS spectrometer is reflected in the signal-to-noise ratio in the spectral region of interest under the conditions of measurement, for example, spectral resolution and time of measurement. Three types of noise can be identified that determine the effective sensitivity of the spectrometer:

(1) Detector noise (P_{DN}) includes all noise originating in the detector, and for our purposes includes noise generated in the first stage of the detector's preamplifier. There are many potential sources of detector noise depending on the type of detector and its mode of operation. These sources of noise, however, are not a function of the incident infrared intensity which is to be measured.

(2) Source limited noise (P_{SN}), also known as shot noise or photon noise, is the fluctuation of power in thermal sources as a result of the random nature of the spontaneous emission process. It is therefore an unavoidable concomitant of the use of broad band thermal sources and is proportional to $P^{1/2}$, where P is the power of the thermal infrared source.

(3) Fluctuation noise (P_{FN}) is defined to include a host of experimentally induced noises which stem from fluctuations in, for example, the sensitivity of detector, source, intensity, or position of optical components (particularly the sample) with time. Further, because one carries out RAIRS by normalizing to, or subtraction of, a base-line (no adsorbate) spectrum, long-term fluctuations can also be troublesome, particularly in the case when the baseline spectrum is heavily structured. These sources of noise can be grouped together because their contribution is proportional to the power of the detected radiation. Signal noise, modulation noise, and scanning noise[31] are all forms of fluctuation noise.

The fluctuation of the photon field is, to a good approximation, equal to the square root of the number of photons.[31] The total flux P_v available for spectroscopy from a thermal black body source at 1500 K (Figure 9b) in the spectral range 250–5000 cm^{-1} is about 4×10^{13} photons s^{-1}, and therefore the superimposed photon noise $P_{PN} \approx 6 \times 10^6$ photons s^{-1}. The photon signal-to-noise limit at this source temperature is therefore over an order of magnitude greater ($P/P_{DN} \approx 7 \times 10^6$) than that at present experimentally achieved in RAIRS ($S/N \approx 10^5$): One must conclude that photon noise is never signal-to-noise limiting, and RAIRS measurements are either detector or fluctuation noise limited, depending on the spectral region of measurement. The long wavelength region is most certainly signal-to-noise limited by P_{DN}, and it is essentially improvements in detector sensitivity that have and will increase the spectral range of RAIRS to lower energies. The main reason for t⊦ dominance of P_{DN} at long wavelengths is the strongly decreasing intensity of thermal emitting infrared sources at longer wavelengths and the difficulties in dispersive monochromation of far infrared radiation, particularly when constrained by the geometric requirement of the reflection experiment. Increasingly sensitive detectors, more intense infrared sources, and FTIR techniques are all potential candidates for the improvement of RAIRS in detector limited signal-to-noise operation, and hence in its long-wavelength limit.

In the spectral region in which the sensitivity of modern detectors is best, and source intensity is high, i.e., the midinfrared range ($\sim$1300–4000 cm^{-1}) and the most accessible region in RAIRs to date, detector noise (P_{DN}) is not signal-to-noise limiting and noise contributions from fluctuation noise (P_{FN}) become dominant. This source of noise is often encountered in measurements in which a small signal change is to be measured on a large signal background. The main contributions to P_{FN} may be difficult to identify and will vary from spectrometer to spectrometer. It is convenient to distinguish two main sources, which are (a) long-term drifts, i.e., changes that occur in, for example, sample position, transmissive medium between source and detector, detector sensitivity, or source power, over the time scale of the measurement, and (b) higher-frequency noises induced by, for example, transmission of vacuum pump vibrations to optical components, or vibrations induced by moving grating or mirror components during scanning. Noises induced by these mechanisms cannot be reduced by increasing source intensity, detector sensitivity (or in some cases measurement time) since $P_{FN} \propto P$, and can be invariably reduced only through improvements in the stability of the optical system. Noise contributions from long-term drifts (a) are critical in systems where the background absorbance spectrum, on which the adsorbate infrared band is superimposed, is heavily structured. This is particularly true if the spectral modulation of the structure is of the same order as the half-width of the adsorbate bands under investigation. Hence atmospheric water bands are particularly insidious, and increased signal to noise in the spectral regions in which they appear can be achieved only be evacuation of the optical system, or polarization modulation (Section 2.4). A summary of the most important experimental criteria for successful RAIRS is therefore as follows:

1. Maximum infrared source intensity in a relatively small solid

angle and aperture (geometric requirement of grazing incidence) with high detector sensitivity.

2. Constant energy resolution over the spectral range of the order of the half-width of the adsorbate band ($\leq 5\ cm^{-1}$).
3. Phase sensitive detection within the response time limits of the detector.
4. Minimization of extraneous absorption bands in the background spectrum. This is most effectively achieved through evacuation of the optical system and/or polarization modulation.
5. A stable optical system with minimization of long-term drifts.

2.1. Optical Configuration

The ideal reflection geometry for RAIRS is determined by the dielectric properties of the substrate (Section 1.2) and for a metal grazing reflection is optimum: only the component of the electric field vector perpendicular to the surface, $E^{\perp}$, can interact with the adsorbate oscillating dipole, and it is exclusively the p component of the incident radiation which contributes to $E^{\perp}$, this being optimum at high angles of incidence ϕ [equation (6)]. It was also shown that a single, rather than multiple reflection is nearly optimum in the detector noise limit due to the balance of loss of intensity during reflection at grazing incidence and the increased absorption by the extra molecules available with each additional reflection. This conclusion is least true for measurements on highly reflecting surfaces or of RAIRS experiments in the midinfrared region where the signal noise may not be detector noise limited. Further, it remains practically much simpler to arrange a stable optical system for a single reflection. Surface preparation and the incorporation of additional surface sensitive techniques is also considerably simplified by using the single reflection geometry.

The incident and reflected beams are characterized by a divergence, defined by the solid angle ψ, and an aperture η which is the size of the image being focused on the sample. The solid angle ψ is directly related [$\sin(\psi'/2) = 1/2f$] to an effective f number of the reflecting optics which must be ideally matched to the f number of the spectrometer. Both the size of aperture and solid angle are restricted by the grazing incidence reflection at the sample. The constraints are greatest in the plane of reflection, ψ' defining beam divergence in this plane. For a small slit width, m, compared to the crystal size, d, the limiting configuration of the reflection optics is given by:

$$(m/d) \approx \sin(\psi'/2) \lesssim \cos \phi \tag{8}$$

The consequence of this limit can be demonstrated by considering reflection at a grazing angle $\phi = 85°$ from a crystal of length $d = 10$ mm. In the limit of a point source ($m \to 0$), one is restricted to $\psi' \lesssim 10°$ (or an effective optics $f > 6$). In the limit $\psi' \to 0$ (parallel incident beam), the slit width is limited to $m \lesssim 1$ mm. Since m combines with m', the slit height, to yield the aperture ($\eta = mm'$), the reflection geometry imposes a restriction on the luminosity or optical throughput of the optical system, given by the product of η and ψ. In the detector noise limit of measurement, one wishes to optimize optical throughput, and hence η and ψ are chosen to be as large as possible under the constraints of grazing incident reflection. Note that any attempt to increase optical throughput at the sample reflection through the use of demagnifying optics is hindered by the concomitant increase in divergence ψ through optically decreasing the aperture η. In the case of dispersive RAIRS systems, an $f/4$ optics has been generally favoured to date, providing a good balance between high angle of incidence ϕ and high luminosity of spectrometer. An optical system $f/4$ corresponds to an angle of divergence $\psi' \approx 14°$, restricting the maximum angle of incidence to $\phi \approx 82°$. Reflecting from a typical 10×10 mm crystal, the limit in slit width is therefore $m < 1$ mm, imposing an overall limit on aperture μ (assuming a maximum slit height $m = 10$ mm) of 1×10^{-5} m^2, and hence luminosity $\psi\mu = 6 \times 10^{-7}$ m^2 Sr. It is the constraint that the angle of incidence and sample size places on the luminosity $\psi\mu$ which makes high brightness sources attractive in cases where signal to noise is detector noise (P_{DN}) limited.

The relatively simple geometry of the single reflection experiment allows RAIRS to be combined with UHV surface, preparative, and other investigative techniques. The geometry of several RAIRS systems are shown in Figure 6. The dispersive RAIRS spectrometer in Figure 6c has a UHV chamber incorporating low-energy electron diffraction (LEED) and temperature programmed desorption (TPD). The UHV chamber is separated from the inferior vacuum ($\sim 10^{-3}$ Torr) of the optical system by UHV compatible ir windows. Several window materials are available which can be either sealed directly to stainless steel flanges or sealed with viton gaskets and differentially pumped[32] if their coefficient of linear expansion varies greatly from that of stainless steel. Table 2 includes a list of the most commonly used infrared transmissive window materials, their range of transmission, and some relevant physical properties. The spectrometer and detector optics are generally fixed in their aligned positions, x, y, and z movement of the sample and its rotation in the reflection plane allowing the final optical adjustment. The infrared windows can be as far away from the sample as the beam divergence and

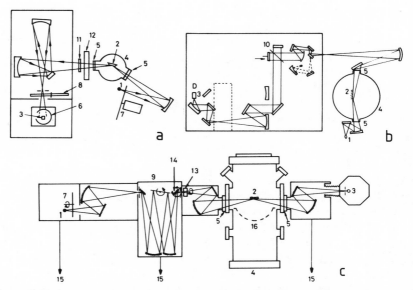

Figure 6. Various types of RAIRS spectrometers incorporating either (a) a rotating filter (after Ref. 51), (b) an interferometer (after Ref. 53), or (c) a dispersive monochromator (the spectrometer used in Refs. 47, 61, 71, 83, 106). 1, Source; 2, sample; 3, detector; 4, UHV chamber; 5, infrared window; 6, lens; 7, chopper; 8, rotating filter; 9, dispersive monochrometer; 10, Michelson interferometer; 11, fixed polarizer; 12, photoelastic modulator; 13, rotating polarizer; 14, vibrating mirror, 15, pump for optics, 16, LEED.

window size will allow, although in the case of *in situ* studies of reactions at high pressures over the surface (Section 4) the optical path may be reduced to prevent gas phase absorption by moving the windows closer to the sample.

2.2. Radiation Sources

As in the case for ir spectroscopy generally, a thermal emitting source is the most commonly used in RAIRS. The power P_λ which can be emitted in a wavelength interval $\lambda - (\lambda + d\lambda)$ from a perfect blackbody source, i.e., the spectral power distribution, is given by Planck's radiation law

$$P_\lambda = \frac{2\pi h c^2 A \, d\lambda}{\lambda^5 [\exp(hc/R\lambda T) - 1]} \tag{9}$$

where A is the source area (or aperture). Since, however, we are interested in a spectrometer with a constant energy $(d\nu)$ rather than

Table 2. Infrared Transmissive Materials for Lenses and Windows

Window material	Low-energy cutoff (cm^{-1})	Melting point (°C)	Thermal expansion $(grad^{-1})$	Comments
LiF	1600	842	35×10^{-6}	
MgF	1400	1266	$13.4(9.2)^a \times 10^{-6}$	
CaF_2	1200	1360	18×10^{-6}	
Silicon	1100^b	1410	3×10^{-6}	
NaF	1000	988	29×10^{-6}	Slightly hydroscopic
BaF_2	900	1280		
NaCl	600	801	39.8×10^{-6}	Slightly hydroscopic
KCl	550	776	36.6×10^{-6}	Slightly hydroscopic
AgCl	500	445	31×10^{-6}	Can corrode metals
Germanium	470	937	6×10^{-6}	
KBr	350	730	38.4×10^{-6}	Hydroscopic
KI	350	682	43×10^{-6}	
CsBr	250	636	59.8×10^{-6}	Very hard
KRS-5^c	250	414	58×10^{-6}	Toxic high v.p.
CsI	180	621	54.9×10^{-6}	Very hard
Diamond	10	>3500	1×10^{-6}	Size restriction, expensive

^{a} Nonisotropic.
^{b} Some transmission below ~500 cm^{-1}.
^{c} Thallous bromide/thallous iodide.

wavelength $(d\lambda)$ window, the energy resolution remaining of the order of the half-width of adsorbate bands over a wide spectral range, it is more meaningful to express the spectral power distribution as a function of energy and constant energy window, $P_v(v, dv)$:

$$P_v = \frac{2\pi h v^3 A \, dv}{c^2[\exp(hv/RT) - 1]} \quad (10)$$

To obtain the amount of energy available in the RAIRS experiment, one can assume a perfectly diffuse (Lambertian) emitter and point source. The fraction of energy available to the spectrometer is given by

$$P_v^\dagger = P_v \sin^2(\psi'/2) = P_v/4f^2 \quad (11)$$

since $\sin \psi'/2 = \frac{1}{2}f$. Figure 7 is a plot of the power available $P_v^\dagger$ from $1 \, mm^2$ of such a black body source at various temperatures in a spectrometer with $f/4$ optics and for constant energy window $dv = 5 \, cm^{-1}$ (the instrumental resolution required for RAIRS). This represents a maximum power expectation since losses due to filters, windows, mirrors and grating or beam splitter will reduce the throughput intensity

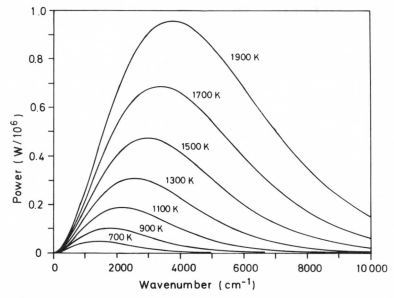

Figure 7. Spectral power distribution for a Lambertian black body at various temperatures [aperture (μ) 1 mm^2] in the solid angle collected by an $f/4$ optical system for a constant spectral window (dv) of 5 cm^{-1}.

as would the reduced emissivity of a nonideal black body. Despite this approximation, it is instructive to calculate[29] using P_v the order of magnitude of sensitivity required for the detection of the available power in RAIRS, i.e., in the detector noise limit. For example, at 2000 cm^{-1}, 4×10^{-7} W of power are available from the source at 1500 K, and assuming one wishes to measure less than one tenth of a monolayer of CO, i.e., an absorption band of ~0.1%, the noise equivalent power ($\equiv P_{DN}$) of the detector should be better than 10^{-10} W. As will be seen in Section 2.5, generally only low-temperature detectors can operate with such intrinsically low noise levels.

The globar appears to be the most popular thermal emitting source in RAIRS, consisting of a silicon carbide rod ohmically heated, generally with direct current. These can be operated in atmosphere at temperatures of up to ~1500 K, but are mostly run at ~1300 K, at which temperature they are relatively stable and long lived. The power requirement of a standard globar (~4 × 15 mm of useful glowing area) is ~150 W, which can easily be provided by a smooth, stabilized, direct current supply to yield a relatively noise free and stable (long-term) infrared source. Air currents can, however, cause cooling of the source when operated in the atmosphere, and this is most effectively avoided by operation in vacuum. A water-cooled source enclosure has also been found to be advantageous with regard to source lifetime and stability.

As has been pointed out, the geometric restriction of reflection at grazing incidence prevents higher intensity being achieved through very low f number optics, and hence alternative sources of high brightness provide a potential improvements in RAIRS in the detector signal-to-noise limit. The two main contenders for high brightness infrared sources are lasers and synchrotron radiation. There are at the present time no infrared lasers that can be tuned over a wide spectral range, and this constitutes their main disadvantage with respect to their possible applications in RAIRS. However, their future potential as collimated, high-power (and highly monochromated) infrared sources should not be underestimated, particularly in the spectral region below $1000 \, \text{cm}^{-1}$ where conventional infrared sources are weakest and some tunable lasers are already commercially available.[34] Of particular note is the diode laser, which can operate in the region 4000–$600 \, \text{cm}^{-1}$, although each laser can only be tuned over a small spectral range (30–$100 \, \text{cm}^{-1}$). The high power ($\sim 100 \, \mu\text{W}$) makes such a laser attractive, however, for RAIRS investigation of a specific absorption band. This is provided that once out of the detector signal-to-noise limit, laser stability is good enough (or can be normalized well enough) to provide a reasonable fluctuation noise limit to allow the detection of the very weak infrared absorption bands in RAIRS. The inconvenience and cost, however, of a series of diode lasers to cover a large spectral range makes it at present an unrealistic infrared source for RAIRS, and while several other tunable lasers are under development that may ultimately become suitable for such experiments (some of which are reviewed by Hoffman),[27] it appears premature to consider their potential use by anyone other than perhaps a laser spectroscopist! The other potential high brightness source of infrared radiation is the synchrotron, which provides a source of continuous radiation from the microwave to the hard x-ray energy region. While the x-ray and uv spectral regions have been the subject of extensive theoretical and experimental characterization (indeed, the proliferation of dedicated synchrotron light sources is a consequence of the excellent intensity, polarization, and spectral characteristics of synchrotron radiation over conventional sources in this spectral range), only relatively recently has the potential of the synchrotron source in the infrared spectral region been considered.[35–40] It appears that the intensity characteristics of the synchrotrons so far studied in the infrared region below $\sim 4000 \, \text{cm}^{-1}$[38–40] are reasonably well predicted by theory.[35–37] It can be seen (Figure 8) that under only the smallest

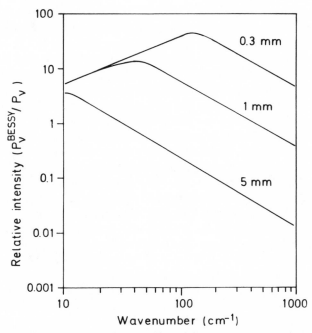

Figure 8. The intensity of infrared synchrotron radiation at BESSY relative to a glowbar at 1500 K for various apertures. The synchrotron radiation is normalized for a ring current of 100 mA and the spectrometer has an acceptance angle of ± 17.5 mrad ($f/13.6$). (After Ref. 40.)

aperture constraints ($0.3\,mm^2$) does the synchrotron appear to afford any real intensity advantage over conventional infrared sources, and then only in the region around $100\,cm^{-1}$ yielding ~ 50 times the intensity of a globar source operating at 1500 K. Since, as has been shown, the aperture constraint is not quite so stringent for RAIRS on typical surfaces, the advantages of the bright and highly collimated synchrotron source ($f/28$, $\mu = 1\,mm^2$ at BESSY[40]) at long wavelengths may not outweigh the problems associated with superimposed noise[40] and normalization of intensity (time structure of synchrotron source) which contribute to the fluctuation noise limit, as well as the cost and complication of working at such a multiuser facility.

2.3. Monochromators and Interferometers

The grating monochromator has been successfully (and most commonly) used for RAIRS[41-48] and is generally of the Czerny–Turner design. It is the combination of the dispersive power of the diffraction grating (the number of lines per millimeter), and the optical path length combined with slit width m (i.e., the range of angles over which the diffracted radiation is collected) that determines the spectral resolution of the instrument. The intensity throughput (or luminosity) is determined by the spectrometer's aperture μ ($=mm'$), optical path, and size of optical components. The latter two are generally combined to yield an optical f number (e.g., for a Czerny–Turner design $f = F/D$, where F is the focal length of the focusing mirrors and D is their diameter[49]) and therefore, together with the aperture, defines the monochromator luminosity and reflects the maximum amount of radiation which can be collected from a black body source for optical throughput are clearly related and operate in opposite directions. Since one wishes to maximize luminosity, and since the potential resolution of which grating spectrometers are capable ($<10^{-3}$) can to some extent be sacrificed, relatively low f number monochromators are favored for RAIRS. An additional advantage of the low f number is the shorter optical paths required, making the monochromator suitable for evacuation. Since the reflection geometry can be considered our limiting optical component ($\psi\mu \approx 6 \times 10^{-7} \text{ m}^2 \text{ Sr}$) it sets the optimum condition (and upper limit) for luminosity for the monochromator and in practice $f/4$ optics are generally favored. The grating is then chosen to provide the required spectral resolution at the maximum slit width (m) of ~ 1 mm (Section 2.1). Since constant energy resolution dv is a function of v for a particular grating and slit width, several gratings may be required to cover the spectral range of interest if maximum intensity throughput (i.e., with optimum slit width m) is to be maintained. Change of grating is also necessitated because of the restricted spectral range over which a grating can be made efficient through blazing with regard to diffracted intensity. Together with the requirement of a cutoff filter to eliminate higher-order diffraction, up to three gratings and filters may be required to cover the experimental range $600–4000 \text{ cm}^{-1}$.

Low intensity of ir sources in the long-wavelength region produces a serious limitation in the use of dispersive monochromators because of the contradictory requirements of luminosity and resolution for both ir spectroscopy generally, and more critically RAIRS: the option of opening the slits to allow greater intensity throughput is restricted in the reflection experiment because of the effective size of the image at grazing incidence [equation (9)]. Adsorbate bands down to $\sim 1000 \text{ cm}^{-1}$ have been measured with dispersive systems,[50] but the limited luminosity at long wavelengths makes any prediction of a RAIRS measurement of submonolayers below $\sim 600 \text{ cm}^{-1}$ with a grating instrument optimistic.

Optical coatings of increasing thickness laid usually on a quartz or germanium substrate can be used to produce a variable wavelength interference filter with a bandpass of relatively small half-width. When manufactured in circular segments on a disk, such films provide a means of monochromatization of transmitted radiation passing through a slit in front of the filter wheel, the energy changed simply by rotating the filter wheel, and a RAIRS spectrometer employing such a filter is shown schematically in Figure 6a. Monochromation in the range 4000–700 cm^{-1} can be achieved with a resolution $(d\lambda/\lambda)$ of $\sim 10^{-2}$ and a slit width $m < 1$ mm. When used for RAIRS, rather low spectral resolutions are typical $(dv = 40$–80 cm^{-1} in the range 1000–3000 cm$^{-1})$[51] and provides a serious drawback in the resolution of close-lying bands in, e.g., isotope experiments (Section 3). Their advantage is short optical paths, ease of operation, and relatively high throughput (>30% transmission).

Vibrational transitions in the range 500–10 cm^{-1}, including many of the stretching and bonding modes of molecules containing heavy atoms, lattice vibrations, and torsional modes, which were generally beyond the effective range of conventional spectrometers, are accessible using Fourier transform infrared spectroscopy[52] (FTIR or FTS). The main advantage of the interferometer over the conventional spectrometer is that high resolution can be obtained without the use of narrow slits, which reduced the luminosity of the conventional instrument. This provides an effective way of making use of the available energy of infrared sources which is of particular importance at long wavelengths. This same advantage (the luminosity or Jacquinot advantage) of FTIR can also be realized in RAIRS, although the restrictive criteria of the reflection optics limit its full exploitation. This inability to make full use of the high luminosity of commercial FTIR instruments for RAIRS was demonstrated by Baker and Chesters,[53] whose instrumental throughput $(7 \times 10^{-6}$ m^2 Sr$)$ was far greater than allowed by the reflection optics at $\phi = 84°$ on a 3×1 cm foil $(\sim 7 \times 10^{-7}$ m^2 Sr$)$, which is similar to that from a single crystal using a dispersive optics of $f/4$ (Section 2.1). A schematic of the FTIR-RAIRS spectrometer used in Ref. 53 is shown in Figure 6c. The second main advantage of FTIR, also in the detector limit signal-to-noise limit, is the multiplex or Fellgett advantage, resulting from the fact that radiation of all wavelengths is being sampled simultaneously at the detector. A signal to noise $N^{1/2}$ times better can be expected for an FTIR over a dispersive instrument for the same spectral range and scanning time, where N is the number of spectral elements in the range.[52] A sensitivity over the midinfrared range compatible with a dispersive RAIRS instrument $(\sim 10^{-5}$ absorbance units) scanning only

$\sim 300\ \text{cm}^{-1}$ is theoretically feasible for an FTIR spectrometer scanning $\sim 400\text{–}4000\ \text{cm}^{-1}$.[53]

However, two experimental problems may hinder the full exploitation of the multiplex advantage of FTIR in the detector noise limit. Because in the FTIR experiment all of the radiation in the measured spectral region falls on the detector, *saturation* may occur (particularly with bolometers). This problem can be tackled by careful choice of detector or to some extent by the use of infrared filters narrowing the spectral region being measured, although the concomitant reduction in the number of spectral elements N reduces the multiplex advantage. The second problem arises from the fact that the large analog detector output signal (the interferogram) must be converted to digital signal for Fourier analysis: The dynamic range of the A/D converter must be sufficient to allow changes due to the very small absorption bands to be digitally resolved. The present state-of-the-art A/D converter places a signal-to-noise ceiling of 3×10^{-3} for a single scan FTIR instrument scanning $400\text{–}4000\ \text{cm}^{-1}$ at $2\ \text{cm}^{-1}$ resolution,[53] although coaddition of interferograms in multiple scanning will improve this signal-to-noise limit.

In the midinfrared spectral region where signal to noise is fluctuation noise limited because of good detectivity and infrared source intensity, the FTIR technique may be at a disadvantage over the dispersive instrument since $P_{FN} \propto P$: as pointed out, the simultaneous monitoring of intensity over the complete spectral range leads to the multiplex advantage in the detector noise limit where $P_{DN} \propto P$, no advantage for photon noise limitation where $P_{DN} \propto P^{1/2}$, and a disadvantage in the case of the fluctuation noise limit $(P_{FN}\alpha P)$[52] in which case, if all the spectral elements of the noise are of comparable intensity, the noise level is $N^{1/2}$ worse for FTIR over a dispersive instrument. One method of improving this situation is again to incorporate filters to restrict the number of spectral units over which the fluctuation noise is observed at the detector. From the theory, at least, one must conclude that while FTIR may be advantageous at long wavelengths (P_{DN} limit), its signal to noise in the midinfrared region (P_{FN} limit) is likely to be inferior. This is to some extent reflected experimentally, since the lowest energy band yet monitored using RAIRS ($\sim 800\ \text{cm}^{-1}$) was achieved using an FTIR instrument,[30] yet the signal to noise achieved in FTIR above $\sim 1500\ \text{cm}^{-1}$[53,54] seems to be below that achievable by conventional dispersive instruments.[42,46,47] The advantages of FTIR in the scanning of large spectral regions, however, must not be underestimated and the incorporation[54,55] of evacuated optics, polarization modulation, and optical filters will undoubtedly lead to improvements in FTIR-RAIRS in both the detector and fluctuation noise limited regions. A recent comparison of dispersive and FTIR instruments for RAIRS[56] recently concluded that both types of instruments will have an important role to play in the future. This indeed appears to be a fair analysis, considering the complementary nature of their strengths.

2.4. Modulation

Their use of signal modulation and phase sensitive detection is commonly used to separate a signal from its background noise spectrum, and in RAIRS there are three main contributions to the improvement in sensitivity that may be expected:

1. The signal to be measured may be separated from the superimposed noise if the latter has a frequency away from the frequency chosen for modulation.
2. Gas phase contributions to infrared absorption in the optical path may be canceled, as can drift in source intensity and detector sensitivity, both of which can make background spectrum normalization difficult.
3. The dynamic range of the signal output of the detector can be reduced, allowing the scanning of larger spectral ranges with high amplifier sensitivities, or more efficient use of A/D converters in FTIR.

The simplest form of modulation is that of total intensity modulation, generally introduced experimentally using a rotating chopper blade and used in combination with both dispersive[41–43,46,47] and filter wheel[51] spectrometers. Demodulation yields I_T (total intensity) with noise rejection due to (1) only above. The signal-to-noise ratio obtained in RAIRS is consequently not particularly good using this method, although surprisingly successful measurements using this technique have indeed been made[43] and intensity modulation is generally sufficient to obtain a dependable measurement of total absorption intensity of strong dipole absorbers[41,42,46–48] and enables a check to be made on instrumental broadening introduced by, for example, wavelength modulation techniques. Total intensity chopping has also been successfully combined with other modulation techniques.[51]

Double beam modulation (applied to RAIRS when conventional spectrometers are adapted for reflection[45]) is the method originally adopted in transmission infrared instruments and involves the alternate sampling of signals (ideally at the same detector) from two beams, one of which passes through the sample and the other via an identical optical path through a blank or reference cell. If the two beams are correctly matched, the demodulated signal yields the absorption bands of the sample, and background absorptions and variations of source intensity and detector sensitivity (with time and wavelength) which are common to both beams are eliminated. All three modulation advantages above are achieved, but it is extremely difficult to apply the dual beam method to RAIRS: since one beam must pass into a UHV chamber and reflect from the single crystal, it is clearly impractical for the reference beam to have an identical path,* and in particular any changes in signal due to the movement of the sample itself cannot be easily compensated. There has also been to date some restriction in the range of modulation frequencies accessible since the movement of an optical component is generally involved.

Wavelength modulation combined with dispersive monochromation has perhaps been the most widely used techniuqe in RAIRS[41,42,46–48] and was originally introduced by Pritchard and co-workers.[41] Essentially a derivative spectroscopy,[57] it has similarities to, for example, the measurement of dN/dE in Auger spectroscopy, where a small Auger peak on a large background of secondary electrons can be easily distinguished. In RAIRS, the demodulated signal $dI/d\lambda$ at the detector yields a nearly constant value for a smoothly changing background, and a clear differential peak over the absorption band. This effect is clearly demonstrated in RAIRS data extracted from Pritchard et al.[58] (Figure 9a) of a monlayer of CO on Cu(100) in a single beam experiment using wavelength modulation. Numerical integration of the differential spectrum provides an absorption spectrum with some reduction in background noise. Clearly the advantages of (1) and (3) above are achieved using this method, but in the single beam experiment wavelength modulation cannot alleviate noise introduced through (2). Wavelength modulation is particularly sensitive to sharp atmospheric absorption bands which have half-widths of the same order as the adsorbate bands (instrumental limited) and introduce strong structure in the derivative spectrum. Evacuation of the optical system is therefore important as is the long-term stability of source and detector. Wavelength modulation also has the disadvantage that it is not as sensitive to broad infrared bands, and there can be some difficulty in deriving absolute integrated absorption intensity and peak half-widths directly from wavelength modulated data, particularly when the phase sensitive detection is

---

* Recently the dual beam technique has been successfully adapted for RAIRS[136] and compensation for gas phase adsorption is possible up to a few torr pressure by use of a reference cell that has the same path length (~10 cm) as that of the UHV chamber.

carried out at twice the modulation frequency yielding the doubly differentiated signal $d^2I/d\lambda^2$.[48,50] Since the movement of an optical component is involved, only relatively low modulation frequencies are accessible, and usually lie in the range 100–400 Hz.

An alternative technique, which most nearly relates to the double beam method, is polarization modulation. Alternate sampling of the $p(I_p)$ and $s(I_s)$ components of the reflected ir beam provides a pseudo-double-beam technique[59]: absorption due to the surface species on a metal occurs only in the p component (Section 1.2) while atmospheric absorption bands etc. together with source intensity and detector sensitivity changes appear in both components, which possess identical optical paths. Sensitivity of measurement through contributions (1), (2), and (3) above are introduced as in the case of the dual beam method, but without the difficulties associated with a double beam spectrometer and with greater flexibility in modulation frequencies. The simplest experimental arrangement consists of a continually rotating polarizer[46,59] and phase sensitive detection of such a modulated beam yields a signal $(I_p - I_s)$. The cancellation of the background gas phase absorption bands is clearly demonstrated in Figure 9b, taken from the work of Hoffmann.[60] The physical rotation of an optical component restricts modulation frequencies to below ~400 Hz. The combination of a photoelastic modulator (which operates at much higher frequencies) and static polarizer also allows demodulation of a signal $(I_p - I_s)$.[51] In a similar way that the dual beam technique is sensitive to intensity imbalances, polarization modulation works most efficiently when I_p and I_s are equal, and if this is not the case the signal $(I_p - I_s)$ becomes sensitive to contributions from (2), and to some extent (3), above. Dispersive instruments transmit unequal intensities I_p and I_s in some spectral regions due to the characteristic of the grating and this type of effect is illustrated in Figure 9c. Near the blaze wavelength of the grating atmospheric absorption is reduced substantially in the signal $I_p - I_s$ since $I_p \approx I_s$, but away from the blaze the background absorption increases as the values of I_p and I_s diverge. One way to minimize the effect of background absorption is to evacuate the optical path.[46,60,61] Alternatively[51] the introduction of a chopper operating at a significantly different frequency allows simultaneous intensity modulation and a combined signal output $(I_p - I_s)/(I_p + I_s)$ which to some extent compensates for differences in the components I_p and I_s. The direct matching of I_p and I_s has also been used in an attempt to alleviate this problem,[51,62] and this method of canceling gas phase contributions to the absorption signal was used to study the adsorption of CO[51] and NO[62] on platinum foils with

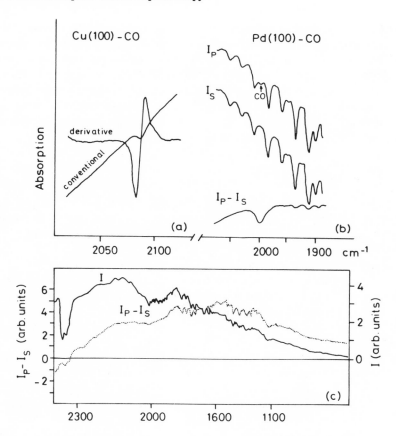

Figure 9. The use of wavelength modulation (a) to produce a clear derivative spectrum from a small absorption peak on a smooth changing background (after Ref. 89). Polarization modulation (b) can be used to eliminate atmospheric absorption bands (which appear in both I_p and I_s) from the adsorbate peak (exclusively in I_p) by subtraction of the two components (after Ref. 60). This is most efficient when I_p and I_s are equal, which in the case of dispersive optics is only the case (c) around the blaze wavelength of the grating (ca. 2300 cm^{-1}).

relatively high ambient pressures (50 Torr) of the reacting gases above the surface.

The technique of polarization modulation in RAIRS is closely related to infrared ellipsometric spectroscopy (IRES).[63] The latter technique yields the real and imaginary parts of R_p/R_s (the ratio of reflection coefficients for p and s polarized light) which can in turn be

separated to yield an ellipsometric absorbance and dispersion (ellip-
sometry yields both intensity and phase information). The absorbance
that is derived from IRES is essentially equal to the reflectance
absorbance for p polarized light obtained in RAIRS using polarization
modulation for experiments on metal substrates in the infrared spectral
region, and for measurements carried out at grazing angles of incidence
ϕ,[63] because of the exclusive absorption of radiation in the p component
(Section 1.2).

2.5. Signal Detection

It has been the introduction of increasingly sensitive ir detectors
operating at low temperature (≤ 78 K) that has made experimental
RAIRS possible, although a conventional thermocouple detector has
indeed been convincingly used by King and co-workers.[43] The high
sensitivity of detector is required owing to the small number of absorbing
molecules giving a maximum absorption peak of only $\sim 1\%$, together
with a restiction in optical throughput imposed by grazing incident
reflection on a small surface, and the inherent low intensity of thermal
black body emitters. It is instructive to recall (Section 2.2) the levels of
detectivity required in RAIRS. Since typically an $f/4$ spectrometer is
found to be optimum with an aperture μ of ca. 1×10 mm at the sample,
the limiting luminosity of the reflection geometry was shown to be
$\sim 6 \times 10^{-7}$ m^2 Sr. If no defocusing optics is used onto a typical detector
of 1×1 mm, the luminosity of the RAIRS spectrometer can be further
reduced to only 6×10^{-8} m^2 Sr, and the maximum power available
(Figure 7) from a black body source under such conditions is $\sim 10^{-7}$ W.
Since we are attempting to measure changes in this power of the order of
10^{-3} or less for typical submonolayer absorptions, the detector must
clearly be sensitive enough to detect changes in the order of 10^{-10} W or
smaller.

Infrared spectrometers are often designed around the characteristics
of the infrared detector. These characteristics are conveniently described
by means of "figures of merit" and while these may be numerous, the
most important[31] generally relate to

1. The minimum radiant power that can be detected above the
   intrinsic noise of the detector.
2. Responsivity, i.e., the output signal (V) per unit radiant power
   (W).
3. Spectral response, i.e., the variation of (1) and (2) above with
   wavelength.

4. The response time.
5. Long-term stability, i.e., changes of (1) or (2) with time.
6. Dynamic range (FTIR measurement).

The large number of possible sources of detector noise that make up P_{DN} and the large number of detector types makes it inappropriate to describe them in detail here and the interested reader is referred to one of several reviews.[31,64,65] A brief summary of the methods used to quantify 1–4 above is included, however, since this can be related to the value D used by manufacturers of infrared detectors.

The noise equivalent power (NEP) of a detector is commonly used to quantity (1) above and is defined as the rms value of sinusoidally modulated radiant power P_0 falling on a detector which will give rise to an rms signal voltage V_s equal to the rms noise voltage V_{DN} from the detector. If the electronic bandwidth of the detector/preamplifier system is Δ, the NEP, and therefore P_{DN} ($\equiv$NEP), is given by

$$P_{DN} = P_0 A^{\alpha} \frac{V_N}{V_S} \Delta^{-1/2} \tag{12}$$

where A is the detector area. For many detectors $\alpha = \frac{1}{2}$[31] and hence they are specified in terms of a quantity D^*, which provides a source-area-independent quantity of merit given by

$$D^* = A^{1/2} P_{DN}^{-1} \tag{13}$$

D^* is known as the detectivity, and characterizes a detector at a given operation temperature and wavelength.

The responsivity R, (2) above, can also be related to D^*:[31]

$$R = D^* V_{DN}/(A\Delta f)^{1/2} \tag{14}$$

The quantity D^* as a function of wavelength, (3) above, therefore provides a measure of the three most important factors of merit of a detector.

The response time τ of a detector (4) depends critically on the physical process being exploited to measure the radiative power. For a semiconductor τ is characterized by the recombination time of majority (photoconductive) or minority (photovoltaic) carriers and is therefore relatively small, i.e., the modulation used for phase sensitive detection can be up to several tens of kilohertz. Low-temperature bolometers, however, have response times governed by the heat capacity of the detector and τ is consequently much larger. The frequency dependence of responsivity R has the form of a low pass filter[31] determining the maximum modulation frequency permissible before the response R [or

D^*—equation (14)] is degraded—i.e., a high-frequency cutoff. The low-frequency limit of all detectors is limited by the onset of $1/f$ type noise in the detector/preamplifier electronics, and the variation of R (or D^*) for a detector with frequency is typically characterized by an intermediate frequency region in which it is most efficiently operated.

In making a choice of detector, one must refer to the manufacturer's curve for D^* as a function of wavelength, keeping in mind the spectral region of interest and the experimentally accessible range of modulation frequencies. Infrared detectors conventionally operated at room temperature (e.g., thermocouple or golay) detect over a wide spectral range but have relatively high noise levels ($D^* \approx 10^{-8}$–$10^{-10}\,\omega\,Hz^{-1/2}$) and relatively long time constants τ. They are therefore not generally suited to RAIRS, although they have been successfully used in the spectral region around 2100 cm^{-1}.[43] The low-temperature detector most commonly used[41,43,46,48,51,52] has been indium antimonide or lead selenide, which have perhaps the highest value of D^* (10^{11}–$10^{12}\,\omega\,Hz^{-1/2}$) of all photoconductive or photovoltaic detectors, but a spectral cutoff at ~1700 cm^{-1} below which one cannot measure. More recently the introduction of increasingly sensitive mercury cadmium telluride detectors, which have D^* values slightly lower than indium antimonide, have enabled RAIRS measurements to be extended to longer wavelengths[50–52] using a detector which again must only be cooled to liquid-nitrogen temperatures. Perhaps the most sensitive infrared detector is the liquid-helium-cooled metal doped germanium bolometer,[47,58,61] which can be operated at high sensitivity continuously down into the far infrared energy region and indeed has been successfully used for RAIRS despite the inconveniences of operating at detector temperatures of ca 1.7 K. The detector noise limit (P_{DN}) is due to background radiation from the surroundings within the detector field of view, and this is reduced using interchangeable helium-cooled filters directly in front of the detector. P_{DN} can be reduced sufficiently to be fluctuation noise limited (P_{FN}) in the accessible range (limit to $\bar{\nu} \gtrsim 1200$ cm^{-1} by CaF$_2$ windows).[58,61] One of the drawbacks of bolometric detection is the severe limitation in detector response time which restricts modulation frequencies to below ca. 150 Hz.

3. Applications

3.1. Selection Rules and the Adsorbate Geometry

A molecule in the gas phase containing N atoms has $3N$ degrees of freedom. These consist of three translation and and three (or two for a linear molecule) rotational degrees of freedom, leaving $3N-6$ (or $3N-5$) normal vibrational modes. While adsorption into a mobile phase could give rise to the loss of only one translational degree of freedom (that perpendicular to the surface), generally localized adsorption takes place, the dynamical constraints leading to the conversion of translational and rotational degrees of freedom into vibrational modes (frustrated translations and rotations). Consequently up to $3N$ vibrational modes can be associated with an adsorbate species. The excitation of any one of these modes is governed by the selection rule for electric dipole transitions with the additional "surface selection rule" for an adsorbate on a metal surface.

The transition probability P_{vv} of a pure vibrational transition is proportional to the square of the transition moment M_{vv}, which is given by[66]

$$M_{vv} = \int_{-\infty}^{+\infty} \psi(v)\mu\psi(v')\,d\tau \tag{15}$$

where $\psi(v)$ and $\psi(v')$ are the vibrational eigenfunctions of the initial and final states and μ is the dipole moment of the molecule. The criterion for dipole activity is that the integral over all space in equation (19) must be nonzero. Since the dipole moment μ can be resolved into three spacial components μ_x, μ_y, and μ_z, the latter integral is nonzero if any one of the component integrals [equation (15)] is nonzero:

$$M_{vv'}^x = \int \psi(v)\mu_x\psi(v')\,dx$$

$$M_{vv'}^y = \int \psi(v)\mu_y\psi(v')\,dy \tag{16}$$

$$M_{vv'}^z = \int \psi(v)\mu_z\omega(v')\,dz$$

The conditions for a nonvanishing integral can be deduced using symmetry arguments and group theory,[66] the requirement being that the product of the symmetry representations Γ of $\psi(v)$, $\psi(v')$, and μ (either μ_x, μ_y, or μ_z) must be totally symmetric. Since one excites generally from the ground state in vibration spectroscopy, $\psi(v)$ is totally symmetric, and the integral is finite only when $\Gamma\psi(v')X\Gamma_\mu$ is totally symmetric, i.e., $\Gamma\psi(v')$ must transform as either x, y, or z. Defining the surface normal as the z direction, the additional imposition of the surface selection rule for metals (Section 1.2) requires that only vibrations in which $\Gamma\psi(v')$ transforms as z will be dipole allowed. Strictly speaking the local

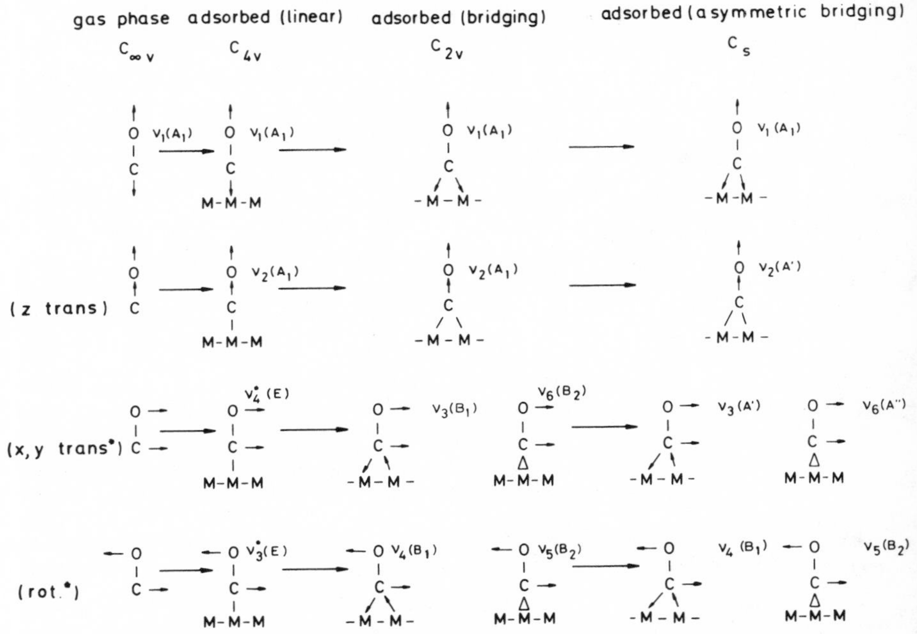

Figure 10. The reduction in local symmetry of CO when adsorbed on the (100) surface of an FCC metal, and the conversion of translational and rotational degrees of freedom into vibrational modes at the surface.

geometry of the adsorption site must be taken into account when determining the symmetry of the adsorbate molecule. For example, a reduction in symmetry from C_v to C_{4v} accompanies the adsorption from the gas phase of CO into a linearly bound configuration of an on-top site on an FCC(100) surface (Figure 10), but adsorption in an on-top site on an FCC(110) surface leads to a lifting of the degeneracy of the v_3 and v_4 modes since its local symmetry is now C_{2v}. In the latter case (the difference between C_{4v} and C_{2v}) no new dipole active modes become allowed as a result of the lowering of symmetry through local substrate geometry, although examples indeed exist that demonstrate the effect of reduced symmetry by the substrate, for example, in the adsorption of benzene on Pt and Ni surfaces.[67] A vibration becoming active due to the lowering of symmetry is related to the strength of the interaction lowering the symmetry, and it appears often sufficient to consider the configuration of the adsorbate and ignore the local symmetry of the substrate site in determining the adsorbates' symmetry species and the appropriate selection rules.[68]

Table 3. Symmetry and Modes of Ethane (Deuterated) on Pt(111)

| | | | Eclipsed confirmation | | | | Staggered confirmation | |
| | | | Gas phase | | Adsorbed phase | | Gas phase | Adsorbed phase |
Mode	Character	Gas phase[a]	D_{3h}	C_{2v}	$C_s(1)$	$C_s(2)$	D_{3d}	C_s
$v_8 + v_{11}$	CH$_3$ def.	2087 cm^{-1}	A_2''	B_1	A''	A'	A_{2u}	A'
v_5	CH stret.	2111 cm^{-1}	A_2''	B_1	A''	A'	A_{2u}	A'
v_7	CH stret.	2236 cm^{-1}	E	$A_1 + B_1$	$2A'$	$A' + A''$	E_u	$A' + A''$
			3	1	2	3	4	3
	Infrared active modes							

[a] Reference 78.

Note that the dipole selection rules alone do not allow the distinction between the two adsorbate geometries C_{4v} and C_{2v} shown in Figure 10: In the absence of any substrate screening, all vibrational modes for the two adsorbate geometries would be allowed, i.e., C_{4v} (A_1, E) and C_{2v} (A_1, B_1, B_2) but the imposition of the surface selection rule reduces the observable modes to v_1 and v_2 (A_1) in C_{4v} or C_{2v}. One must therefore resort to a "fingerprint" analysis (Section 3.2) of the observed bands to distinguish the absorbate species, i.e, linear or bridged. A reduction in symmetry to $C_s(1)$ through the formation of an asymmetric bridging complex, on the other hand, produces two additional dipole allowed modes $v_3(A')$ and $v_4(A')$, while a tilting of the bridged molecule reduces the symmetry to $C_s(2)$ with $v_5(A')$ and $v_6(A')$ becoming active.*

The adsorbate symmetry can therefore be deduced through an analysis of the number of observed and expected vibrational bands in the measured spectral range. In the case of RAIRS, the surface selection rule for adsorbates on metals must be applied rigorously since no short-range or resonant mechanism is available for energy transfer as in the case of EELS. The use of the strict dipole selection rule in RAIRS is illustrated by adsorption studies of ethane on platinum[69] and of the methoxy[70] and formate[71] species on copper surfaces.

Deuterated ethane produces a single absorption band in RAIRS in the range 2000–2300 cm^{-1} when adsorbed on Pt(111),[69] while three active vibrational modes are observed in the gas phase (Table 3). In the light of the small energy difference between the two possible gas phase conformations, eclipsed (D_{3h}) and staggered (D_{3d}), the effect of lowering the symmetry by the surface of both species is considered by reference to symmetry correlation tables. Adsorption of the eclipsed form (D_{3h}) with

---

* $C_s(1)$ and $C_s(2)$ correspond to reflection planes σ_{xz} and σ_{yz} in the C_s point group, and must be defined with respect to the adsorbate configuration in some cases.

302

Brian E. Hayden

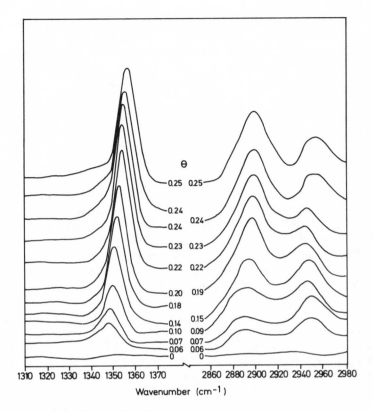

Figure 11. The RAIRS spectra of formate on Cu(110) as a function of coverage (after Ref. 71).

the C–C axis parallel to the surface produces a C_{2v} configuration with a single active mode (A_1) because of the imposition of the surface selection rule. A further reduction in symmetry, e.g., tilting in the zx $[C_s(1)]$ or yz $[C_s(2)]$ planes leads to an additional mode becoming active. The minimal lowering of symmetry for the D_{3d} symmetry species on adsorption is C_s, i.e., there is only one possible reflection plane for the adsorbed staggering confirmation, leading to three active modes (A'). The observation of a single band may be used to conclude that ethane adsorbs in the eclipsed conformation with the $c–c$ axis parallel to the surface.*

* This conclusion differs slightly from that originally drawn from the results in Ref. 70 but has subsequently been confirmed by the author.[71]

A bridged formate species with C_{2v} or $C_s(1)$ symmetry on Cu(110) was also characterized in RAIRS[71] measuring in the spectral region 1200–3000 cm^{-1} and spectra are shown for various coverages of formate in Figure 11. The observation of the symmetric stretch vibration $\nu_s(COO)$ at 1348–1358 cm^{-1} and the $\nu(C–H)$ stretch at 2891–2900 cm^{-1} in the absence of the asymmetric stretch $\nu_a(COO)$ excludes the possibility that the formate symmetry is further reduced to $C_s(2)$ or C_1 on the surface (Table 4), which would be the case for a monodentate type adsorbate species. In addition to the fundamental modes expected of a formate species in C_{2v} or $C_s(1)$ symmetry (either a bridging or tilted bridging configuration, respectively), a peak at ~2950 cm^{-1} was observed. This band had previously been observed in EELS and assigned to an unspecified C–H stretch mode which was active because of the less stringent selection rules of the electron spectroscopy.[74] The rigidity of the dipole selection rule in RAIRS required its reassignment[71] to a dipole allowed combination band ν_{comb}^1 made up to two otherwise inactive modes at the surface, $\nu_a(COO) + \sigma(CH)$ (Table 4): The symmetry of ν_{comb}^1 is correct for dipole activity and is also correct for it to gain additional intensity through Fermi resonance with $\nu(CH)$.[71] Note from Table 4 that the observation of only one combination band should allow the deduced symmetry to be narrowed to C_{2v}. The inability to observe ν_{comb}^3, however, which may be very weak in the absence of intensity gain through Fermi resonance, was considered insufficient grounds for discounting the possibility that the formate species has $C_s(1)$ symmetry.

Table 4. Symmetry and Modes of Formate on Cu(110)

Mode	Na salt[a]	Gas phase		Adsorbed phase		
		C_{2v}	C_{2v}	$C_s(1)$	$C_s(2)$	C_1
$\nu_s(COO)$	1366 cm^{-1}	A_1	A_1	A'	A'	A
$\sigma(CH)$	1377 cm^{-1}	B_1	B_1	A''	A'	A
$\nu_a(COO)$	1567 cm^{-1}	B_1	B_1	A''	A'	A
$\nu(CH)$	2841 cm^{-1}	A_1	A_1	A'	A'	A
Infrared active (fundamental) modes		4	2	2	3	4
$\nu_{comb}^1 = \nu(COO) + \sigma(CH)$	2953 cm^{-1}	A_1	A_1	A'	A'	A
$\nu_{comb}^2 = \nu_s(COO) + \sigma(CH)$	2720 cm^{-1}	B_1	B_1	A''	A'	A
$\nu_{comb}^2 = \nu_s(COO) + \pi(CH)$	2435 cm^{-1}	B_2	B_2	A'	A'	A
Infrared active (combination) modes		3	1	2	2	3

[a] Reference 75.

Table 5. Symmetry and Modes of Methoxy on Cu(100)

Mode	Gas phase C_{3v}	Adsorbed phase C_{3v}	C_3	C_1
$\nu_s(CH_3)$	A_1	A_1	A'	A
$\nu_a(CH_3)$	E	E	$A' + A''$	A
Infrared active modes	2	1	2	2

In a recent RAIRS study of methanol decomposition to form the methoxy species on Cu(100),[70] the observation of both the symmetric $\nu_s(CH)$ and asymmetric $\nu_a(CH)$ stretch vibrations of the $-CH_3$ moiety in the adsorbed methoxy species (2795 cm^{-1} and 2870 cm^{-1}, respectively) was used to demonstrate a reduction in adsorbate symmetry from C_{3v} to C_s or C_1. This provided evidence that the methoxy species was tilted on the surface, since only the symmetric stretch would have been observed in the molecule in C_{3v} symmetry orientated perpendicular to the surface (Table 5). Note that the influence of the substrate symmetry, and the fourfold symmetry of the surrounding methoxy species in a $c(2 \times 2)$ overlayer,[70] are presumed alone to be too small a perturbation to reduce the C_{3v} symmetry of a vertically orientated methoxy species.

3.2. Vibrational Frequencies as Fingerprints

The use of vibrational frequencies alone as a means of determining the nature of a particular adsorbate species is often termed "fingerprinting," and as pointed out by Sheppard et al.[76] provides the basis of interpretation in many of the infrared studies of adsorbed species on supported metal catalysts pioneered by Eischens.[7] Generally the first step in interpreting vibrational spectroscopic data in surface studies is made by comparison with infrared spectra obtained from compounds containing similar molecular entities: such spectra can be of the liquid, solid, gaseous, or matrix isolated phases of the molecule itself, or of a related molecule or compound (invariably an ionic salt or organometallic complex) containing the suspected surface moiety. There also exists a substantial literature concerning the infrared vibrational spectra of adsorbed molecules on supported metal catalysts providing considerable scope for comparison with single-crystal data from RAIRS and EELS.[78] As soon as a molecule adsorbs on the metal surface, in the absence of interaction with any other molecule, it experiences a shift from the gas phase due to mechanical renormalization, its dipole image and its chemical bonding to the surface. All of these effects essentially determine the singleton frequency ω_0 of the adsorbate mode, although to a first approximation, renormalization and dipole image shifts are assumed relatively insensitive to the metal substrate, vibrational "fingerprinting" therefore allowing a relatively general characterization based on chemical effects. Frequency shifts that are chemical in nature include those arising from bonding configuration (generally giving rise to the largest frequency shifts), the coordination of the substrate atom(s), and local chemical environment. The latter contribution includes effects due to coadsorbed species on the surface (e.g., often adatoms preadsorbed to modify the catalytic or adsorption behavior of the surface) and the influence of neighboring molecules of the same species at finite coverage. Both dipole image and chemical bonding influence not only coverage-dependent frequency shifts but also the halfwidth, line shape, and intensities of vibrational peaks (Sections 3.4 and 3.5). The chemical forms of frequency perturbation manifest themselves on single-crystal and supported metal surfaces alike, but can be conveniently separated in RAIRS studies of adsorption on well-characterized single-crystal surfaces, particularly when combined with complementary techniques such as low-energy electron diffraction (LEED), temperature programmed desorption (TPD), or work function measurements ($\Delta\phi$). The use of vibrational frequencies as fingerprints, and frequency modification due to chemical environment, is illustrated using results obtained in RAIRS studies of mainly CO on Pd, Pt, Ru, and Cu single-crystal surfaces.

The adsorption of CO on supported Pd investigated by Eischens et al.[77] using infrared transmission techniques (Figure 12a) is a good illustration of the influence of both bonding configuration and site geometry on the $v(CO)$ stretch vibration of chemisorbed CO. The increasing number of vibrational bands with increasing coverage was interpreted as reflecting the "heterogeneity of the surface,"[78] the latter also being responsible for the large halfwidth of the adsorption peaks. It was also realized that several configurations of chemisorbed CO [linearly $v(CO) \approx 2080$ cm^{-1} and bridge species $v(CO) \approx 1923, 1835$] were present on the surface.[7] The latter assignment originated from comparison of the observed band frequencies with those obtained in vibrational studies of carbonyl compounds, and the use of this fingerprint technique continues to provide a satisfactory method of empirically deducing the configuration of CO chemisorbed on metal surface[76,78]:

Linear	ca. 2130–2000 cm^{-1}
Bridge	ca. 2000–1860 cm^{-1}
Threefold	ca. 1920–1800 cm^{-1}
Fourfold	ca. 1800–1700 cm^{-1}

While CO may be found in a number of configurations on a single-crystal

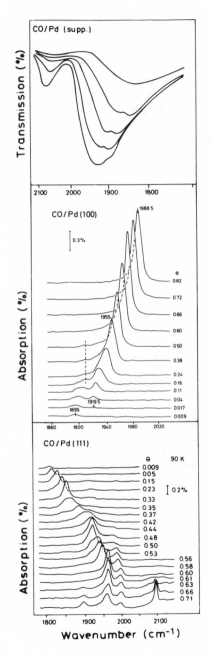

Figure 12. A comparison of the infrared spectra at various coverages of adsorbed CO on supported Pd (after Ref. 7) at 300 K, Pd(100) at 300 K (after Ref. 79) and Pd(111) at 90 K (after Ref. 80) single crystal surfaces.

surface, the coordination of substrate atoms is well defined. Further, if the adsorbed overlayer is ordered and CO adsorption is site specific (i.e., it prefers one of the above configurations) the LEED experiment can be used as a complementary technique in determining the bonding configuration of the adsorbate. Figures 12b and 12c are RAIRS spectra of CO adsorbed at various coverages on Pd(100) and Pd(111) single-crystal surfaces.[79,80] The adsorption of CO on Pd(100) provides a convenient adsorption system to study configuration specific internal frequencies of CO:[79] only twofold bridge sites can be occupied in the observed ordered overlayer [LEED results show a $c(4 \times 2)$ R45°[81] or $(2\sqrt{2} \times 2\sqrt{2})$R45°[85] structure at $\theta = 0.5$] if equivalent adsorption sites are assumed for all molecules, and indeed only a single band is observed in RAIRS, which shifts as a function of coverage (Figure 12b): $\nu(CO) \sim$ 1900–2000 cm^{-1}. Of the bands appearing for CO on supported palladium (Figure 13a) with increasing coverage (1835–1923–2061 cm^{-2}), only that which is associated with bridged CO (1923 cm^{-1}) is observed on Pd(100), while there are equivalents of all three bands in RAIRS results for CO adsorbed on Pd(111) at various coverages (1808–1951–2097 cm^{-1}) (Figure 12c). These results also can be interpreted through frequency fingerprints and with the aid of the series of observed LEED structures.[46,80] The $(\sqrt{3} \times \sqrt{3})R30°$ LEED structure observed at lowest coverages requires that all CO molecules in the overlayer are in equivalent sites, i.e., exclusively linear, bridge, or threefold configurations. The single RAIRS band at 1808 cm^{-1} in this range indicates that all

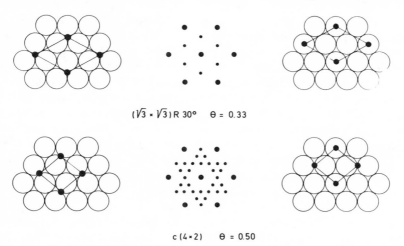

$(\sqrt{3} \times \sqrt{3})R\,30°$ $\theta = 0.33$

$c\,(4 \times 2)$ $\theta = 0.50$

Figure 13. The LEED patterns and some possible real space models of the $(\sqrt{3} \times \sqrt{3})R30°$ and $c(4 \times 2)$ structures of CO on Pt(111) and Pd(111).

molecules are in threefold sites (Figure 13). An increase in coverage to $\theta = 0.5$ is accompanied by the appearance of a $c(4 \times 2)$ LEED pattern: if site specific adsorption of CO takes place, the only two acceptable real space models correspond to CO adsorbed half in bridging and half in linear sites [as is the case on Pt(111)] or exclusively in bridging positions (Figure 13). The latter is concluded to be the case for Pd(111) with a single RAIRS peak at 1951 cm^{-1}. Spectra for coverages $\theta > 0.5$ are characterized by the reappearance of a peak in the spectral region associated with the threefold species $v(CO) = 1900 \text{ cm}^{-1}$, and a peak which can be assigned to CO in an on-top site or linear configuration at 2097 cm^{-1}. The presence of these species as determined in RAIRS can be reconciled[81] with the series of hexagonal LEED structures observed at high coverages: the relatively sharp and clearly assignable peaks in RAIRS illustrates also the site specific nature of CO adsorption in overlayers which appear to be incommensurate with the substrate, an observation that has also been made for CO adsorbed on Pt(111)[47] and Cu(111) surfaces.[83]

CO adsorption on Pt(111)[84] shows similarities to its adsorption on Pd(111) in that the initial adsorbate-induced LEED structures observed are $(\sqrt{3} \times \sqrt{3})R30°$ (although rather weak on Pt(111)) at $\theta = \frac{1}{3}$ followed by a $c(4 \times 2)$ at $\theta = \frac{1}{2}$. The CO adsorbate configuration, however, in the two ordered layers on Pt(111) differ from those on Pd(111) (described above) as revealed in the vibrational spectroscopic data.[85,86] Linearly bound CO, $v(CO) = 2094\text{--}2110 \text{ cm}^{-1}$ is exclusively adsorbed in the $(\sqrt{3} \times \sqrt{3})R30°$ structure on Pt(111), while linear and bridging $v(CO) = 1840\text{--}1857 \text{ cm}^{-1}$ configurations are observed at $\theta = 0.5$ in the $c(4 \times 2)$ overlayer. The real space structures compatible with both the LEED and vibrational spectroscopic data are shown in Figure 13. High-resolution RAIRS spectra (Figure 14) reveal, however, rather than a single band associated with the bridging species in the $c(4 \times 2)$ overlayer (Figure 13) as observed in EELS,[85,86] a doublet in this spectral region separated by $\sim 40 \text{ cm}^{-1}$.[47] It was also found that this doublet structure appeared to be temperature dependent (Figure 14b) while the peak associated with linear bound CO remains unchanged, and the only dependence of the $c(4 \times 2)$ LEED structure with temperature in the range $95 < T(K) < 300$ was that it becomes slightly diffuse at higher temperatures. In accordance with the general rule that increasing the site coordination in the bonding configuration of CO decreases the $v(CO)$ frequency, the band $v(CO) = 1840\text{--}1857 \text{ cm}^{-1}$ was attributed to bridging CO, and the band $v(CO) = 1810 \text{ cm}^{-1}$ to CO in a threefold hollow site. The temperature dependence of the spectra in the absence of any structural change in LEED was accounted for in terms of facile interconversion of

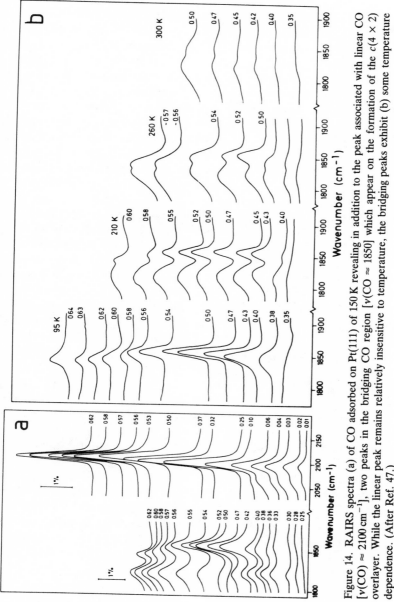

Figure 14. RAIRS spectra (a) of CO adsorbed on Pt(111) of 150 K revealing in addition to the peak associated with linear CO [v(CO) ≈ 2100 cm^{-1}], two peaks in the bridging CO region [v(CO ≈ 1850] which appear on the formation of the c(4 × 2) overlayer. While the linear peak remains relatively insensitive to temperature, the bridging peaks exhibit (b) some temperature dependence. (After Ref. 47.)

bridging to threefold sites (which are present on either side of the bridging position) via the excited states of a low-frequency frustrated translation of the molecule. At low temperature the more stable bridging species is "frozen out," while at higher temperatures the threefold sites are also populated via the reaction coordinate x (see Figure 24), the mean position of the CO molecule remaining constant. This surface phenomenon is analogous to the rapid interconversion of linear and bridging CO ligands on polynuclear metal carbonyls [e.g., $Rh_4(CO)_{12}$[87]] giving rise to two or more possible molecular isomers, and is an example of stereochemical nonrigidity, also known as scrambling or fluxionality.[88] It has also been shown that this process occurs on Cu(111),[83] where the interconversion of bridging to threefold configurations in the overlayer (Figure 23b) is even more facile (lower adsorption energy on copper) and the bridging species can be frozen out only at temperatures as low as ~7 K.

Clearly in the case of CO adsorption on palladium, platinum, and copper surfaces, intermolecular interactions can lead to a series of adsorbate configurations as a function of coverage and temperature, and it is likely that both this effect and site inhomogeneity contribute to the coverage dependence of infrared spectra of CO adsorbed on the supported catalyst.* The effects of surface heterogeneity on CO adsorption on supported catalyst and polycrystalline surfaces was illustrated by Pritchard and coworkers[24,89] by comparing RAIRS data for linearly adsorbed CO on various crystal surfaces of copper (i.e., various substrate atom coordinations) with infrared spectra of CO on supported- or polycrystalline copper surface (Figure 15). It was concluded that the polycrystalline surface (and the silica or alumina supported Cu surface which gives rise to similar spectra to the polycrystalline film for adsorbed CO[89]) is dominated by higher index crystal planes of copper. Conversely, the dominance in the infrared spectrum of CO on MgO supported copper[90] of a band at 2080 cm^{-1} indicates an abundance of close packed crystal faces on such samples.

The presence of relatively high concentrations of defect sites on small supported metal catalysts has long been regarded as having an important influence on the adsorption behavior and catalytic activity of such surfaces. An example of the effect of these "special surface sites" is found in infrared spectra of dinitrogen which adsorbs only on particles of Ni, Pd, and Pt in the range $15 \text{ Å} < d < 70 \text{ Å}$ where these defect sites

---

* For example, an interconversion of CO adsorption configurations as a function of coverage on the supported Pd surface has been demonstrated by selective isotope adsorption studies.[137]

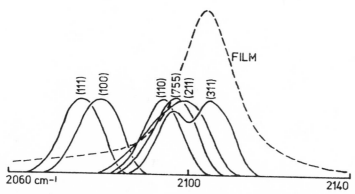

Figure 15. A comparison of RAIRS spectra for linearly adsorbed CO on a variety of cooper single crystal surfaces and a polycrystalline surface (after Ref. 89).

dominate.[91] RAIRS studies of CO adsorption on stepped and kinked stepped surfaces of platinum, chosen so as to model adsorbate sites available on supported metal catalysts, has been carried out[92,93] and the frequencies of CO adsorbed at well-characterized defect sites have been measured. These results and those obtained on the close-packed Pt(111) surface[46] have been subsequently[94] compared to infrared measurements of CO adsorbed on well-characterized supported platinum surfaces[95] to demonstrate the feasibility of modeling catalytic systems using well-characterized single-crystal surfaces. The v(CO) stretch frequency of linearly bound CO provides a fingerprint for its adsorption site coordination on the platinum surfaces.[93,94] Figure 16 contains the RAIRS spectra of linear bound CO on Pt(111), Pt(533), and Pt(432) as a function of coverage. The latter two surfaces are stepped and kink-stepped faces of platinum chosen so as to model sites available on supported platinum particles (Figure 17). The band observed at lowest coverages on Pt(533) and Pt(432) is shown to be associated with CO adsorbed at the step or kink step site on the surface and gives rise to the high-temperature desorption peak in thermal desorption experiments.[92,93] The peak at highest coverages on these surfaces is similar in frequency to that found on Pt(111) and is associated also with CO which is adsorbed on the (111) terraces of Pt(533) and Pt(432). The frequencies of the observed bands from these data for linear CO at step, kink step, and terrace coordinated platinum atoms is summarized in Table 6. Adsorption of CO on supported platinum particles of various carefully characterized sizes studied by infrared spectroscopy[95] allowed the assignment of three observed v(CO) bands (Table 6) to linearly bound

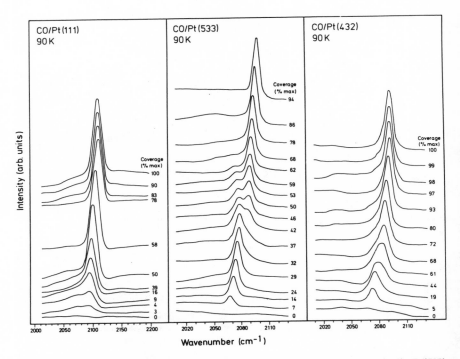

Figure 16. A comparison of RAIRS spectra of linearly adsorbed CO on Pt(111), Pt(533), and Pr(432) surfaces (after Ref. 93).

CO at edge, corner, and face atoms of the supported catalyst (Figure 17): These peaks were found to vary over the range of particle sizes studied with intensity in a similar fashion to the concentration of edge, corner, and face atoms with which they were associated. The similarity in frequency between independently identified species of CO at sites of various coordination (Table 6) demonstrated a successful modeling of the catalyst particles using stepped single crystals.[93–95] It is interesting to note how close the frequencies of CO associated with the defect (step or kink) and terrace species lie (Figure 16) despite their differences in sticking probability and desorption temperatures. A RAIRS study of CO adsorbed on a vicinal copper surface[96] also showed relatively small perturbations of $\nu(CO)$ for CO species adsorbed at defects from the value observed for CO adsorbed on the terrace.

The local chemical environment experienced by an adsorbate molecule can also clearly influence the frequency of internal vibrational modes of adsorbed molecules. The modification of adsorption behavior and

Figure 17. Models of small supported platinum clusters with corner and edge atoms highlighted together with their mimicked counterparts on the (a) stepped Pt(533) and (b) kink-stepped Pt(432) surfaces.

surface reactions through coadsorption of other molecules on single-crystal surfaces provides an increasingly important method of studying mechanisms of catalytic poisoning and promotion, and vibrational spectroscopy provides an ideal tool for investigating the concomitant perturbations in local chemical environment. The high spectral resolution of RAIRS is particularly advantageous where the chemically induced vibrational shifts are relatively small or when the mechanism of the chemical perturbation, e.g., the range over which it is felt, are being investigated. The effect of preadsorbed oxygen on $v(NO)$ of linearly adsorbed NO on Ru(001)[60] is shown in Figure 18: NO adsorbed on an oxygen-covered $[p(2 \times 2)Ru(001)]$ surface has a stretch frequency $v(NO) = 1846 \text{ cm}^{-1}$, shifted from $v(NO) = 1819 \text{ cm}^{-1}$ on the clean

Table 6. CO Adsorbed on Pt(111), Pt(533), and Pt(432)[a]

Surface	Surface atom densities (cm^{-1})			CO coverages from TPD[b]		CO vibrational frequencies (cm^{-1})		
	Terrace	Step	Kink	Terrace	Step/kink	Terrace[c]	Step	Step/kink
Pt(111)	1.51×10^{15}			>0.95		2093–2110		($\sim$2068 cm^{-1})[d]
Pt(533) Pt 4(111) × (100)	1.22×10^{15}	4.06×10^{14}		0.73	0.27	2085–2097	2067–2078	
Pt(432) Pt 4(111) × (31$\bar{1}$)	1.19×10^{15}	2.39×10^{14}	2.39×10^{14}	0.70	0.30	2080–2094		2072–2077

[a] After Ref. 93.
[b] Expressed as a fraction of maximum coverage (0.05) for Pt(533) and Pt(432).
[c] In phase mode of the vibrationally coupled system (see text).
[d] Frequency of CO molecules assumed to be associated with a low concentration of defect sites.

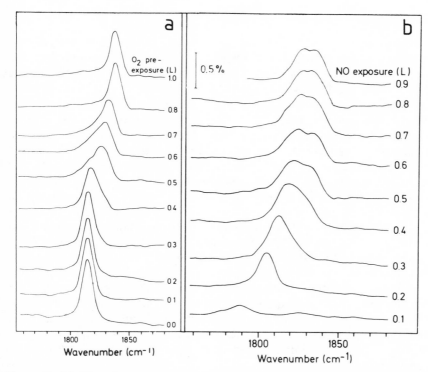

Figure 18. (a) The effect of preadsorbed oxygen on the RAIRS spectrum of linearly adsorbed NO on Ru(001). (b) The sequential adsorption of No on clean and oxygen covered areas of a partial oxygen covered surface. (After Ref. 61).

surface. Note the coexistence of two peaks in RAIRS at intermediate oxygen coverages (Figure 18a) due to NO adsorbed on clean and oxygen-covered regions of the surface. The consecutive adsorption of NO in clean and then oxidized areas on such a partially oxygen-covered surface is demonstrated in the RAIRS spectra of Figure 18. Similar frequency shifts due to coadsorbed oxygen are observed on the supported Ru catalyst: $v(NO) = 1810 \text{ cm}^{-1}$ on the fully reduced sample and the frequency shifted upwards to $v(NO) = 1880 \text{ cm}^{-1}$ on the oxidized surface.[97] The effect of coadsorbed oxygen on the $v(CO)$ vibrational frequency of linearly bound CO has also been the subject of a RAIRS investigation on Cu(111) and Cu(110) surfaces.[98] As in the above case, relatively small perturbations of $v(CO)$ were observed ($\sim 40 \text{ cm}^{-1}$) and a close comparison with similar effects for CO adsorbed on clean and oxidized supported Cu catalysts could be made.

3.3. Coverage-Dependent Frequency Shifts

Assuming that the adsorbed molecule is otherwise structurally or electronically unperturbed with respect to its gas phase counterpart, the mere fact that one or more of the adsorbate molecules' atoms becomes attached to a rigid (or vibrating) substrate atom results in a mechanical shift of one or more of its internal normal mode frequencies to higher energy. The simplest model[26] yields a value for the upward shift in $v(C–O)$ of ~50 cm^{-1} from its gas phase value of 2143 cm^{-1}. Vibrational frequencies of adsorbates on metals are invariably, however, shifted downward in frequency from their gas phase values owing to the effect of the image dipole and the various chemical effects, which allow, for example, vibrational frequencies to be used as vibrational fingerprints. In the absence of additional adsorbate–adsorbate interactions at finite coverage θ, mechanical renormalization, image dipole, and chemical effects determine the singleton frequency ω_0 of the adsorbate mode (i.e., the frequency of the adsorbate peak as $\theta \to 0$). The latter two contributions also influence the coverage-dependent frequency shift, halfwidth, line shape, and intensity of adsorbate bands. These phenomena have been the subject of some theoretical effort, and while they are discussed in more detail elsewhere,* a brief background is included insofar as it relates closely to high-resolution RAIRS data in investigations of dipole coupling, chemical shifts, vibrational line shapes, and absorption intensities.

The shift in adsorbate normal mode frequency from ω' (shifted already from the gas phase as a result of mechanical renormalization and effects chemical in origin) as a consequence of its image dipole to the singleton frequency ω_0 is given, in the absence of vibrational damping, by[99]

$$\omega_0 = \omega'\left(1 - \frac{\alpha_v/4d^3}{1 - \alpha_e/4d^3}\right)^{1/2} \tag{17}$$

where α_v and α_e are the molecular and electronic polarizabilities and d is the distance of the dipole to its image plane. The metal surface is regarded as a sharp discontinuity in the dielectric response between metal and vacuum, the metal screening a point dipole. Despite the approximations and sensitivity of the resulting equation to the value of d, the effect of the dipole image will be to lower the adsorbate normal mode frequency ω', and therefore to some extent offset the upward frequency shift of mechanical renormalization.

---

* The reader is referred to Chapter 1 and the recent review by Willis et al.[26] and Hofmann.[27]

In addition to changing the resonant frequency of the adsorbed dipole, the effect of the self-image field on the mechanical dipole is to increase the oscillation strength through an enhancement in polarizability $\bar{\alpha}(\omega)$ which can be estimated in the limit of a perfect metal, or imperfect screening using the Thomas–Fermi approximation.[26] The effective polarizability is expressed as a constant term $\bar{\alpha}_e$ plus the contribution at a single resonance $(\bar{\alpha}_v)$ which occurs at the renormalized frequency ω_0 in equation (16). A value for $\bar{\alpha}_v$ can be obtained directly from absolute intensity measurements since $\bar{\alpha}(\omega)$ determines the intensity of the excitation process.

The vibrational frequencies of adsorbate modes measured experimentally in RAIRS are invariably found to be a function of coverage even when there is no change in adsorbate configuration. There are two components to this generally smooth coverage-dependent frequency shift which can be distinguished. The first is due to vibrational coupling between adsorbed species as a result of the electrostatic dipole interaction, or as suggested by Moskovits and Hulse[101] via short range, through metal bonding interactions. The second is known as the static shift, and comprises both of coverage-dependent chemical and static dipole[102] contributions. The frequency of a vibrational band at any finite coverage, $\omega(\theta)$, is shifted by an amount $\Delta\omega(\theta)$ from the singleton frequency ω_0 by a contribution $\Delta\omega_D(\theta)$ due to dipole coupling, and an amount $\Delta\omega_S(\theta)$ due to the static shift:

$$\Delta\omega(\theta) = \omega(\theta) - \omega_0 = \Delta\omega_D(\theta) + \Delta\omega_S(\theta) \qquad (18)$$

Fortunately, both the static and dipole contributions to $\Delta\omega(\theta)$ can be separated experimentally in RAIRS, as can the singleton frequency ω_0, which is the frequency of the band as $\theta \to 0$. The origin and nature of the two components $\Delta\omega(\theta)$ and $\Delta\omega_D(\theta)$ have been the subject of some theoretical effort, and their measurement in RAIRS provides a source of potentially valuable information concerning the nature of the chemisorption bond and adsorbate interactions generally.

3.3.1. Dipole Coupling

The electric field experienced by the adsorbate molecule, influenced already by the image dipole field [equation (17)], is also modified by the field induced by neighboring molecules and their dipole images. Of the two normal modes of the coupled ensemble of identical oscillators, only the in-phase collective mode is dipole active (the antiphase mode is inactive) and hence observed in RAIRS. Within the wavelength of the external electric field, a large number of molecular dipoles oscillating in

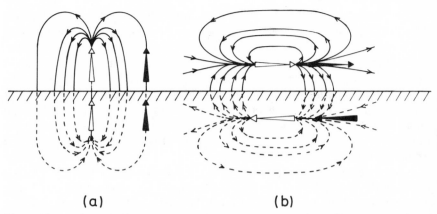

(a)   (b)

Figure 19. A schematic illustrating the influence of the electrostatic field of a neighboring dipole (and its image) on an identical dipoles' local environment causing either (a) depolarization or (b) polarization.

phase are absorbing radiation. Figure 19 shows two important orientations of in phase oscillating dipoles at the surface, and the effect of the electric field of a neighboring dipole in each case. The neighboring oscillator may stiffen (shift to higher frequency) in case (a), or soften (shift to lower frequency) in case (b), the force constant (frequency) of the vibrational mode. Since the surface selection rule requires that only dipoles oscillating perpendicular to the surface are observable in RAIRS, i.e., dipole orientated as in Figure 19a, nearly exclusively upward dipole coupling shifts with increasing coverage have been observed experimentally in RAIRS, although it will be seen that the downward shift that can be expected from the dominance of "end to end" coupling depicted in Figure 19b has also been observed in RAIRS indirectly.[71] The extent of the coupling will depend on the effective molecular ($\bar{\alpha}_v$) and electronic ($\bar{\alpha}_e$) polarizabilities of the molecule in the direction of the vibration, the number of nearby oscillators, and their distance. The dipole coupling on metals also involves the interaction with the image dipole of the neighboring dipoles. For identical oscillating dipoles arranged on a regular two-dimensional mesh described using Cartesian coordinates x_m and y_m defined relative to some arbitrary origin on the mesh, the singleton frequency ω_0 shifts to a value $\omega_D(\theta)$ given by[26]

$$\omega_D(\theta) = \omega_0\left(\frac{1 + c\bar{\alpha}_v U_0}{1 + c\bar{\alpha}_e U_0}\right)^{1/2} \tag{19}$$

where $\Delta\omega_D(\theta) = \omega_D(\theta) - \omega_0$ and U_0 is the two-dimensional lattice sum which, unlike the three-dimensional case, converges in the absolute limit.

c is the partial coverage and is equal to unity when the mesh defined by U_0 is complete. In this analysis [equation (19)] the self–image frequency shift is included in the singleton frequency ω_0 [equation (17)] while any other self-image effects are included as the effective $\bar{\alpha}_v$ and $\bar{\alpha}_e$ values introduced earlier. In the case of dipoles oriented perpendicular to the surface (Figure 19a) the lattice sum is given by

$$U_0^a = \sum_m \left\{ \frac{1}{(x_n^2 + y_m^2)^{3/2}} + \left[\frac{1}{(x_n^2 + y_m^2 + 4d^2)^{3/2}} - \frac{12\,d^2}{(x_n^2 + y_m^2 + 4d^2)^{5/2}} \right] \right\}$$

(20)

including direct dipole–dipole interactions (first term) and interaction with other image dipoles (second bracketed term). Since U_0^a is positive [equation (20)] ω_0 increases with increasing coverage c [equation (19)] and is consistent with the upward dipole frequency shift of the majority of dipole frequency shifts measured in RAIRS due to interaction of perpendicular dipoles (Figure 19a). In the case of dipoles lying parallel to the surface (Figure 20b) the lattice sum has the form[71]

$$U_0^b = \sum_m \left\{ \left[\frac{1}{(x_m^2 + x_m^2)^{3/2}} - \frac{3x_m^2}{(x_m^2 + x_m^2)^{5/2}} \right] - \left[\frac{1}{(x_m^2 + x_m^2 + 4d^2)^{3/2}} - \frac{3x_m^2}{(x_m^2 + y_m^2 + 4d^2)^{5/2}} \right] \right\}$$

(21)

with the two bracketed terms being due to direct and image interactions, respectively: U_0^b can be negative or positive [equation (21)] depending on the possible dominance of either "end to end" dipole coupling (Figure 19b) or coupling between parallel dipoles lying parallel to the surface

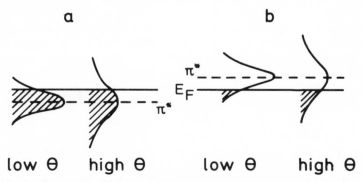

Figure 20. A schematic showing the variation in occupation of a broadened $2\pi_b^*$ state for the cases in which the state lies (a) below and (b) above the Fermi level of the metal (after Ref. 106).

(out of the plane of Figure 20b). In this case, therefore, the value of d and the choice of mesh will determine whether the coverage-dependent frequency shift will be positive or negative.

The difficulty in determining d, the approximation of a point dipole, the method of incorporating the image dipole shift, and the correct choice of $\bar{\alpha}_v$ and $\bar{\alpha}_e$ in such an analysis have all been the subject of some discussion, and vibrational coupling via the substrate has also been suggested[101] to occur. While a conclusive outcome of such deliberations requires perhaps a greater base of experimental RAIRS data for comparison with theory, many of the measurements already made and described below appear consistent with the type of dipole coupling analysis described, although frequency perturbations via the through metal coupling would appear to be difficult to distinguish from those of the dipole interaction.[26]

3.3.2. The Static Shift

A change in the electronic density distribution between the adsorbate and the metal can accompany a change in coverage, and these in turn can give rise to changes in the vibrational frequency of internal and adsorbate–metal modes. Such coverage-dependent shifts, consequently, depend in a detailed fashion on the nature of the adsorbate–metal bonding. In the case of a molecule such as CO on a metal the adsorbate metal bond is formed by charge transfer from the 5σ molecular orbital of CO to the metal, with backdonation of the metal d electrons into the unoccupied $2\pi^*$ orbital of CO, i.e., a backbonding scheme that will be similar to adsorbed NO and analogous to that used to explain the bonding in organometallic carbonyls (and nitrosyls). Since the CO 5σ orbital is only weakly bonding with respect to the C–O bond, and $2\pi^*$ is strongly antibonding, it is generally believed that changes in the extent of backdonation from the metal as a function of coverage dominates the chemical contribution to the observed coverage dependent static frequency shift invariably measured in RAIRS. The original model for CO along these lines developed to explain upward vibrational shifts for CO observed on supported metal catalysts was proposed by Blyholder,[103] who argued that as coverage increased, the increasing competition for metal d electrons would lead to a reduced backdonation per molecule, an increased C–O force constant, and a concomitant increase in vibrational frequency. This rationale for an upward frequency shift with coverage is less appropriate for noble metals such as copper on which backdonation would be expected to be far less important. In this case Pritchard and co-workers have offered an alternative model to explain the downward chemical shift observed in RAIRS experiments of isotopic mixtures on copper surfaces,[104,105] arguing the importance of bonding effects involving the 5σ CO orbital when backbonding is unimportant due to the energetic separation of the d band. An alternative explanation[106] that has much in common with the arguments of Blyholder was developed to explain the positive, zero, and negative chemical shifts observed in RAIRS. It is based on the premise that the existence, or otherwise, of π^* backbonding depends on the relative energies of the Fermi level and the CO $2\pi^*$ level, allowing backbonding from sp metal density of states as well as from the d band when the latter is not available. Following cluster calculations[107] it is assumed that interaction with the metal splits the $2\pi^*$ orbital into two states, $2\pi_a^*$ and $2\pi_b^*$, which are antibonding and bonding with respect to the metal. While the $2\pi_a^*$ level is expected to lie well above the Fermi level, the $2\pi_b^*$ overlaps with E_F, and because of its finite broadness two situations can be distinguished: In the case of strong backbonding, the center of gravity of $2\pi_b^*$ lies below E_F (Figure 20a) while in the case of weak backbonding it lies above E_F (Figure 20b). It was then argued that an increase in broadening of $2\pi_b^*$ with increasing coverage as a result of increasing overlap of the orbitals (band formation) causes $2\pi_b^*$ to become less filled if its center lies below E_F (a) and increasingly filled if above E_F (b) causing a concomitant frequency increase or decrease, respectively. The model depicted in Figure 20 for transition (a) and noble (b) metals is therefore consistent with the conventional wisdom that backdonation is important in the former case and relatively unimportant in the latter. Most attention has been given to the mechanism giving rise to chemical shifts in adsorbed CO and NO because they have been studied most extensively through the use of mixed isotope experiments in RAIRS. The detailed information concerning adsorbate–metal bonding from such experiments, however, can clearly be obtained for other molecules with the extension of the mixed isotope method to a wider range of adsorption systems. Only through the extension of experiments involving isotopic mixtures of NO and CO to other surfaces will a better understanding be possible of such chemical shifts, and perhaps also of any additional contribution to the static shift from the modification in the throughspace static dipole interaction with changes in the internuclear separation.[102]

The dipole and static contribution to the coverage dependent frequency shifts can be separated experimentally in RAIRS by the use of isotopic mixtures, after similar experiments on supported metal particles.[108] While neighboring molecules with identical frequencies experience dipole coupling, molecules with sufficiently different frequencies can be considered vibrationally decoupled. The frequency difference between any pairs of isotopically substituted molecules, for example, $^{13}C^{16}O/^{12}C^{16}O$, $^{12}C^{18}O/^{12}C^{16}O$ and $^{14}N^{16}O/^{15}N^{16}O$, is sufficient to assume complete vibrational decoupling of the internal mode. Therefore, an adsorbed molecule, say $^{12}C^{16}O$, surrounded entirely by molecules of another isotope, say $^{13}C^{16}O$, at any finite coverage θ experiences no frequency perturbation due to vibrational coupling: This situation is realized in practice as the relative concentration of $^{12}C^{16}O$ (in $^{13}C^{16}O$) tends to zero, and is known as the dilution limit. This effect is illustrated in Figure 21, where for CO adsorbed on Pd(100) at constant coverage θ the "switching off" of the dipole coupling between ^{12}CO molecules causes the frequency decrease with reducing partial coverage. At the dilution limit, however, the $^{12}C^{16}O$ molecules remain under the influence of any

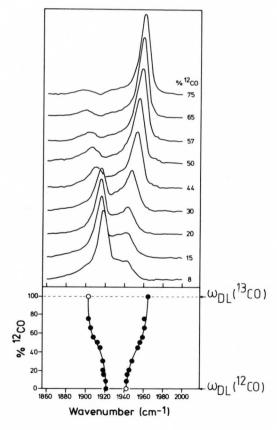

Figure 21. The experimental separation of the static and dipole contributions to the coverage-dependent chemical shift using isotopic mixtures illustrated for the case of CO on Pd(100) (after Ref. 79.) Coverage is kept constant and the frequencies of both ^{11}CO and ^{13}CO plotted as a function of isotopic composition to obtain their frequencies at the dilution limit $\omega_{DL}(\theta)$.

static frequency perturbation due to neighboring $^{13}C^{16}O$ molecules since they are in any static dipole or chemical sense indistinguishable from $^{12}C^{16}O$. In other words, at any coverage θ, the frequency shift $\Delta\omega_1(\theta)$ away from the singleton value ω_0 experienced by a molecule in its *dilution limit*, where its frequency is $\omega_L(\theta)$, is only due to the static contribution:

$$\Delta\omega_L(\theta) = \omega_L(\theta) - \omega_0 = \Delta\omega_s(\theta) \qquad (22)$$

Since the singleton frequency ω_0 and total frequency shift $\Delta\omega(\theta)$ can be measured, both static ($\Delta\omega_s$) and dipole ($\Delta\omega_0$) contributions to $\omega(\theta)$ can be identified [equations (18), (22)]. The simple treatment of dipole interactions of coupled oscillators[108] has been extended using self-consistent models of the local oscillatory electric field by several authors and are reviewed elsewhere.[26] There are some dangers in the use of isotopes with frequencies as close as $\sim 50\ \mathrm{cm}^{-1}$ in some cases,[106] but the error induced due to cross coupling in the experimental separation of $\Delta\omega_0(\theta)$ is fortunately only of the order of $\sim 2\ \mathrm{cm}^{-1}$.

It was the use of isotopic mixtures in infrared experiments on supported platinum that enabled Hammaker *et al.*[108] to qualitatively demonstrate the importance of dipole coupling on the coverage-dependent frequency of adsorbed CO. In the RAIRS experiments of King and co-workers[43,110] it was quantitatively shown that the upward frequency shift $\Delta\omega$ ($\theta = 0.25$) $\approx 35\ \mathrm{cm}^{-1}$ [43] of $\nu(C–O)$ from the singleton value ($\omega_0 = 2063\ \mathrm{cm}^{-1}$) which accompanies the adsorption of linearly bound $^{12}C^{16}O$ on platinum to saturation coverage ($\theta_{LINEAR} = 0.25$, $\theta_{BRIDGE} = 0.25$) at 300 K, could be completely accounted for in terms of dipole coupling.[110] The frequency of the ^{12}CO band in the dilution limit (ω_1) was found to be the same as ω_0, and hence [equation (22)] $\Delta\omega_s = 0$ and [equation (18)] $\Delta\omega_D \approx 35\ \mathrm{cm}^{-1}$. The isotopic method of separation has subsequently been used on a variety of single-crystal surfaces, but nearly exclusively for adsorbed CO (Table 7), and in most of these studies $\Delta\omega_0(\theta)$ and $\Delta\omega_s(\theta)$ have been determined at a series of coverages yielding their functional dependence. As alluded to earlier, both negative and positive changes in static shift for CO on various metals have been measured, leading to some discussion as to the position of the $2\pi_b$ level with respect to E_F, assuming $\Delta\omega_s(\theta)$ due exclusively to chemical contributions. Although no isotopic measurements have been made, strong frequency shifts with coverage observed in, for example, adsorbed formate have also been suggested as being due to the crossing of an affinity level of the adsorbate with E_F due to lateral interactions at high coverages.[71] Dipole shifts are always positive for modes perpendicular to the surface (Table 7) as one would expect from equations (19)

Table 7. Dipole and Static Shifts Separated Using Isotopic Mixtures

Adsorbate system	Adsorbate configuration	Coverage θ	Dipole shift $\Delta\omega_D(\theta)$ (cm^{-1})	Static shift $\Delta\omega_s(\theta)$ (cm^{-1})	Reference
CO/Cu(100)	Linear	0.5	43	−34	48
CO/Cu(111)	Linear	0.33(0.52)^a	25 (53)	−27 (−63)	42
CO/Cu(110)	Linear	0.5	50	−44	106
CO/Pt(100)	Linear	0.75^a	35	0	112
CO/Pt(111)	Linear	0.25^a	36	0	110,112
NO/Pt(111)	Bridge	~0.1^c	5	18	109
	Linear	0.25	33	−13	109
CO/Pd(100)	Bridge	0.8^a	35	60	79

^{a} Not identical absorption sites.
^{b} Total coverage = 0.5 (linear 0.25, bridge 0.25).
^{c} No accompanying LEED pattern.

and (20), but the indirect measurement of dipoles parallel to the metal surface through a combination mode in adsorbed formate[71] (see also Section 3.1) revealed a downward frequency shift at low coverages suggested as due to a dominance of "end to end" coupling [equation (21)] while the remaining modes simultaneously shifted to higher frequency as one would expect since their interacting dipoles were perpendicular to the surface.

The static and dipole shifts of two configurations of an adsorbate on the same surface has also been measured, namely, bridge and linear species of NO on Pt(111).[109] In the single isotope experiment (Figure 22b) it was demonstrated that initial adsorption of ^{14}NO produces exclusively the bridge species [$v(N-O) = 1476$–1498 cm^{-1}] converting at higher coverages to yield an ordered overlayer containing only linear adsorbed NO[$v(N-O) = 1700$–1725 cm^{-1}]. The two $v(N-O)$ peaks both show coverage-dependent frequency shifts (Figure 22a) which could be separated, using isotopic mixtures of ^{14}N^{16}O/^{15}N^{16}O, into static and dipole components (Figure 22a). Both linear and bridge species exhibit positive dipole shifts, with the shift exhibited by the bridge species unexpectedly small although it predictably fell to zero as its partial coverage was reduced to zero during its conversion to the linear species ($0.12 > \theta > 0.16$). A residual static contribution to the frequency shift of bridge NO in this coverage range (Figure 22a) was ascribed to a chemical interaction between the linear and bridged NO species, and labeled an intermolecular (as opposed to an intramolecular) static shift.[109] More importantly, perhaps, it was shown that the two adsorbed configurations exhibited chemical shifts in opposite directions, the most strongly

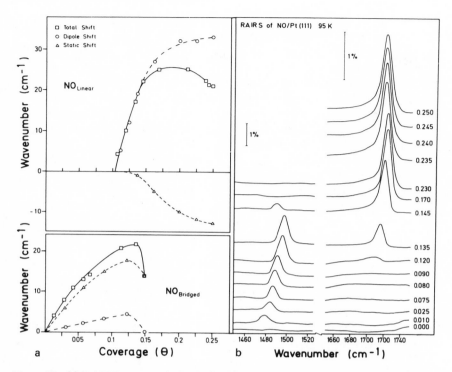

Figure 22. (a) RAIRS spectra of adsorbed NO on Pt(111) showing a bridged species at low coverage converging to a linear species at high coverages. The coverage-dependent frequency shift (b) for both species has been separated into its static and dipole components.

bridging species showing a positive shift (associated with strong back-bonding) and the linear a negative shift.

While the frequency separation between isotopes may be sufficient to assume little frequency perturbation through dipole coupling between neighboring isotopes, sufficient coupling invariably arises to cause an effective transfer of intensity from the low to high frequency component band, and at high coverages (where coupling is greatest) this can even give rise to intensity inversion. This effect is understandable within the framework of both dipole and through-metal coupling mechanisms[26] and is expected and observed not only in cases of random mixtures of isotopes, but wherever two species with similar singleton frequencies are intermixed on the surface. The observation or otherwise of intensity borrowing in such cases can be used as evidence of complete intermixing of the adsorbate phase,[98] or for the existence of separate domains.[61]

Vibrational coupling, which was shown to dominate the frequency behavior of CO on platinum surfaces, also gives rise to the frequency changes and intensity borrowing observed in RAIRS (Figure 16) for CO on stepped (and kink-stepped) surfaces of platinum.[92,93] Despite the fact that two distinct species of CO are desorbed in the TPD spectrum at high coverages which can be associated with the occupation of step and terrace sites, only a single peak is observed in RAIRS. At low coverages, a single peak associated with adsorption at the step shifts to higher frequency because of vibrational coupling in a one-dimensional domain (Figure 16). At intermediate coverages, a second peak appears as terrace sites are occupied, forming immediately two-dimensional domains consisting of molecules adsorbed at both step and terrace sites. The increased vibrational coupling in a two-dimensional over a one-dimensional lattice sum [equation (20)] gives rise to an increase in the degree of vibrational coupling and the concomitant increase in frequency. The observed bands from the adlayer in this midcoverage range (Figure 16) are simply the in-phase (high-frequency) and out-of-phase (low-frequency) modes of an ensemble of molecules containing both sorts of CO, or viewed alternatively, of an ensemble containing both one-dimensional and two-dimensional subdomains. Intensity is then increasingly transferred to the high-frequency component as coverage is increased until it is ultimately all in the high-frequency mode. The identical effect is also discussed by Hollins et al.[95] in connection with CO adsorption on defect sites of Cu(110).

The dependence of frequency on lateral interactions has also been used to study order–disorder transitions and the phase boundaries between lattice gas, island, and condensed phase. For example, the vibrational frequency of the adsorbate band shifted by the adsorbate–adsorbate interactions experienced in an ordered island, relative to the singleton value ω_0, has been used to estimate island size,[111–114] enabling a thermodynamic study of the equilibrium between the island and lattice gas on the surface.[113] In addition to the information concerning chemical bonding from coverage-dependent static frequency shifts in RAIRS, order–disorder phenomena as probed through the dipole shift will be increasingly a subject of investigation in RAIRS in combination with one or more complementary surface sensitive techniques.

3.4. Natural Half-Widths and Line Shapes

RAIRS is an ideal vibrational spectroscopy for the study of vibrational line shapes and half-widths of adsorbate modes, not only because of its high spectral resolution and sensitivity, but also because surface temperature, coverage, and ambient pressure above the surface are relatively easily varied during measurements on highly ordered, well-characterized surfaces. The half-widths (FWHM) of vibrational bands of gas phase molecules lie typically in the range $10^{-3} <$ FWHM $(cm^{-1}) < 10^{-1}$ while the internal modes of adsorbed molecules on metals (the adsorbate/substrate mode is not yet accessible in RAIRS) have half-widths in the range $5 <$ FWHM $(cm^{-1}) < 50$. The many mechanisms that have been identified as potentially giving rise to this increased half-width can be subdivided into one of two classes, namely, homogeneous and inhomogeneous line broadening mechanisms, and a more detailed description of their origin and importance can be found in Chapter 1; only a brief outline is included below.

Inhomogeneous line broadening, as its name implies, is a result of an inhomogeneity in the distribution of the oscillator frequencies being measured. This is commonly due to a lack of order in the adsorbate overlayer, giving rise to an inhomogeneous distribution of distances between interacting oscillators, or can be due to the occupation of nonequivalent adsorption sites. Both of these effects give rise to a heterogeneity in the local environment of the oscillators and can give rise also to an asymmetry in line shape. The random adsorption of molecules into a lattice gas,[113] in which adsorbed molecules maximize their intermolecular distances through repulsive interaction, gives rise to a broadened, often symmetric peak at low coverages. Molecules adsorbed in islands lacking a high degree of order at low coverage can also give rise to line broadening. Where adsorbate island and lattice gas are both present in equilibrium, two separate (but often overlapping) peaks can invariably be distinguished.[111-114] Decreasing island size increases the relative weight of oscillator frequencies from molecules at the island's edge, and their difference (with respect to molecules inside the island) in local environment manifests itself in an asymmetrically broadened peak. This phenomenon has been treated semiquantitatively by Pfnür et al.[111] using a modified or effective dipole lattice sum. Vibrational coupling in a randomly adsorbed immobile layer can also give rise to an asymmetric RAIRS peak, and, following theoretical calculations on random oscillators,[115] one may expect it to manifest itself in a low-frequency absorption tail. The above inhomogeneous broadening effects, all of which can be associated with heterogeneous dipole coupling, are most pronounced at low adsorbate coverages. Broadening as a result of adsorption at surface defects or impurities provides a mechanism of chemically induced inhomogeneity and is also most pronounced at low coverages. The sharpest bands observed in RAIRS occur at intermediate and high coverage on the formation of well-ordered commensurate overlayers, and linewidths down to $4.5\,\text{cm}^{-1}$ have been reported,[48] although they are more generally of the order of $6\text{--}8\,\text{cm}^{-1}$. In the study of homogeneous line broadening, one must be able to assume that under such conditions the linewidth is no longer limited by inhomogeneous line broadening effects. At high adsorbate coverages an additional inhomogeneous broadening may occur if the still well-ordered adsorbate layer becomes incommensurate with the substrate periodicity, leading to the occupation of a range of inequivalent adsorbate sites. The presence, or otherwise, of such broadening under these conditions in RAIRS experiments, together with evidence extracted from coverage-dependent intensity changes (Section 3.5), has been exploited[47,79,80,83,111,114] in discussions concerning the interpretation of LEED patterns of densely packed overlayers. These can often be interpreted in terms of a site specific rather than an incommensurate real space model of adsorption (see, for example, Ref. 117). Through the measurement of half-widths and line shapes in the inhomogeneous line-broadening limit, RAIRS provides a valuable tool for the detailed study of adlayer growth and order–disorder phenomena, and is nicely illustrated in the results of Pfnür et al.[111] for the adsorption of CO on Ru(001). Figure 23 shows the change in half-width of $\nu(\text{C--O})$ as a function of coverage for adsorption at 300 K. In the coverage region $0.2 < \theta < 0.33$, islands of a commensurate $(\sqrt{3} \times \sqrt{3})R30°$ ordered overlayer are present, and the natural halfwidth of the band (i.e., the measured peak width corrected for instrumental broadening) is $8\,\text{cm}^{-1}$. At lower coverages, the inhomogeneity of dipole coupling in small islands or the influence of some

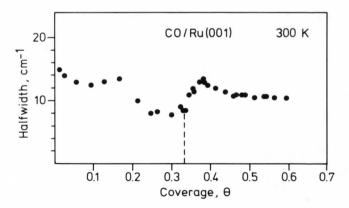

Figure 23. The half-width of $\nu(\text{CO})$ as a function of CO coverage on Ru(001) (after Ref. 111).

surface (chemical) heterogeneity produces a half-width of up to $14\,\text{cm}^{-1}$. At high coverage, the gradual disappearance of the extra LEED features associated with the $(\sqrt{3} \times \sqrt{3})R30°$ structure indicates a decrease of order in the adlayer at 300 K which is reflected in an increased halfwidth of $10\text{--}12\,\text{cm}^{-1}$. A similar order–disorder transition and concomitant change in half-width is observed at lower adsorption temperatures as the coverage is increased beyond that corresponding to a $(2\sqrt{3} \times 2\sqrt{3})R30°$ ordered overlayer.[111] Inhomogeneous broadening has also been used as a monitor of such phenomena for CO adsorbed on palladium surfaces, and these observations are described in detail in Ref. 27. The inhomogeneity of vibrational coupling in the random mixtures of adsorbed isotopes has also been the subject of some theoretical and experimental effort[118] and is invariably observed in experiments in which isotopic mixtures are used to separate dipole and static contributions to vibrational frequency shifts (Section 3.3).

Homogeneous line broadening is due either a shortening of the lifetime τ of the vibrationally excited state, which requires a mechanism allowing dissipative decay, or is due to pure dephasing which requires no net energy transfer. Mechanisms of dissipative decay and pure dephasing, and their effect on half-width and line shape, are discussed in more detail in Chapter 1 and in Refs. 119, 120. A mechanical coupling of the adsorbate mode to the substrate phonons constitutes a potential mechanism for dissipative decay when the frequency of the adsorbate mode lies below the phonon band edge. When the frequency of the adsorbate mode lies well above that of the substrate phonon modes, as is the case for most internal modes of adsorbates and those at present accessible in RAIRS, anharmonic coupling is weak and cannot easily explain the observed half-widths.[121] The quantum mechanical description of the classical image dipole effect (Section 3.3) reveals a second potential mechanism of dissipative decay through electron–hole pair creation or plasma oscillations which more effectively reduce the final state lifetime of high-frequency vibrational modes and hence can produce a greater degree of broadening.[119] The value of the final state lifetime one can expect through the creation of electron–hole pairs through such a mechanism appears to depend upon the model applied and varies from $\tau \approx 10^{-10}\,\text{s}$ (infinite barrier model[122]), $\tau \approx 10^{-10}\,\text{s}$ (finite step barrier model[123]) and $\tau \approx 5 \times 10^{-12}\,\text{s}$ (Lang–Kohn density profile[124]). The latter model is indeed capable of explaining the lowest limiting natural half-width of RAIRS peaks in ordered overlayer. An alternative mechanism for electron–hole pair creation through a more short-range interaction was suggested[125] to explain the linewidths of adsorbed CO. Vibrational damping via electron–hole pair excitations is suggested to occur through charge oscillations between a partly filled adsorbate induced resonance (namely, $2\pi^*$ in adsorbed CO) and the metal. Since the final state lifetime is shown to be inversely proportional to the square of the charge fluctuation, clearly the degree of line broadening will be sensitive to the exact placing of the affinity level with respect to the Fermi level. Estimates of the final state lifetime using this model ($\sim 10^{-12}$ s) have been used to explain the observed half-widths in RAIRS peaks of CO on Cu(100),[48] the change in FWHM as a function of the extent of backbonding to the metal $2\pi^*$ level as a function of adsorption site and coverage for CO on Pd and Ru surfaces,[27] and has been extended to explain RAIRS data for molecules other than CO.[14] In the latter study the RAIRS technique was employed to investigate the mechanisms of dissipative decay in adsorbed layers where it is explicitly assumed that the linewidth is limited only by mechanisms of homogeneous broadening.[126] Persson and Ryberg report the vibrational damping due to final state lifetime of the v(C–H) and v(C–D) modes (symmetric v_s and asymmetric v_a) of the adsorbed methoxy species and its deuterated isotope on Cu(100). Assuming that the observed natural widths are determined by the lifetime of the final state in dissipative decay, its value τ was calculated from the natural halfwidth at $\theta = 0.2$ for each band v_s(C–H), v_s(C–D), v_a(C–H), and v_a(C–D). The ratio τ_s(C–D)$/\tau_s$(C–H) and τ_a(C–D)$/\tau_a$(C–H) was found to be equal within the limits of experimental error to the ratio of the reduced masses m^*(D)$/m^*$(H), as one would expect[125] for vibrational damping caused by electron–hole pair creation. The authors go on further to explain the preferred short-range model of electron–hole pair creation by periodic fluctuation of charge between the metal and the adsorbate $2\sigma^*$ affinity level [the antibonding component of the molecular orbital combination $C(2sp^3)$ and $H(1_s)$], a resonance that is assumed to straddle the metal Fermi level. Recently, temperature-dependent broadening of internal modes has been observed in RAIRS[47,83] associated with a so-called exchange coupling, which similarly determines vibrational line shapes in molecular crystals.[127] The latter line broadening mechanisms is homogeneous, and is a pure dephasing process.[120] The temperature dependence of the bridge bands of CO on Pt(111)[47] and Cu(111)[83] is described briefly in Section 3.2 and ascribed to the facile interconversion of bridge and threefold CO species in a process similar to that of fluxionality in organometallic chemistry. It was also noted[47,83] that at constant coverage this process appeared to be reversible and accompanied by an increase in linewidth and a shift in frequency. This is demonstrated in Figure 24a for the $c(4 \times 2)$ ordered overlayer of CO on Pt(111). Note the coalescence of broadening bands at higher temperatures. This was ascribed[47] to

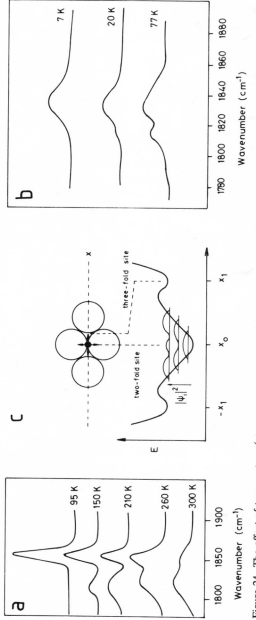

Figure 24. The effect of temperature (at a constant coverage of $\theta \sim 0.5$) on the RAIRS spectrum of CO in the bridging $\nu(CO)$ region on (a) Pt(111) and (b) Cu(111). A threefold species [$\nu(CO) \sim 1810\,cm^{-1}$] is populated in addition to a bridged species [$\nu(CO) \sim 1850\,cm^{-1}$] at higher temperatures via a large-amplitude frustrated translation along the direction x (c). The latter mode also couples with $\nu(CO)$ causing the broadening and shifting of the observed peaks through a pure dephasing mechanism. (After Refs. 147, 83).

coupling of the higher frequency $v(C-O)$ band being measured to the high-amplitude frustrated translation, responsible also for the surface fluxionality. As the temperature is raised, excited vibrational states of the frustrated translation will become increasingly occupied, with the result that the probability of finding the molecule away from the equilibrium position rises steeply (Figure 24b). The resultant averaging of the resultant force constant of the C–O bond over the potential curve of the frustrated translation gives rise to the observed peak shift, the anharmonic coupling between the two modes [$v(CO)$ and the frustrated translation] also accounting for the strong temperature-dependent broadening. There were clearly similarities[47] between the temperature-dependent shift and broadening on the surface and that experienced in the coupling of low-frequency librations to higher-frequency modes in molecular crystals.[127] The two ingredients necessary for exchange-coupling dephasing, i.e., the existence of a large amplitude mode and anharmonic coupling,[120] were clearly present. Pure dephasing mechanisms provide another means of homogeneous line broadening for adsorbate modes, and has indeed also been offered as an explanation for the temperature dependence of the $v(CO)$ linewidth for adsorbed CO on Ni(111).[100]

3.5. Intensities

Like the frequency, half-width, and line shape of the adsorbate band, the integrated intensity is also influenced by the interaction of the adsorbate dipole with the substrate, and, at finite coverage, by the interaction with the surrounding adsorbate molecules. In the absence of adsorbate–adsorbate interactions, one expects a linear increase in integrated absorption intensity with coverage, the value of which is conveniently expressed in terms of the effective dynamic charge (e^*) of the molecule[128]:

$$\int \Delta R/R \, d\omega \approx 32\pi^3 \sec \phi \, n(e^*)^2/cm_r \tag{23}$$

where n and m_r are the density of adsorbate oscillators and their reduced mass, respectively. The effective dynamic charge is simply related to the vibrational polarizability for the gas (α_v) or adsorbed ($\bar{\alpha}_v$) phase:

$$\bar{\alpha}_v = (e^*)^2/m_r\omega_0^2 \tag{24}$$

and can also be related to the integrated molar intensity $\bar{A}$ of the gas phase molecule[128]:

$$\bar{A} = 3.87 \times 10^{-5}(e^*)/m_r \tag{25}$$

Values of e^* obtained using absolute RAIRS intensity data for adsorbed CO on Pt(111)[112] appear to correspond closely with those obtained in EELS measurements, and to the gas phase value of $0.65e$.[128] Similar conclusions have been drawn from surveys of e^* values of CO adsorbed on other metal surfaces,[101,129] which suggest a small deviation of $\bar{\alpha}_v$ from α_v. There appears to be, however, some controversy remaining as to realistic values of $\bar{\alpha}_v$ and $\bar{\alpha}_e$ even for the case of adsorbed CO. Average gas phase values ($\alpha_e \approx 1.9\,\text{Å}^3$, $\alpha_v = 0.06\,\text{Å}^3$)[99] as well as the value corresponding to the C–O bond axis ($\alpha_e \approx 2.5\,\text{Å}^3$, $\alpha_v = 0.06\,\text{Å}^3$)[112,130] have been applied, and recently Persson and Ryberg[118] found reasonable agreement between experiment and theory for coverage dependent frequency [equation (19)] and intensity [equation (26)] changes with values $\bar{\alpha}_e \approx 3\,\text{Å}^3$ and $\bar{\alpha} \approx 0.3\,\text{Å}^3$ for CO on Cu(100), and the results[111] for CO on Ru(001). The modification of adsorption intensity as a result of the dielectric screening due to the polarizability of neighboring molecules is given by[118]

$$\int \Delta R/R \, d\omega \sim \frac{\bar{\alpha}_v c U_0}{1 + \bar{\alpha}_e c U_0} \tag{26}$$

The increased polarizabilities obtained by Persson and Ryberg[118] were interpreted in terms of increased charge density in the $2\pi^*$ orbital of the adsorbed CO due to backbonding. The latter observation provides an additional complication in choosing $\bar{\alpha}_v$ and $\bar{\alpha}_e$ which should now reflect the coverage dependence of backbonding, as evidenced by the static contribution to chemical shift (Section 3.3). The dependence of $\bar{\alpha}_e$ and $\bar{\alpha}_v$ on chemical and physical surface environment is not well understood at present, and careful measurement and intercomparison of absolute intensities and their coverage dependencies provides a potential means of clarifying the situation.

Relative integrated absorption intensities can be used in a semiquantitative fashion to successfully monitor relative coverages in adsorption processes, obtaining thermodynamic quantities and allowing a comparison of absorption cross sections of various adsorbate species. In such experiments, no attempt is being made to obtain absolute quantities [equation (23)] and the effect of adsorbate–adsorbate interactions [equation (26)] is being ignored. The latter approximation is clearly most valid at low coverages and is reflected experimentally in a linear increase in integrated intensity with concentration of the adsorbate on the surface. Figure 25b shows the change in integrated absorption intensity of CO as a function of coverage on Pt(111).[47] The ratio of integrated absorption intensity of $v(\text{CO})$ of the linear CO species at $\theta = \frac{1}{2}$ [corresponding to a $(\sqrt{3} \times \sqrt{3})R30°$ structure of exclusively linear bound species] to that at

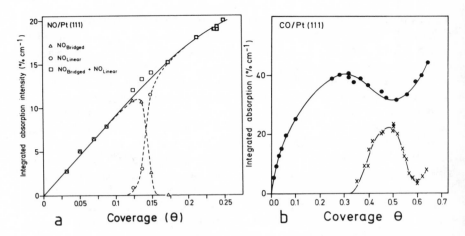

Figure 25. The integrated absorption intensities of the linear and bridged species of (a) NO and (b) CO as a function of coverage on Pt(111) (after Refs. 147, 109).

$\theta = \frac{1}{2}$ [where both linear and bridge species are present to equal concentrations in a $c(4 \times 2)$ layer] is close to that expected $(\frac{1}{3}:\frac{1}{4})$ on the basis of the LEED data assuming intensity is proportional to the CO coverage. Increasing coverage beyoud $\theta = \frac{1}{2}$ is characterized by an increase in the intensity of the linear band and provided evidence to support a site specific mode[86] of adsorption over an incommensurate model.[85] The similarity in intensity of the bridge and linear bands in the $c(4 \times 2)$ structure at $\theta = 0.5$ was also used as evidence for the similarity in absorption coefficient of the two species, a conclusion also supported in recent data for the bridged and linear species of CO on Cu(111).[83]. Further, the linear and bridged species of NO on Pt(111) were found to have nearly identical absorption coefficients.[109] Figure 25a shows the intensity of the bridged and linear species of NO (and their sum) as a function of coverage, and nicely demonstrates the complete interconversion of the two species in a rather small coverage range, the smooth transition indicating very similar absorption coefficients. Note also the relative linear change in absorption intensity with coverage of NO.[109]

The relative intensities of the threefold and bridge CO species[47] on Pt(111) as a function to temperature (Figures 14, 24) were used to estimate a difference in their adsorption enthalpy of ~4 kJ mol⁻¹, again assuming to the first approximation that intensity was proportional to partial coverage. The intensity of the O–H stretch vibration in RAIRS experiments of water adsorption on Ru(001)[137] was used to follow the formation of hydrogen-bonded clusters on the surface. The intensity of the O–H band (3400 cm^{-1}) was measured dynamically while adsorbing water at constant pressure onto clean, and increasingly oxygen-covered surfaces (Figure 26). On the clean surface there is a linear relationship between integrated absorption intensity and exposure, while the presence of small quantities of preadsorbed oxygen delays the onset of absorption. It was concluded that the oxygen atoms "bind" the water molecules (reflected also in the thermal desorption spectra), this preventing cluster formation and in turn eliminating the intensity enhancement due to hydrogen bonding.

Since chemical bonding and the distance of the oscillator dipole from the metal surface may both play an important role in determining

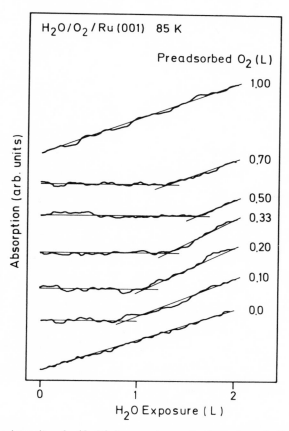

Figure 26. The intensity of v(O–H) from water on Ru(001) dynamically measured as a function of water exposure. The experiment is carried out on the clean and increasingly oxygen covered surface (after Ref. 131).

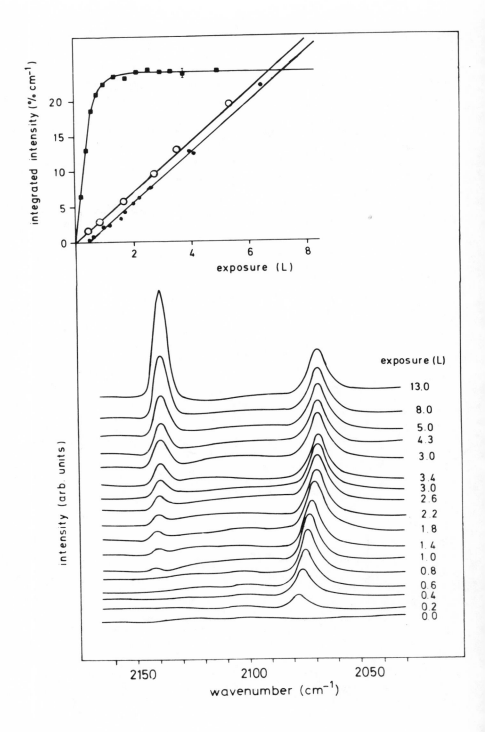

absorption intensity, it is instructive to compare the relative integrated absorption intensities of chemisorbed and physisorbed species of the same molecule on the same surface. Such an experiment[83] was attempted for the adsorption of CO on Cu(111) at low temperatures (~7 K), where not only could the band associated with the stereochemically nonridged bridge bound species of CO be "frozen out" (Figure 24c), but also physisorbed layers of CO could be adsorbed in addition to chemisorbed CO, and their RAIRS spectra measured (Figure 27). The intensity of the linear chemisorbed species and the physisorbed layers growing on top are plotted as a function of exposure. Assuming similar sticking probabilities, the chemisorbed molecule clearly has a significantly higher absorption coefficient than the physisorbed species, even taking into account the random orientation of the latter and the metal's screening. There was also no detectable difference in intensity between the adsorption in the first and subsequent layers of the physisorbed species, and if chemisorption was prevented by preadsorption of a monolayer of Ar the adsorption coefficient of the physisorbed species remains the same. The results demonstrate a significant difference between the absorption cross section of chemisorbed and physisorbed CO on Cu(111), and hence differences [equations (23), (24)] between their vibrational polarizabilities.

4. Outlook

There has been some success in the past decade in improving the sensitivity of RAIRS in both the fluctuation and detector noise limits of operation. In the midinfrared region, where fluctuation noise generally dominates, signal-to-noise ratios of the order of 10^5 are achievable, allowing the detection of down to 10^{-3} of a monolayer of a strong dipole absorber such as CO. The spectral range has been extended through improvements in the detector noise limit at low energies, and reaction intermediates are now accessible. It is clearly difficult to predict how quickly the long-wavelength limit will be further extended, although the rewards in being able to measure adsorbate-substrate modes with the resolution and sensitivity at present afforded by RAIRS in the study of internal modes must surely give some experimental impetus to this end.

Figure 27. The RAIRS spectrum of linearly chemisorbed and randomly physisorbed CO adsorbed on Cu(110) at 7 K. The integrated absorption intensity of the bands is plotted in the inset ■-chemisorbed CO, ●-physisorbed CO, ○-physisorbed CO on an Ar monolayer (after Ref. 83).

Increases in detector sensitivity and the use of more intense infrared
sources are both potential candidates for the extension of the present
spectral range through improvements in the detector noise limit. The
incorporation, for example, of liquid-helium-cooled band pass filters will
reduce the noise equivalent power of background noise limited detectors,
such as liquid-helium-cooled bolometers, at long wavelengths. This
approach is being pursued in combination with slow scan interferometry
and vacuum optics.[132] An alternative infrared source being used is a
Nd:YAG-pumped dye laser combined with three stages of nonlinear
optics which will be used to provide radiation tunable in the region
1–25 µm for RAIRS experiments.[133] Whichever approach is pursued to
extend the low-energy limit in RAIRS, some attention must be given to
the level of fluctuation noise that may be introduced by the detector,
spectrometer and source. The increased spectral range and sensitivity
already achieved will allow a wide variety of vibrational studies to be
carried out with the simultaneous application of other surface sensitive
techniques. Such studies are likely to make full use of perhaps two of the
greatest advantages of RAIRS, namely, its intrinsically high resolution
and its insensitivity to ambient pressure above the surface under
investigation.

4.1. High-Resolution Vibrational Studies

RAIRS will continue to provide a high-resolution vibrational spectroscopic technique for the study of submonolayer concentrations of
adsorbates, furnishing information that is not available from techniques
with lower spectral resolution. The natural linewidth of internal adsorbate modes is of the order of $6\,cm^{-1}$, and in many cases bands are
separated in energy by less than $\sim50\,cm^{-1}$, and many correspond to
different adsorbate configurations, adsorption site, or many arise from
differing lateral interactions in adsorbed phases. Similarly, the absorption
bands of a wide variety of molecular isotope mixtures can be resolved in
RAIRS. It has been shown that the careful measurement of absorption
bands, particularly their integrated absorption intensity and frequency,
yields detailed information concerning the adsorbate, with the experimentalist free to vary coverage, temperature, or ambient pressure
during the RAIRS measurement, and simultaneously combine the
technique with complementary surface sensitive techniques such as
LEED and $\Delta\phi$. As a high resolution surface vibrational spectroscopy,
one is also able to measure natural half-widths and line shapes of the
adsorbate, yielding data directly relating to adsorbate disorder in the
inhomogeneous line-broadening limit. In the homogeneous broadening

limit in RAIRS studies of order overlayers, the final state vibrational lifetime or a measure of adsorbate dynamical motion is available in the dissipative decay and pure dephasing regimes, respectively. RAIRS is perhaps also well suited for the study of the commensurate–incommensurate phase transition at high adsorbate coverages, distinguishing cases in which site specific adsorption may dominate, other forms of order–disorder phenomena and island growth, and forms of surface stereochemical nonrigidity. The use of isotopic mixtures in RAIRS to separate static from dipole coupling contributions to coverage-dependent frequency shifts may also receive increasing attention as the relationship between adsorbate bonding and the chemical shift is better understood. The spectroscopic identification of isotopes also will allow an investigation of isotopic exchange on surfaces. The mechanisms of thermal accommodation in adsorption and reaction, and the local freedom of adsorbate molecules are directly related to the surface dynamic behavior of molecules, and this aspect may give some impetus to the study of adsorbate half-widths and line shapes in the homogeneous line broadening limit.

4.2. Catalysis on Single Crystals

There has been an increasing awareness of the potential shortcomings in drawing conclusions based on adsorbate studies carried out on single crystals in the pressure regime $10^{-11} \lesssim P$ (Torr) $\lesssim 10^{-6}$ pertaining to the mechanisms of heterogeneous catalysis over commercial catalysts at pressures $P > 10^2$ Torr. Recently it has been convincingly shown that metal single-crystal surfaces can indeed mimic the characteristics of supported metal catalysts at higher pressures by studies of reaction rates and turnover numbers,[134] and much attention has been given to the modification of the single crystal in such a way as to reproduce the surface of the commerical catalyst through coadsorption of promotors and poisons, and the introduction of well-characterized defects (e.g., steps and kinks). *In situ* studies on single-crystal surfaces of adsorption and reaction at higher pressures, which have in the past been successfully carried out on the supported catalyst, have been unfeasible because of the sensitivity of the majority of single-crystal investigative techniques to pressure. The majority of such techniques entail the use of hot filaments and require a high mean free path for electrons above the surface, and hence are restricted to pressures below $\sim 10^{-6}$ Torr. Purely optical, i.e., photon in/photon out, techniques provide a powerful means of circum-navigating this problem because of their insensitivity to ambient pressure in the optical path. Hence RAIRS, which can deliver high-resolution vibrational spectra and hence detailed chemical information concerning the adsorbed species and reaction intermediates, is an ideal spectroscopy for the *in situ* study of catalytic reactions on clean and "modified" single-crystal metal surfaces. It is not surprising, therefore, that several groups are already constructing "high" pressure cells for such *in situ* RAIRS studies and the resulting vibrational spectroscopic data will undoubtedly appear in the catalytic and surface science literature in the near future. The potential of optical techniques generally has been increasingly recognized mainly in regard to bridging the gap between data obtained on single crystal and in studies of heterogeneous catalysis,[135] and the wealth of vibrational spectroscopic data already accumulated through infrared studies on supported and oxide catalysts makes an optical vibrational spectroscopic technique even more attractive. It should now be possible using RAIRS to bridge the gap between conventional low- and high-pressure studies on single-crystal metal surfaces, and *in situ* studies of heterogeneous catalysis on supported metal catalysts.

References

1. A. M. Buswell, K. Krebs, and W. H. Rodebush, *J. Am. Chem. Soc.* **59,** 2603 (1937).
2. N. G. Yaroslavski, and A. V. Karyakin, *Dokl. Acad. Nauk SSR* **85,** 1103 (1952).
3. N. Sheppard and D. J. C. Yates, *Proc. R. Soc. London A* **238,** 69 (1956).
4. R. P. Eischens, S. A. Francis, and W. A. Pliskin, *J. Phys. Chem.* **60,** 194 (1956).
5. L. H. Little, *Infrared Spectra of Adsorbed Species,* Academic, New York (1966).
6. M. L. Hair, *Infrared Spectroscopy in Surface Chemistry,* Marcel Dekker, New York (1967).
7. R. P. Eischens and W. A. Pliskin, *Adv. Catal.* **10,** 1 (1958).
8. J. F. Harrod, R. W. Roberts, and E. F. Rissmann, *J. Chem. Phys.* **71,** 343 (1967).
9. H. L. Pickering and H. C. Eckstrom, *J. Phys. Chem.* **63,** 512 (1959).
10. S. A. Francis and A. H. Ellison, *J. Opt. Soc. Am.* **49,** 131 (1959).
11. M. L. Kottke, R. G. Greenler, and H. G. Tompkins, *Surf. Sci.* **32,** 231 (1972).
12. R. G. Greenler, *J. Chem. Phys.* **50,** 1963 (1969).
13. M. D. J. Low and J. C. McManus, *Chem. Commun.* 1166 (1967).
14. A. M. Bradshaw, J. Pritchard and M. L. Sims, *Chem. Commun.* 1519 (1968).
15. H. C. Eckstrom, G. G. Possley, S. E. Hannum, and W. H. Smith, *J. Chem. Phys.* **52,** 5435 (1970).
16. R. G. Greenler, *J. Chem. Phys.* **44,** 310 (1966).
17. R. G. Greenler, *Jpn. J. Appl. Phys. Suppl.* **2**(2), 265 (1974).
18. R. G. Greenler, *J. Vac. Sci. Technol.* **12,** 1410 (1975).
19. J. Pritchard and M. L. Sims, *Trans. Faraday Soc.* **66,** 427 (1970).
20. G. W. Poling, *J. Colloid Interface Sci.* **34,** 365 (1970).
21. R. G. Greenler, R. R. Rahn, and J. P. Schwartz, *J. Catal.* **23,** 42 (1971).
22. J. D. E. McIntyre and D. E. Aspnes, *Surf. Sci.* **24,** 417 (1971).
23. R. W. Ditchburn, *Light,* Blackie, London (1952); E. A. Stratton, *Electromagnetic*

Theory, McGraw-Hill, New York (1941); O. S. Heavens, *Optical Properties of Thin Solid Films*. Butterworths, London (1955).

24. P. Hollins and J. Pritchard, in *Vibrational Spectroscopy of Adsorbates* (R. F. Willis, ed.), Springer Series in Chemical Physics, Vol. 15, Springer, Berlin (1980).
25. M. Ito and W. Suetaka, *J. Phys. Chem.* **79**, 1190 (1975).
26. R. F. Willis, A. A. Lucas, and G. D. Mahan, in *The Chemical Physics of Solid Surfaces and Heterogeneous Catalysis*, Vol. 2 (D. A. King and D. P. Woodruff, eds.), Elsevier, Amsterdam (1982), p. 59.
27. F. M. Hoffmann, Infrared reflection absorption spectroscopy of adsorbed molecules, *Surf. Sci. Reports,* **3**, 2/3 (1983).
28. J. D. Jackson, *Classical Electrodynamics*, Wiley, New York (1962), p. 110.
29. J. F. Blanke, S. E. Vincent, and J. Overend, *Spectrochim. Acta*, **32A**, 163 (1976).
30. N. D. S. Canning and M. A. Chesters, *J. Electron. Spectrosc. Relat. Phenom.* **19**, 69 (1983).
31. P. W. Kruse, L. D. McGlauchlin, and R. B. McQuistan, *Elements of Infrared Technology*, Wiley, New York (1962).
32. P. Hollins and J. Pritchard, *J. Vac. Sci. Technol.* **17**, 665 (1980).
33. *American Institute of Physics Handbook,* McGraw-Hill, New York (1963); D. E. McCarthy *Appl. Opt.* **2**, 591 (1963); **4**, 317, 507, 878 (1965); **6**, 1896 (1967).
34. J. J. Turner, in *Vibrational Spectroscopy—Modern Trends* (A. J. Barnes and W. J. Orville-Thomas, eds.), Elsevier, New York (1977); R. L. Byer, in *Tunable Lasers and Applications*, (A. Mooradian, T. Jaeger and P. Stokseth, eds.), Springer-Verlag, Berlin (1976).
35. J. R. Stevenson, H. Ellis, and R. Bartlett, *Appl. Opt.* **12**, 2884 (1973).
36. J. R. Stevenson and J. J. Latheart, *Nucl. Instrum. Methods.* **172**, 367 (1980).
37. P. Lagarde, *Infrared Phys.* **18**, 395 (1979).
38. G. P. Williams, Brookhaven National Laboratory (BNL) Report 26947 (1979); G. P. Williams and R. N. Brawer, BNL 28265 (1980).
39. W. D. Duncan and J. Yarwood, Technical Memorandum, Daresbury.
40. J. Nagel and E. Schweizer, Technical Report BESSY, TB 63/84 (1984).
41. K. Horn and J. Pritchard, *Surf. Sci.* **52**, 437 (1975); *J. Phys. (Paris)* **38**, C4, 164 (1977).
42. P. Hollins and J. Pritchard, *Surf. Sci.* 456 (1979).
43. R. A. Shigeishi and D. A. King, *Surf. Sci.* **58**, 379 (1976).
44. F. J. Krebs and H. Lüth, *Appl. Phys.* **14**, 337 (1977).
45. J. C. Campuzano and R. G. Greenler, *Surf. Sci.* **83**, 301 (1979).
46. A. M. Bradshaw and F. M. Hoffmann, *Surf. Sci.* **72**, 513 (1978).
47. B. E. Hayden and A. M. Bradshaw, *Surf. Sci.* **125**, 787 (1983).
48. R. Ryberg, *Surf. Sci.* **114**, 627 (1982).
49. K. D. Möller and W. G. Rotschild, *Far Infrared Spectroscopy*, Wiley-Interscience, New York (1971).
50. R. Ryberg, *Phys. Rev. Lett.* **49**, 1579 (1982).
51. W. G. Golden, D. S. Dunn, and J. Overend, *J. Phys. Chem.* **82**, 843 (1978); *J. Catal.* **71**, 395 (1981).
52. J. Chamberlain, *The Principles of Interferometric Spectroscopy*, Wiley, New York (1979).
53. M. D. Baker and M. A. Chesters, in *Vibrations at Surfaces* (R. Caudona, J. M. Gilles, and A. A. Lucas, eds.), Plenum Press, New York (1982).
54. W. G. Golden and D. A. Saperstein, *J. Electron Spectrosc. Relat. Phenom.* **30**, 43 (1983).

55. M. J. Dignam and M. D. Baker, *J. Vac. Sci. Technol.* **21**, 80 (1982).
56. W. G. Golden, D. D. Saperstein, M. W. Stevenson, and J. Overend, *J. Phys. Chem.*, in press.
57. M. Cardona, Modulation spectroscopy, *Sol. State Phys. Suppl.* **11**, 105 (1969).
58. J. Pritchard, in *Modern Methods of Surface Analysis*, DECHEMA Monograph No. 78, DECHEMA Frankfurt (1975).
59. A. M. Bradshaw and F. M. Hoffmann, *Surf. Sci.* **52**, 449 (1975).
60. F. M. Hoffmann, Ph.D. thesis, University of Munich (1977).
61. B. E. Hayden, K. Kretzschmar, and A. M. Bradshaw, *Surf. Sci.* **123**, 366 (1983).
62. D. S. Dunn, W. G. Golden, M. W. Severson, and J. Overend, *J. Phys. Chem.* **84**, 336 (1980).
63. J. D. Fedyk, M. Mahaffy, and M. J. Dignam, *Surf. Sci.* **89**, 404 (1979).
64. R. D. Hudson Jr. and J. W. Hudson (ed.), *Infrared Detectors*, Benchmark Papers in Optics 2, Dowden, Hutchinson and Ross (1975).
65. R. J. Keyes (ed.), *Optical and Infrared Detectors*, Topics in Applied Physics, Vol. 19, Springer-Verlag, New York (1977).
66. F. A. Cotton, *Chemical Applications of Group Theory*, Wiley-Interscience, New York (1971).
67. S. Lehwald, H. Ibach, and J. E. Demuth, *Surf. Sci.* **78**, 577 (1978).
68. N. V. Richardson and A. M. Bradshaw, in *Electron Spectroscopy: Theory, Techniques and Applications* (B. Baker and R. Brundle, eds.), Vol. 4, p. 153, Academic, London (1980).
69. K. Horn, in *Vibrations in Adsorbed Layers*, Conf. Proc. KFA Jülich (H. Ibach and S. Lehwald, eds.), p. 141 (1978).
70. R. Ryberg, *Chem. Phys. Lett.* **83**, 423 (1981).
71. B. E. Hayden, K. C. Prince, D. P. Woodruff, and A. M. Bradshaw, *Phys. Rev. Lett.* **51**, 475 (1983); *Surf. Sci.* **133**, 589 (1983).
72. K. Horn, private communication.
73. G. Hertzberg, *Molecular Spectra and Molecular Structure*, Vol. 2, Van Nostrand Reinhold, New York (1945).
74. B. A. Sexton, *Surf. Sci.* **88**, 319 (1979).
75. K. Ito and H. J. Bernstein, *Can. J. Chem.* **34**, 1970 (1956).
76. N. Sheppard and T. T. Nguyen, in *Vibrations in Adsorbed Layers*, Conf. Proc. KFA Jülich (H. Ibach and S. Lehwald, eds.), p. 146 (1978).
77. R. P. Eischens, S. A. Francis, and W. A. Pliskin, *J. Phys. Chem.* **60**, 194 (1956).
78. N. Sheppard and T. T. Nguyen, in *Advances in Infrared and Raman Spectroscopy*, Vol. 5 (R. E. Hester and R. J. H. Clark, eds.), Heyden and Son, London (1978).
79. A. Ortega, F. M. Hoffmann, and A. M. Bradshaw, *Surf. Sci.* **119**, 79 (1982).
80. P. Hofmann, R. Ortega, F. M. Hoffmann, K. Kretzschmar, and A. M. Bradshaw, *Surf. Sci.*, to be submitted.
81. J. C. Tracy and P. W. Palmberg, *J. Chem. Phys.* **51**, 4852 (1969).
82. R. J. Behm, K. Christmann, and G. Ertl, *J. Chem. Phys.* **73**, 2984 (1980).
83. B. E. Hayden, K. Kretzschmar, and A. M. Bradshaw, *Surf. Sci.* **155**, 553 (1985).
84. G. Ertl, M. Neumann, and k. M. Streit, *Surf. Sci.* **64**, 393 (1977).
85. A. M. Baro and H. Ibach, *J. Chem. Phys.* **71**, 4812 (1979).
86. N. R. Avery, *J. Chem. Phys.* **74**, 4202 (1981).
87. F. A. Cotton, L. Kruczynski, B. L. Shapiro, and L. F. Johnson, *J. Am. Chem. Soc.* **94**, 6191 (1972).
88. R. D. Adams and F. A. Cotton, in *Dynamic N.M.R. Spectroscopy* (L. M. Jackman, M. Lloyd, and F. A. Cotton, eds.), Academic, New York, p. 489 (1975).

89. J. Pritchard, in *Vibrations in Adsorbed Layers,* Conference Records Series of KFA (H. Ibach and S. Lehwald, eds.), Jülich (1978), p. 114.
90. J. Pritchard, T. Catterick, and R. K. Gupta, *Surf. Sci.* **53**, 1 (1975).
91. R. Van Hardeveld and A. Van Montfoort, *Surf. Sci.* **4**, 396 (1966).
92. B. E. Hayden, K. Kretzschmar, A. M. Bradshaw, and R. G. Greenler, *Surf. Sci.* **149**, 1394 (1985).
93. R. G. Greenler, B. E. Hayden, K. Kretzschmar, R. Klauser, and A. M. Bradshaw, Proc. 9th Intern. Congress on Catalysis, Berlin IV-197, Verlag Chemie, Weinheim (1984).
94. R. G. Greenler, K. D. Burch, K. Kretzschmar, R. Klauser, A. M. Bradshaw, and B. E. Hayden, *Surf. Sci.* **152**, 338 (1985).
95. R. G. Greenler and K. D. Burch, to be published.
96. P. Hollins, K. J. Davis, and J. Pritchard, *Surf. Sci.* **138**, 74 (1984).
97. A. A. Davydov and A. T. Bell, *J. Catal.* **49**, 332 (1977).
98. P. Hollins and J. Pritchard, *Surf. Sci.* **134**, 91 (1983).
99. G. D. Mahan and A. A. Lucas, *J. Chem. Phys.* **68**, 1344 (1978).
100. M. Trenary, K. J. Uram, F. Bozos, and J. T. Yates Jr., *Surf. Sci.* **146**, 269 (1984).
101. M. Moskovits and J. E. Hulse, *Surf. Sci.* **78**, 397 (1978).
102. S. Efrima and H. Metiu, *Surf. Sci.* **109**, 109 (1981).
103. G. Blyholder, *J. Phys. Chem.* **68**, 2772 (1964).
104. M. A. Chester, J. Pritchard, and M. L. Sims in *Adsorption–Desorption Phenomena,* (F. Ricca, ed.), Academic, London (1972), p. 277.
105. P. Hollins and J. Pritchard, in Vibrational Spectroscopies for Adsrobed Species ACS Symposium Series No. 137, Am. Chem. Soc., Washington (1980).
106. D. P. Woodruff, B. E. Hayden, K. C. Prince, and A. M. Bradshaw, *Surf. Sci.* **123**, 397 (1982).
107. For example C. R. Brundle, P. S. Bagus, D. Mensel, and K. Hermann, *Phys. Rev. B* **24**, 7041 (1981); R. P. Messner, S. H. Larmson, and D. R. Salahub, *Phys. Rev. B* **25**, 3576 (1982).
108. R. M. Hammaker, S. A. Francis, and R. P. Eischens, *Spectrochim. Acta* **21**, 1295 (1965).
109. B. E. Hayden, *Surf. Sci.* **131**, 419 (1983).
110. A. Crossley and D. A. King, *Surf. Sci.* **68**, 528 (1977).
111. H. Pfnür, D. Mensel, F. M. Hoffmann, A. Ortega, and A. M. Bradshaw, *Surf. Sci.* **93**, 431 (1980).
112. A. Crossley and D. A. King, *Surf. Sci.* **95**, 131 (1980).
113. D. A. King, *J. Electron. Spectrosc. Relat. Phenom.* **29**, 11 (1983).
114. P. Hollins, *Surf. Sci.* **107**, 75 (1981).
115. G. K. T. Conn and D. G. Avery, *Infrared Methods,* Academic, New York (1960).
116. J. Pritchard, *Surf. Sci.* **79**, 231 (1979).
117. J. P. Briberian and M. A. van Hove, *Surf. Sci.* **118**, 443 (1982).
118. B. N. J. Persson and R. Ryberg, *Phys. Rev. B* **24**, 6954 (1981).
119. B. N. J. Persson, *J. Phys. Chem.* **17**, 4741 (1984).
120. J. W. Gadzuk and A. C. Luntz, *Surf. Sci.* **144**, 429 (1984).
121. M. Metiu and W. E. Palke, *J. Chem. Phys.* **69**, 2574 (1978).
122. B. N. J. Persson and M. Persson, *Surf. Sci.* **97**, 609 (1980).
123. M. A. Kozkusknev, V. G. Kistalev, and B. R. Shub, *Surf. Sci.* **81**, 261 (1979).
124. P. Apell, *Phys. Scr.* **29**, 146 (1984).
125. B. N. J. Persson and M. Persson, *Solid State Commun.* **36**, 175 (1980).
126. B. N. J. Persson and R. Ryberg, *Phys. Rev. Lett.* **48**, 549 (1982).

127. R. M. Corn and H. L. Strauss, *J. Chem. Phys.* **76,** 4834 (1982).
128. H. Ibach, *Surf. Sci.* **66,** 56 (1977).
129. S. Anderson and J. Davenport, *Solid State Commun.* **28,** 677 (1978).
130. M. Scheffler, *Surf. Sci.* **81,** 562 (1979).
131. K. Kretzschmar, J. K. Sass, A. M. Bradshaw, and S. Holloway, *Surf. Sci.* **115,** 183 (1982).
132. E. Schweizer and A. M. Bradshaw, private communication.
133. A. Luntz, private communication.
134. R. D. Kelley and D. W. Goodman, in *The Chemical Physics of Solid Surfaces and Heterogeneous Catalysis* (D. A. King and D. P. Woodruff, eds.), Vol. 4, p. 427, Elsevier, Amsterdam (1981).
135. J. M. White, *Science* **218,** 429 (1982); J. Demuth and P. Avouris, *Phys. Today* **36**(11), 62 (1983).
136. M. Trenary, K. F. Uram, and J. T. Yates Jr., *Surf. Sci.* **157,** 512 (1984).
137. P. Gelin and J. T. Yates Jr., *Surf. Sci.* **136,** 21 (1984); P. Gelin, A. R. Siedle, and J. T. Yates Jr., *J. Phys. Chem.* **88,** 2978 (1984).

8

Raman Spectroscopy

Alan Campion

1. Introduction

Raman spectroscopy as a probe of surface structure and dynamics is still developing its power and potential. The Raman process itself—inelastic light scattering by molecular vibrations—is inherently weak. Only a small fraction, 10^{-6} or so, of the photons incident on a sample are Raman scattered. Thus it might appear that the application of such an intrinsically inefficient process to the detection of a small number of molecules would not be very productive.

Two developments over the past six years, however, have created great interest in Raman spectroscopy as a surface analytical tool: surface-enhanced Raman scattering (SERS) and multichannel surface Raman spectroscopy without enhancement. The discovery of SERS conferred unprecedented sensitivity and selectivity to surface Raman spectroscopy.[1-5] The Raman signals from molecules adsorbed on the roughened surfaces of certain metals are over a millionfold stronger than expected from their gas phase cross sections and surface densities. Much of the research in SERS has been devoted to elucidating the origin of this giant enhancement, an understanding of which has now been reached, at least in broad terms. SERS has proven to be a sensitive and powerful probe of surface, and interfacial processes and interesting applications are being reported regularly.

Alan Campion • Department of Chemistry, University of Texas at Austin, Austin, Texas 78712.

Unfortunately, however, SERS has not been as universally applicable as had been hoped originally. The effect is strongest for roughened surfaces of only a few, highly reflective metals. Prompted by the excitement of the discovery of SERS, and motivated by its restricted applicability, we began to try and develop a method of surface Raman spectroscopy that did not require either surface or resonance enhancement. If successful our approach would be much more universal than SERS or resonance Raman spectroscopy as it would require no special properties of either the substrate or the adsorbate. We have developed and applied multichannel optical techniques to surface Raman spectroscopy. The multiplex advantage of 10^2–10^3 is sufficient to obtain unenhanced Raman spectra of molecules adsorbed on planar surfaces at submonolayer coverages in a few minutes.[6] This technological advance means that Raman spectroscopy can be applied to a wide range of important problems in surface physics and chemistry.

Raman spectroscopy offers a number of advantages as a surface vibrational spectroscopy. As an optical technique it has high resolution ($1 \, cm^{-1}$) and immunity to the presence of an ambient gas phase, at least to moderate pressures (1 atm). The relatively low scattering cross section compared with that for infrared absorption, say, is somewhat compensated for by the intensity of visible laser sources and the availability of true quantum detectors for visible photons. The recent application of multichannel techniques, which provide the same multiplex advantage gained in Fourier transform infrared spectroscopy (FTIR), has resulted in a sensitivity of better than 5% of a monolayer of a modest scatterer adsorbed on a planar surface, ca. 10^{13} molecules cm^{-2}. While it is difficult to compare the sensitivities of the various spectroscopies because the "best" Raman scatterers are weak infrared absorbers and vice versa, it is safe to say that the sensitivity of Raman spectroscopy is at least comparable to other methods. Technological advances can be expected to result in an improvement in sensitivity of at least one order of magnitude in the near future. Finally, the visible wavelengths used place minimal constraints on the window material used and allow the entire infrared region of the spectrum to be studied.

Raman spectroscopy can be used to study a wide variety of surfaces over a wide range of conditions. Molecules adsorbed on planar substrates, either metals or semiconductors, can be examined under either ultrahigh vacuum or elevated pressure. High surface area materials, which include supported catalysts, are also amenable to study, again under either vacuum or elevated pressure. It is hoped that a technique that is sufficiently versatile to be used in all of the regimes of modern surface science will help bridge the gap between surface physics and catalytic chemistry.

On the more fundamental side, Raman spectroscopy provides spectra that are complementary to those obtained by either infrared absorption or electron energy loss spectroscopy (EELS, dipolar loss mechanism). Raman scattering depends upon the change in the molecular polarizability during a molecular vibration, while infrared absorption depends upon the change in the dipole moment. In centrosymmetric molecules, no mode is both infrared and Raman active. In molecules of reasonably high symmetry, there is still a partial exclusion. Thus both spectroscopies are necessary to obtain the complete vibrational spectrum of an adsorbed molecule.

Raman spectroscopy also is preferred for acquiring the spectra of molecules adsorbed on oxides. Since these materials absorb infrared radiation strongly (they are in fact opaque below 1000 cm^{-1}), it is nearly impossible to take infrared spectra of molecules adsorbed on them. In contrast, ionic materials like oxides are very weak Raman scatterers, and the Raman spectrum of the adsorbate can be taken with very little interference from the substrate.

Finally, since laser beams can be focused to very small spot sizes (10 μm) a high spatial resolution map of the molecular composition of the surface can be made. Entirely analogous to the scanning Auger microprobe (which provides a two-dimensional map of the surface elemental composition), the Raman microprobe can map out the locations of different compounds.

Clearly Raman spectroscopy has much to offer in surface analysis. Its high resolution, wide spectral range, high sensitivity, complementary selection rules, and high spatial resolution will undoubtedly find many exciting applications in years to come.

I have tried to organize this chapter so that it will provide a more or less complete introduction to the field, beginning with theory and ending with applications. The field itself is becoming much broader and it is impossible to cover all topics of interest in one article. Accordingly, I have chosen to focus on the fundamental aspects in almost a tutorial fashion so that the interested reader could conduct surface Raman experiments if desired. Section 2 reviews the basic theory necessary to understand Raman scattering from isolated molecules. Section 3 outlines general experimental considerations applicable to both single-crystal studies and experiments with high surface area materials. Estimates of sensitivity and selectivity are given. A detailed discussion of the properties of the electromagnetic field near a conducting surface is given.

An understanding of this subject is absolutely essential for designing and interpreting surface Raman experiments. These considerations are important for both planar surfaces and supported metal particles. A complete discussion of the angular dependence of the scattered intensities and the modifications of the selection rules for Raman scattering imposed by the presence of a conducting surface is presented. Experimental conditions for studying high surface area materials are also discussed.

In Section 4 I discuss the instrumentation required for surface Raman spectroscopy, emphasizing new developments in multichannel optical methods that result in an enormous increase in sensitivity over conventional detection. Section 5 presents a few illustrative examples and applications. In Section 6 I review the essential physics necessary to understand the mechanisms leading to SERS, and show how the proposed mechanisms explain most of the key observations in that field. The section closes with some very interesting applications of the SERS technique.

2. Theory

2.1. Classical: The Polarizability Tensor

When a molecule is irradiated with light, a small portion of it may be scattered either elastically or inelastically. The elastically scattered light is called Rayleigh scattering and the inelastically scattered light is either Raman or Brillouin scattering. Light scattering arises from dipole moments induced in atoms or molecules by the incident field, through the polarizability of the electrons. The static polarizability leads to Rayleigh scattering, while modulation of the polarizability by electronic, vibrational or rotational motion leads to Raman scattering. Brillouin scattering arises from modulation of the incident light by acoustical waves and will not be discussed further here. I will also limit the discussion to vibrational Raman scattering at this point. The treatment presented here follows closely that of Long.[7]

A simple classical model illustrates most of the essential features of Raman scattering. A laser illuminates the sample at frequency v_0; the amplitude of the electric field of the light is

$$\mathbf{E} = \mathbf{E}_0 \cos(2\pi v_0 t) \tag{1}$$

Suppose the polarizability of the molecule is modulated at the vibrational frequency v_1,

$$\alpha = \alpha_0 + \alpha_0 \cos(2\pi v_1 t) \tag{2}$$

where α_0 is the polarizability at the equilibrium nuclear geometry. Since the electric field induces a dipole moment in the molecule according to

$$\mu = \alpha E \tag{3}$$

the time dependence of μ is given by

$$\begin{aligned}
\mu &= \alpha_0 E_0 (1 + \cos 2\pi v_1 t)(\cos 2\pi v_0 t) \\
&= \alpha_0 E_0 \cos 2\pi v_0 t + (1/2)\alpha_0 E_0 \cos 2\pi(v_0 + v_1)t \\
&\quad + (1/2)\alpha_0 E_0 \cos 2\pi(v_0 - v_1)t
\end{aligned} \tag{4}$$

The scattered light has been modulated by the oscillating polarizability and sidebands appear in the spectrum. The first term is the Rayleigh scattering at the unshifted frequency v_0, the second term is anti-Stokes Raman scattering, and the third term is the Stokes Raman scattering. The spectrum is sketched in Figure 1. This simple classical model provides a selection rule for Raman scattering: the molecular polarizability must change during a vibration for that vibrational mode to be Raman active. The classical picture does not, however, give any information about intensities, which I will discuss later.

An oscillating electric field may induce a dipole moment in a molecule whose axis is not parallel to the applied field, because of the particular symmetry of the molecular electronic distribution. Thus, in general the induced dipole is related to the applied field through the polarizability tensor:

$$\begin{matrix} \mu_x \\ \mu_y = \\ \mu_z \end{matrix} \begin{bmatrix} \alpha_{xx} & \alpha_{xy} & \alpha_{xz} \\ \alpha_{yx} & \alpha_{yy} & \alpha_{yz} \\ \alpha_{zx} & \alpha_{zy} & \alpha_{zz} \end{bmatrix} \begin{bmatrix} E_x \\ E_y \\ E_z \end{bmatrix} \tag{5}$$

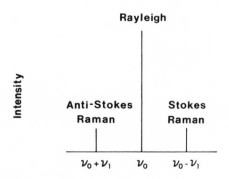

Figure 1. Schematic of a Raman spectrum showing the frequency relationship between Rayleigh scattering and Stokes and anti-Stokes Raman scattering.

For nonresonant Raman scattering, the polarizability tensor is real and symmetric. (This follows from consideration of the potential energy of the system when two orthogonal fields are applied, $U = E_x \cdot \alpha_{xy} \cdot E_y = E_y \cdot \alpha_{yx} \cdot E_x$, so $\alpha_{xy} = \alpha_{yx}$.) For resonant Raman scattering, however, the polarizability tensor may have imaginary components. The real polarizability tensor can always be diagonalized by choosing a principal axis system. In this system,

$$\begin{matrix} \mu_x \\ \mu_y = \\ \mu_z \end{matrix} \begin{bmatrix} \alpha_{xx} & & \\ & \alpha_{yy} & \\ & & \alpha_{zz} \end{bmatrix} \begin{bmatrix} E_x \\ E_y \\ E_z \end{bmatrix} \tag{6}$$

and the magnitude and orientation of the induced dipole may be readily computed.

The polarizability and ellipsoid provides a convenient visualization of the molecular polarizability. In the principal axis system the equation

$$\alpha_{xx}x^2 + \alpha_{yy}y^2 + \alpha_{zz}z^2 = 1 \tag{7}$$

describes an ellipsoid with semiaxes $\alpha_{xx}^{-1/2}$, $\alpha_{yy}^{-1/2}$, $\alpha_{zz}^{-1/2}$. Perhaps a more useful alternative definition is

$$(x^2/\alpha_{xx}) + (y^2/\alpha_{yy}) + (z^2/\alpha_{zz}) = 1 \tag{8}$$

because the more polarizable directions are now represented by the longer axes. For molecules with cylindrical symmetry (e.g., diatomics) two of the principal values of the polarizability tensor must be equal and the ellipsoid is one of revolution. For molecules of tetrahedral or higher symmetry all principal values of the polarizability tensor are the same and the ellipsoid becomes a sphere.

Qualitative insight into the Raman activity of vibrational modes can be gained by an examination of the changes in the polarizability ellipsoid during a molecular vibration. CO_2 provides a simple example; its three vibrational modes and associated changes in the shape of the polarizability tensor are shown in Figure 2.[8] For the symmetric stretch v_1, the polarizability clearly changes during the vibration, the ellipsoid being stretched along the bond axes. In general it is difficult to guess the sign of the change, however, and calculations must be carried out if the sign is needed. For the antisymmetric stretch and the bend, no change is observed in the size or shape of the polarizability ellipsoid. If this is not intuitively obvious, consider the following argument. Since bond polarizabilities are additive, compression of one $C{=}O$ bond results in a polarizability change that is exactly offset by that due to expansion of the other $C{=}O$ bond. For a nonlinear triatomic such as SO_2, the situation is

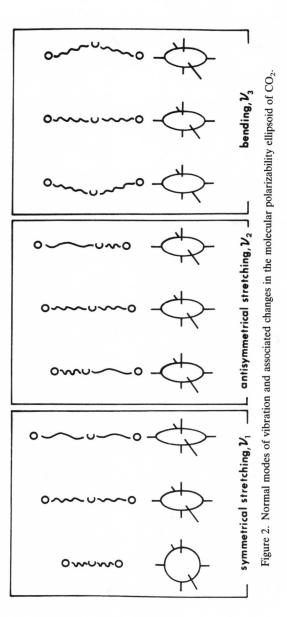

Figure 2. Normal modes of vibration and associated changes in the molecular polarizability ellipsoid of CO_2.

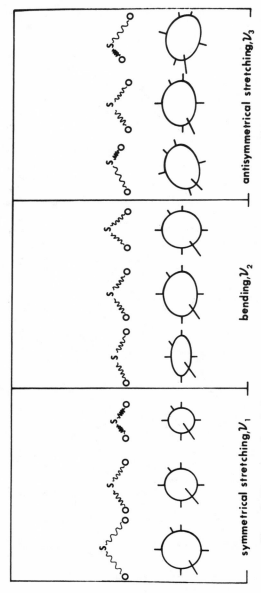

Figure 3. Normal modes of vibration and associated changes in the molecular polarizability ellipsoid of SO_2.

more interesting, as shown in Figure 3. Since the symmetry of the molecule does not change during the symmetric stretching and bending vibrations v_1 and v_2, the polarizability ellipsoid simply expands and contracts with the vibrations as for v_1 in CO_2. For the nontotally symmetric vibration v_3, however, the polarizability ellipsoid retains its shape but rotates with the vibration. In the space-fixed coordinate system determined by the equilibrium geometry this corresponds to the introduction of off-diagonal elements in the polarizability tensor. At the equilibrium position, their values are zero and they have the same magnitudes with opposite signs at the extremes of the vibration. Thus the polarizability is a function of the normal coordinate and v_3 is Raman active. There are two ways, then, that a mode can be Raman active. If it is totally symmetric, the polarizability ellipsoid retains its shape but must change its size during vibration. For nontotally symmetric vibrations the polarizability ellipsoid may keep its shape but change its orientation in space.

Since the Raman tensor is real and symmetric it has two invariants. Thus, only two quantities can be measured in liquids or randomly oriented systems. These are conventionally called the mean polarizability (trace) and the anisotropy is given by

$$a = 1/3(\alpha_{xx} + \alpha_{yy} + \alpha_{zz}) \tag{9}$$

$$\gamma^2 = 1/2(\alpha_{xx} - \alpha_{yy})^2 + (\alpha_{yy} - \alpha_{zz})^2 + (\alpha_{zz} - \alpha_{xx})^2 \tag{10}$$

in the principal axis system.

For Raman scattering by bulk samples the depolarization ratio can be expressed in terms of the tensor invariants and yields information about molecular symmetries. The most common depolarization ratio measured today is the ratio of intensities of light scattered with its electric vector parallel and perpendicular to the incident electric vector, when the laser is linearly polarized and the scattering is observed at 90°. The depolarization ratio is then given by

$$\rho^{\perp}(\pi/2) = (3\gamma^2)/(45a^2 + 4\gamma^2) \tag{11}$$

and range from 0 (polarized) to 3/4 (completely depolarized). Intermediate values are said to be partially polarized. Normal mode symmetries can be determined from the depolarization ratio. For example, a totally symmetric mode of a molecule with an isotropic polarizability has $\gamma = 0$ and $\rho^{\perp}(\pi/2) = 0$. For non-totally-symmetric modes, the diagonal elements vanish, $a = 0$, and $\rho^{\perp}(\pi/2) = \frac{3}{4}$. Quite different considerations are important for oriented molecules, and they will be discussed later.

354

Alan Campion

The classical picture has taken us as far as it can, and we must now resort to quantum and statistical mechanics and group theory for further progress.

2.2. Quantum Mechanical: Selection Rules and Intensities

The molecular polarizability can be calculated from second-order perturbation theory (Kramers–Heisenberg equation) as

$$\alpha = \sum_i \frac{\langle f |H| i\rangle\langle i |H| g\rangle}{(E_i - E_g) - h\nu + i\Gamma_i} + \sum_i \frac{\langle f |H| i\rangle\langle i |H| g\rangle}{(E_i - E_g) + h\nu} \tag{12}$$

where g, i and f are the ground, intermediate, and final states of the molecule, respectively, H is the molecule-radiation interaction Hamiltonian (often taken to be the electric dipole operator), $h\nu$ is the photon energy, and Γ_i is the natural linewidth of the intermediate state, which keeps the expression from blowing up if resonance is achieved. The Raman process is represented pictorially in Figure 4 and can be seen to involve absorption and emission from an intermediate state. The intermediate state is a time-dependent superposition of molecular eigenstates and is often called a virtual state. As the excitation frequency is brought into resonance with a real molecular eigenstate, the Raman scattered intensity increases dramatically owing to the vanishing denominator in equation (12). Resonance Raman scattering is very important for increasing the sensitivity of the Raman technique, for understanding the electronic and vibronic structure of molecules, and for localizing the Raman probe to a particular portion of a molecule in large molecules that contain chromophoric groups. One of the current proposed mechanisms of surface-enhanced Raman scattering is analogous to resonance Raman scattering (see Section 6.3). As the frequency of the exciting laser is brought into exact resonance with a molecular electronic transition, the

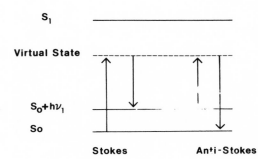

Figure 4. Schematic of the Raman process.

distinction between resonance scattering and resonance fluorescence becomes blurred.

We can now address the relative intensities of the Stokes and anti-Stokes bands by reference to Figure 4. The Stokes bands arise from molecules in the ground vibrational state and their intensities can be calculated from the classical formula for the intensity of radiation emitted by an oscillating dipole

$$I = (16\pi^4 v^4/3c^3)\mu^2 \tag{13}$$

μ is the dipole strength, which can be calculated from the Kramers–Heisenberg dispersion equation by recalling that $\mathbf{\mu} = \alpha\mathbf{E}$. The anti-Stokes Raman lines arise from molecules originating in thermally populated higher-lying vibrational levels of the ground electronic state, and thus their intensity must be weighted by the Boltzmann factors that describe the populations of those levels. Noting that the matrix elements for a given transition are the same whether it is Stokes or anti-Stokes, and recalling that the frequencies are different, we arrive at an expression for the intensities of the anti-Stokes bands relative to the Stokes bands:

$$I(\text{anti-Stokes})/I(\text{Stokes}) = [(v_0 + v_1)^4/(v_0 - v_1)^4]\exp(-hv_1/kT) \tag{14}$$

Low-frequency vibrations can be observed on the anti-Stokes side but high frequency vibrations are generally too weak to be observed. For example, at a frequency shift of $1000\ \text{cm}^{-1}$, the anti-Stokes scattering is about a hundred times weaker than the Stokes scattering. It is not generally practical, therefore, to scan the anti-Stokes side of the spectrum for higher-frequency modes using spontaneous Raman scattering.

We have so far considered only the static molecular polarizability. The Raman polarizability is the variation in the polarizability with normal coordinate and is usually expressed by expanding α in a Taylor series

$$\alpha = \alpha_0 + \sum_i (\partial\alpha/\partial Q_i)_{Q_i=0} + \cdots \tag{15}$$

The second term is the derived polarizability and is often denoted α'. All of the expressions and considerations of the classical treatment, which were derived for α, hold for α' as well, which is why we did not introduce this detail earlier in the discussion.

Selection rules for Raman scattering can be deduced by invoking the Born–Oppenheimer approximation for the separation of electronic and nuclear motion, and the harmonic approximation for the vibrations. Thus a term in the expansion of the derived tensor would be

$$\alpha_k^{fi} = (\partial\alpha/\partial Q_k)_{Q_k=0}\langle\phi_{v_k^f}|Q_k|\phi_{v_k^i}\rangle \tag{16}$$

where the $\phi_{v_k^f}$ and $\phi_{v_k^i}$ are the vibrational wave functions of the kth normal mode in the final and initial states respectively. From the properties of harmonic oscillator wavefunctions we know that

$$\langle \phi_{v_k^f} | Q_k | \phi_{v_k^i} \rangle = 0 \qquad \text{for } v_k^f = v_k^i$$
$$= (v_k^i + 1)^{1/2} b_{vk} \qquad \text{for } v_k^f = v_k^i + 1 \qquad (17)$$
$$= (v_k^i)^{1/2} b_{vk} \qquad \text{for } v_k^f = v_k^i - 1$$

where

$$b_{vk}^2 = h/(8\pi^2 v_k) \qquad (18)$$

Thus, the leading term in the Taylor series expansion of the polarizability leads to the selection rule $\Delta v = \pm 1$, in the harmonic approximation. Overtones and combinations are allowed in higher-level approximations.

Symmetry considerations can be used to determine which vibrational modes are infrared and Raman active. The rate of spectral transitions is given by Fermi's Golden Rule

$$W_{f \leftarrow i} = (2\pi/\hbar)\langle f | H | i \rangle^2 \rho(E_f) \qquad (19)$$

where H is the interaction Hamiltonian and $\rho(E_f)$ is the density of final states. The Herzberg–Teller expansion (Taylor series in Q) gives

$$H = \mu_0 + (\partial \mu / \partial Q_k)_{Q_k=0} Q_k \qquad (20)$$

for infrared absorption and

$$H = \alpha_0 + (\partial \alpha / \partial Q_k)_{Q_k=0} Q_k \qquad (21)$$

for Raman scattering.

Since each term in a Hamiltonian must have the full symmetry of the molecule, each must transform as the totally symmetric representation in the molecular point group. Thus, both $(\partial \mu / \partial Q_k)$ and $(\partial \alpha / \partial Q_k)$ must transform in the same way as Q_k itself. This leads to the conclusion that, for infrared activity, Q_k must transform as the electric dipole operator, a vector, while for Raman activity it must transform as one element of the polarizability tensor. The activity of a given vibrational mode can easily be determined, therefore, from character tables. A mode transforming as one of the Cartesian components of a vector is infrared active while one transforming as one of the nine Cartesian tensor components is Raman active. Herein lies the origin of the complementary nature of the techniques. In fact there is an exclusion rule in centrosymmetric molecules. Since the vector components are antisymmetric on inversion of coordinates and the tensor components are symmetric, no mode can be both Raman and infrared active in centrosymmetric molecules.

3. Experimental Considerations

3.1. Sensitivity and Surface Selectivity

Raman spectroscopy of molecules adsorbed at submonolayer coverage on single-crystal surfaces without enhancement presents the greatest experimental challenge of the situations encountered here and will be discussed first. The intensity of Raman scattered radiation is given generally by

$$I_R = nF(\partial\sigma/\partial\Omega)\Omega \tag{22}$$

where n is the number of molecules in the illuminated volume, F is the laser flux in photons $cm^{-2} s^{-1}$, $\partial\sigma/\partial\Omega$ is the differential Raman scattering cross section, and Ω is the solid angle over which the scattered photons are collected. For surface Raman spectroscopy, the relation can be simplified by substituting $n = \sigma'A$, where σ' is the density of adsorbed species and A is the illuminated area and $F = PA^{-1}$, where P is the laser power in photons sec^{-1}, to give

$$I_R = \sigma'P(\partial\sigma/\partial\Omega)\Omega \tag{23}$$

The illuminated area drops out of the expression for the intensity. The focal diameter can therefore be chosen for other reasons, e.g., as a compromise between increased brightness for efficient monochromator illumination and reduced brightness for minimal substrate heating. To estimate the scattered intensity, substitute some typical values into equation (23). There are ca. 10^{15} adsorption sites per square centimeter on transition metal surfaces, Raman cross sections are ca. $10^{-30} cm^2 molecule^{-1} sr^{-1}$ for moderately strong scatterers, 100 mW of continuous laser power in the green yields 2.5×10^{17} photons sec^{-1}, and with fast ($f/1$) collection optics about one steradian is collected. Thus we expect a total Raman scattered intensity of *ca.* 2.5×10^2 counts per second. If the throughput of the monochromator is 10% and the quantum efficiency of the photomultiplier is 10%, then the signal from a monolayer in this example would be 2–3 counts per second. This is a very weak signal, one that could be recorded using conventional instrumentation, but not easily. For example, if a signal-to-noise ratio of 10:1 is desired, 30 s of integration are necessary in the absence of any background. To record $4000 cm^{-1}$ of the spectrum at a resolution of $5 cm^{-1}$ requires more than six hours! This is an unacceptably long time for surface studies, so alternative approaches must be considered. Examination of equation (23) suggests three possibilities. The number of scatterers can be increased by increasing the surface area of the substrate, for example, by roughening a metal surface or by studying metals dispersed on high area supports. Factors of 10^1-10^3 in signal can be realized using this approach, which is also, of course, extremely relevant to the study of technical catalysts. The second possibility is to try to increase the laser power at the sample. The practical upper limit is determined by the absorption of light and dissipation of the concomitant heat. In our experience, power levels around 1 W are the maximum practical before desorption or sample damage becomes a problem. The laser power felt by the molecule can be effectively increased without a great increase in the overall power by working with substrates whose structure and dielectric constant support electromagnetic resonances at the laser frequencies used. Enhancement of the local electromagnetic field at the surfaces of such structures is responsible for a large part of the total SERS effect, about which more will be said later. The next possibility is to increase the Raman scattering cross section. This can be accomplished in two ways: molecules can be chosen with resonances in the visible so that resonance enhancement of the cross section occurs with the commonly used ion lasers or new deep ultraviolet lasers can be employed to excite resonances in virtually any molecule of interest. The most general approach to overcome the limitation of low count rates is to improve upon the efficiency of the detection system. In our laboratory we have used multichannel detection to decrease the data acquisition time by more than a factor of 10^2. Thus, the six hour experiment mentioned above can be completed in a few minutes.

The presence of any background in the spectrum can increase the time needed to acquire it. For example, in our experiments on single-crystal metal surfaces we observe an inelastic continuum of around ten counts per second, on which the Raman signal of say one count per second is superimposed. Recalling that the noise is equal to the square root of the total number of photons counted in the shot noise limit, the signal-to-noise ratio is given by

$$S/N = I_R/(I_R + I_B)^{1/2}$$

where I_R and I_B refer to the Raman and background intensities, respectively. Thus for our typical example $S/N = 0.3$ for a one-second integration and $S/N = 10$ can be achieved in a 15-min integration, which is certainly a reasonable period.

As important as sensitivity is selectivity, the ability to discriminate between adsorbed species and those in an ambient gas or liquid phase. Again, the worst situation is provided by unenhanced spectroscopy on low-area surfaces. In this case the number of molecules in the ambient gas or liquid phase sampled is of the order of the laser beam diameter cubed times the density of the gas or liquid, while the number of surface molecules sampled is of the order of the square of the beam diameter times the surface density. Thus the ratio of surface scattering to bulk scattering is simply $\sigma'/\rho d$, where σ' and ρ are the surface and bulk densities, respectively, and d is the laser beam diameter. Note that smaller beam diameters emphasize surface scattering, which can be used to advantage in designing an experiment. For a laser spot focus of 100 μm, the ratio of surface to bulk scattering becomes 1 at 10 torr or so, which represents an upper limit in the worst case, namely, identical vibrational frequencies for both the bulk and adsorbed species. It is rarely the case that both species have the same vibrational frequency so that the problem then becomes one of dynamic range. That is, the spectral features of the ambient must not saturate the detector before adequate surface signal is collected. By judicious choice of frequency region sampled, and by background subtraction if necessary, we are confident that experiments can be conducted in the presence of one atmosphere of an ambient gas phase. For example, we are studying intermediates in the Fischer–Tropsch synthesis of hydrocarbons over single-crystal model catalysts using *in situ* Raman spectroscopy. In this case, the reactants (H_2 and CO) have vibrational frequencies at 4161 and 2145 cm^{-1}, respectively, regions where no surface intermediates have bands. For this reaction, we expect to be able to use very high pressures with no interference from the gas phase. Polarization modulation can be employed to discriminate against the bulk scattering, much as it is used in the infrared to discriminate against bulk absorption (see the chapter by Hayden in this volume). As will be discussed in detail later, only p-polarized radiation is effective in exciting surface Raman scattering while both polarizations excite the bulk scattering. Thus, subtraction of the s spectrum from the p spectrum yields only the surface spectrum.

For high surface area materials, or SERS, the selectivity improves in much the same way as the sensitivity. The high surface area-to-volume ratio in the former case and the confinement of the SERS effect to the first few monolayers near a surface in the latter case guarantee selectivity.

To recapitulate briefly, then, Raman spectroscopy of molecules adsorbed at low coverage in single-crystal surfaces is a challenging experiment. The most general approach is simply to use multichannel detection, so that there are no restrictions on the nature of the substrate or the adsorbate. Other interesting experiments may involve high surface area substrates, or adsorbates which show either resonance-enhanced Raman cross sections or SERS or both. In these cases conventional instrumentation is adequate but multichannel detection opens up exciting new possibilities, for example time-resolved measurements.

3.2. Single-Crystal Surfaces

3.2.1. Electromagnetic Fields at Surfaces

No surface spectroscopic experiment involving electromagnetic radiation can be designed without an understanding of the behavior of the electromagnetic field near a surface. The problem is an old one, and can be treated simply by solving Maxwell's equations with the appropriate boundary conditions. For Raman scattering these considerations apply to both the incident and scattered fields and can be used to optimize the experimental design, and determine vibrational symmetries and molecular orientations, which I will discuss in subsequent sections. Greenler[9] was the first to consider Raman scattering by a molecule adsorbed on a planar surface as a problem in classical electrodynamics.[10] His model is simple. The primary field felt by the molecule is the sum of the incident and reflected fields, whose amplitudes may be calculated from Maxwell's equations. The primary field induces an oscillating dipole in the molecule which radiates. The sum of the directly emitted field and that suffering a single reflection from the surface is the secondary or scattered field. The incident angle and input polarization determine the intensity of the primary field. The strength and orientation of the induced dipole depend upon the mode symmetry and the scattering cross section. Finally, the orientation of the induced dipole determines the variation in scattered intensity with observation angle.

3.2.1a. The Primary Field. The boundary conditions on Maxwell's equations require that the tangential components of the electric and

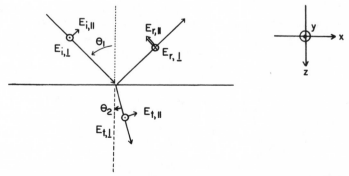

Figure 5. Geometry for light impinging on a surface. E_i, E_r, and E_t are the incident, reflected, and transmitted fields, respectively. The associated angles of incidence, reflection, and transmission are also indicated.

magnetic fields and the normal component of the displacement and induction fields be continuous across the interface between two media of finite conductivity.[10] Consider the geometry pictured in Figure 5. Light is incident from the left making an angle of θ_1 with the surface normal. A portion of the light is reflected at an angle θ_1 and transmitted at an angle θ_2. The electric fields of each ray are denoted by E_i, E_r, E_t, respectively. Solving Maxwell's equations with the boundary conditions mentioned above (assuming the materials are nonmagnetic) results in the Fresnel equations for the reflected and transmitted fields:

$$r^\perp = E_r^\perp/E_i^\perp = \frac{n_1 \cos\theta_i - (n_2^2 - n_1^2 \sin^2\theta_i)^{1/2}}{n_1 \cos\theta_i + (n_2^2 - n_1^2 \sin^2\theta_i)^{1/2}}$$

$$t^\perp = E_t^\perp/E_i^\perp = \frac{2n_1 \cos\theta_i}{n_1 \cos\theta_i + (n_2^2 - n_1^2 \sin^2\theta_i)^{1/2}}$$

$$r^\| = E_r^\|E_i^\| = \frac{-n_2^2 \cos\theta_i + n_1(n_2^2 - n_1^2 \sin^2\theta_i)^{1/2}}{n_2^2 \cos\theta_i + (n_1(n_2^2 - n_1^2 \sin^2\theta_i)^{1/2}}$$

$$t^\| = E_t^\|/E_i^\| = \frac{2n_1 n_2 \cos\theta_i}{n_2^2 \cos\theta_i + n_2(n_2^2 - n_1^2 \sin^2\theta_i)^{1/2}}$$

(24)

where the electric field amplitude is broken down into components parallel ($E^\|$) and perpendicular ($E^\perp$) to the plane of incidence, and n_1 and n_2 are the refractive indices of the two media, n_1 referring to the medium containing the incident and reflected rays.

In polar form the reflection coefficients become

$$r^\perp = R^{1/2} \exp(i\delta_r^\perp)$$
$$r^\| = R^{1/2} \exp(i\delta_r^\|)$$

(25)

where $R^{1/2}$ is the amplitude of the reflected field relative to the incident field. R is known as the reflectivity. δ, the reflection phase shift, is given by

$$\delta = \tan^{-1}[\text{Im}(r)/\text{Re}(r)]$$

(26)

The total field at the surface is the sum of the incident and reflected fields. More important is the intensity which is the mean square total electric field, components of which are given by

$$\langle E_y^2 \rangle = \langle E_\perp^2 \rangle(1 + R_\perp + 2R_\perp^{1/2} \cos\delta_\perp)$$
$$\langle E_x^2 \rangle = \langle E_\|^2 \rangle(1 + R_\| - 2R_\|^{1/2} \cos\delta_\|)(\cos^2\theta_i)$$
$$\langle E_z^2 \rangle = \langle E_\|^2 \rangle(1 + R_\| + 2R_\|^{1/2} \cos\delta_\|)(\sin^2\theta_i)$$

(27)

where $\delta_{\perp,\parallel} = \delta_r + (4\pi z/\lambda)n_1 \cos \theta_i$ determines how the phase shift varies with distance from the surface. Since we are usually concerned with molecules absorbed at submonolayer coverage we can ignore the distance dependence of the phase shift.

The surface fields are most easily evaluated using a computer. Through these relations, the total field intensity is limited to between zero and four times the incident field intensity depending on the phase shift, reflectivity, and angle of incidence. The limiting case is referred to as the "perfect mirror" in which $R_\perp = R_\parallel = 1$ and $\delta_\perp = 180°$ and $\delta_\parallel = 0°$. This situation is very nearly satisfied for metals at infrared wavelengths (Figure 6). Note how this results in the so-called "surface selection rule" for surface infrared spectroscopy. No field tangential to the surface exists at any incident angle. The field normal to the surface

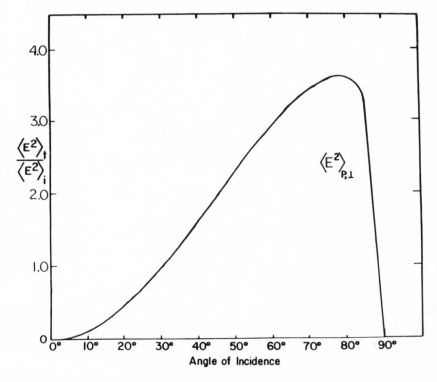

Figure 6. The total electric field intensity at a silver surface in vacuum relative to the incident field intensity as a function of incident angle and polarization. $\lambda = 5 \, \mu m$ for which the bulk silver optical constants are $n = 3.50$ and $k = 30.40$.[11]

follows essentially a $4 \sin^2 \theta_i$ dependence dictating an experimental geometry near glancing incidence.

The dielectric properties of the surface material are strongly wavelength dependent. The same material at visible wavelengths has very different properties than in the infrared. The reflectivity is often less than one and phase shifts are not equal to 0° or 180°. For silver, the reflectivity remains fairly constant from the infrared to the visible. At 0°, however, the phase change moves from 3° to 30°. Tangential fields are therefore possible (Figure 7). At high angles of incidence, the phase change for p-polarized light shifts from close to 0° to close to 180°, resulting in the maximum intensity occurring near 65° rather than at higher angles. From the shape and breadth of the curve it is clear that incident angles from 50° to 70° are optimum for excitation of surface Raman spectra.

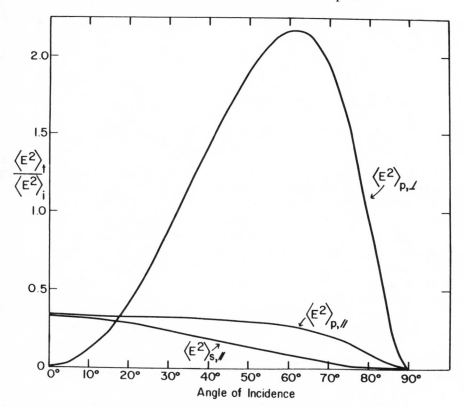

Figure 7. The total electric field intensity at a silver surface in vacuum relative to the incident field intensity as a function of incident angle and polarization. $\lambda = 520$ nm, for which the bulk silver optical constants are $n = 0.0427$ and $k = 3.3988$.[11]

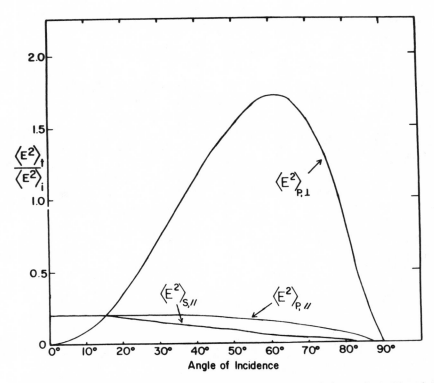

Figure 8. The total field intensity at a nickel surface in vacuum relative to the incident field intensity as a function of incident angle and polarization. $\lambda = 520$ nm, for which the bulk nickel optical constants are $n = 1.8796$ and $k = 3.5061$.[11]

Figures 8 and 9 demonstrate the effect of changing materials at the same wavelength of light. Upon shifting from silver to nickel to silicon the reflectivity steadily decreases. The phase shift progressively decreases as well, however. The result is that all the fields are diminished on the less reflective surfaces.

Finally, if the ambient medium is changed (from vacuum to water, for instance), the curves again shift away from the perfect mirror limits (Figure 10). Although the reflectivity is virtually unchanged, the phase shift has increased. The tangential fields therefore increase at the expense of the perpendicular fields. This result suggests that a different configuration may be useful in electrochemical experiments. Normal incidence, normal collection may be a more convenient geometry as it minimizes the volume of electrolyte sampled while not giving up much sensitivity for a water ambient. Note that in all cases using visible light, the peak

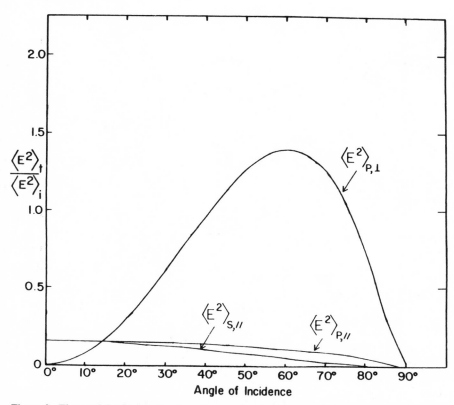

Figure 9. The total field intensity at a silicon surface in vacuum relative to the incident field intensity as a function of incident angle and polarization. $\lambda = 515\,nm$, for which the bulk silicon optical constants are $n = 4.16$ and $k = 0.1$.[11]

positions remain roughly the same while the levels may vary. Thus, for most materials, optimum excitation is achieved with p-polarized light incident at 50°–70°.

The presence of an adsorbed layer and possible deviations from classical behavior have been ignored above. Some authors have explored three-phase models in which two interfaces and multiple reflections are involved.[12] The two-phase model for Raman scattering can be justified on several grounds. The adsorbate is very thin and generally nonadsorbing so that it has very little interaction with the light. The Raman effect is also a scattering process where light is analyzed at a shifted wavelength and in a direction away from the specular direction. Multiphase models appear to be much more important for spectroscopies such as infrared

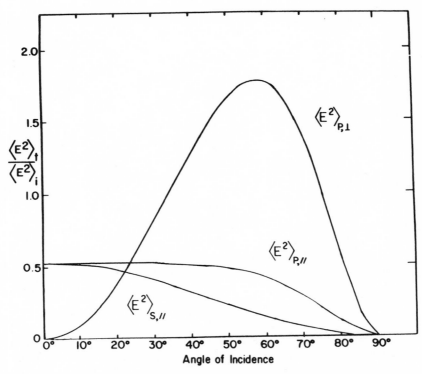

Figure 10. The total field intensity at a silver surface in water ($n = 1.33$) relative to the incident field intensity as a function of incident angle and polarization. $\lambda = 520$ nm, for which the bulk silver optical constants are $n = 0.0427$ and $k = 3.3988$.[11]

and differential reflectance where the total effect is a change of at most a few percent in the specular beam. The two-phase model should be adequate for Raman and possibly luminescence spectroscopy. Nonclassical corrections to the surface field intensity have been studied by others as well.[13] They can be considered if future experimental results warrant it.

3.2.1b. The Secondary Field. The intensity of the scattered light can also be calculated from Greenler's model.

The oscillating dipole can be considered a point source emitting radiation above the surface plane. The dipole direction may be described in terms of three possible orientations relative to the observation point (Figure 11). The observation point lies at an angle θ_3 with respect to the surface normal. The observer will see light scattered directly from the dipole and light scattered toward the surface and reflected toward the

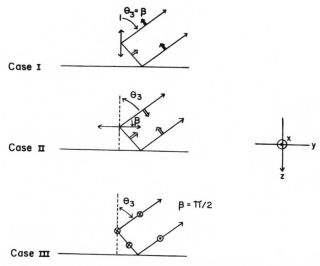

Figure 11. The three possible orientations for an oscillating dipole on a surface relative to the observation direction. The direct and reflected fields sum constructively in case I and destructively in cases II and III.

observation point. The angle at which the indirect light strikes the surface is also θ_3 since the angle of incidence equals the angle of reflection.

The light is p-polarized in cases I and II and is s-polarized in case III. The magnitude of the directly scattered light is proportional to $\sin^2 \theta_3$ for case I and $\cos^2 \theta_3$ for case II. It is constant for case III. Considering the geometry of the rays in Figure 11 and comparing them to the geometry in Figure 5, the instantaneous field strengths headed toward the observer are

$$\bar{E}_{\text{I}} = (E_d^{z,\text{I}} + E_r^{z,\text{I}} + E_d^{x,\text{I}} - E_r^{x,\text{I}})$$
$$\bar{E}_{\text{II}} = (E_d^{z,\text{II}} - E_r^{z,\text{II}} + E_d^{x,\text{II}} + E_r^{x,\text{II}}) \qquad (28)$$
$$\bar{E}_{\text{III}} = (E_d^{y,\text{III}} + E_r^{y,\text{III}})$$

where $E^{i,j}$ is the amplitude of the field in direction i due to the dipole oriented as in case j and d, r refer to directly detected and reflected rays, respectively. The total mean square field strength observed in the direction θ_3 is therefore

$$\langle E_{\text{I}}^2 \rangle_{\text{tot}} = \langle E_{\text{I}}^2 \rangle \sin^2 \theta_3 (1 + R_{\|} + R_{\|} + 2R_{\|}^{1/2} \cos \delta_{\|})$$
$$\langle E_{\text{II}}^2 \rangle_{\text{tot}} = \langle E_{\text{II}}^2 \rangle \cos^2 \theta_3 (1 + R_{\|} - 2R_{\|}^{1/2} \cos \delta_{\|}) \qquad (29)$$
$$\langle E_{\text{III}}^2 \rangle_{\text{tot}} = \langle E_{\text{III}}^2 \rangle (1 + R_{\perp} + 2R_{\perp}^{1/2} \cos \delta_{\perp})$$

where $\langle E_j^2 \rangle$ is the mean square field emitted normal to oscillator j. The fields described by equations (29) have the same form as equations (27). For an observer located infinitely far from the surface such that only one observation angle is detected, the variation in scattered intensity with observation angle for a dipole on a surface is the same as the variation in surface field intensity with incident angle. Case I corresponds to $\langle E_z^2 \rangle$, case II to $\langle E_x^2 \rangle$, and case III to $\langle E_y^2 \rangle$.

These equations are not useful in practice because the collection lens is never infinitely far from the observer. Thus these equations for scattered intensity must be integrated over the solid angle of the collection optics. This is a complicated problem since contributions from case II and case III dipoles become mixed. The problem has been solved numerically[14] and the solution presented here in Figure 12. These calculations were carried out for an $f/0.95$ lens that we use routinely in our experiments ($\Omega = 0.72\,\mathrm{sr}$) for silver at 5400 Å. The results are in

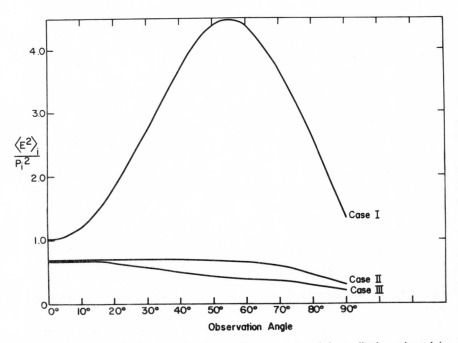

Figure 12. Angular distribution of radiation Raman scattered from dipoles oriented in different directions. Case I is a dipole oriented normal to the surface; case II is oriented parallel to the surface and is contained in the observation plane; case III is oriented parallel to the surface and perpendicular to the scattering plane. The calculations integrate Fresnel's equations over the finite angular range of the collection optics.

accord with expectations, namely, that the angular profiles become broader, but retain their general shape. Note that the optimum detection angle for case I is 55°, the observation angle around which our UHV chamber was designed. Details on the numerical procedure may be found in Ref. 14.

It is clear from the foregoing discussion that the optimum angle of incidence is near 60° for most metals in the visible, that dipoles oriented normal to the surface radiate most efficiently, and that the radiation from those dipoles peak near 60° from the surface normal because the physics is identical in both cases. Relative intensities of vibrational modes excited and observed under different conditions can be used to determine mode symmetries and molecular orientations. In the next section I will describe how the angular distributions of the scattered radiation are measured and what information they may provide.

3.2.2. Angle-Resolved Measurements

Recall that the orientation of the induced dipole depends upon the orientation of the driving electric field and on the symmetry of the vibration. Totally symmetric modes are usually the most intense in Raman spectra; for these modes the induced dipole is parallel to the incident electric field. Thus, for the totally symmetric modes the geometry described in the last section—p-polarized light incident at 60°, observation at 60°—will yield the most intense spectra. These excitation conditions produce an electric field normal to the surface which induces a normal dipole whose radiation peaks near 60°. Although the totally symmetric modes can provide information about molecular orientation and symmetry and are essential in providing a complete picture of the spectrum. It is clear that the nontotally symmetric modes require different excitation and observation conditions, which can be determined by the methods of the last section.

We have designed a surface Raman spectrometer that is capable of measuring the angular distribution of the scattered radiation with two goals in mind. First, we wanted to check the adequacy of the classical electrodynamic analysis of surface Raman scattering by comparing the experimental distribution with the theoretical predictions. Second, we wanted to be able to use the technique to determine mode symmetries and molecular orientations.

We built an instrument that allows the observation angle to be scanned while keeping the incident angle and polarization constant. Figure 13 shows the manipulator used. The laser beam enters the UHV chamber horizontally and then is deflected by a 45° mirror so that it is

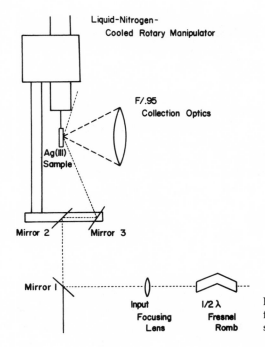

Figure 13. Experimental arrangement for angle-resolved surface Raman spectroscopy.

coaxial with the manipulator axis. Thus the position of the incident beam on the second mirror is invariant to the rotational setting of the manipulator. The second and third mirrors deflect the beam onto the crystal surface at a fixed angle of incidence. Not only must the incident angle be fixed, but the input polarization must remain constant. This was accomplished by using dielectric (rather than metal) mirrors so that the linear polarization state is preserved upon reflection. The polarization was rotated synchronously with the change in observation angle by a $\lambda/2$ Fresnel rhomb mounted externally to the chamber.

As a test case, we chose benzene physically adsorbed at half monolayer coverage on a Ag(111) surface. It is known that benzene lies flat on this surface[15] so that, under these excitation conditions, the induced dipole is given by $\mu_z = \alpha_{zz}E_z$. Note that α_{zz} is the least polarizable component of the scattering tensor, but we are still able to obtain spectra quite easily. Points were taken between $+80°$ and $-80°$ and the results for the two quadrants averaged. Figure 14 shows the experimental points compared to the theoretical prediction. The curve was generated from equation (29) using tabulated values of the silver dielectric constant at the Stokes shifted Raman frequency, and integrat-

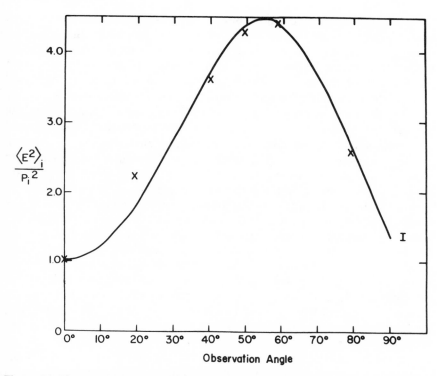

Figure 14. The angular distribution of Raman scattered radiation from v_1 of benzene (992 cm^{-1}) adsorbed at half-monolayer coverage on Ag(111) at 80 K. The solid line is the theoretical curve generated by integrating Fresnel's equations over the finite range of observation angles subtended by the collection optics. The crosses are the experimental points averaged over the two quadrants and normalized to the intensity at 0°.

ing trajectories over the solid angle of our collection optics. Apart from normalization there are no adjustable parameters; the excellent fit suggests strongly that classical electrodynamics is adequate to understand surface Raman spectra.

We tried to observe scattering from dipoles oriented parallel to the surface but have so far been unsuccessful. We are planning to design a second generation angle-resolved spectrometer that will overcome some of the limitations of our first generation instrument. In particular, we plan to be able to switch from high angles of incidence to normal incidence, for more effective excitation of tangential dipoles, and will use different collection optics for greater efficiency.

The results presented here suggest a standard set of surface Raman experiments to determine the symmetries of the bands observed as well as the principal values of the molecular polarizability tensor in the surface frame of reference. For example, a totally symmetric mode will show maximum intensity for p-polarized excitation at a high angle of incidence when observed off-normal. The angular scan of the scattered light will have a minimum along the surface normal. A mode transforming as xz or yz will also be more intense for p-polarized high-angle excitation, but the angular distribution of the scattered light will be relatively flat. Distinctions between xz and yz modes can be made by analyzing the polarization of the scattered light. Finally, xy modes will be very weak. They are preferentially excited by s-polarized light at normal incidence and are observed most efficiently along the surface normal. A polarizer must be used to distinguish between the totally symmetric modes which will scatter with the same polarization as the incident laser, and the xy modes which will scatter with the orthogonal polarization. Table 1 shows the difference in intensities of various modes assuming that their cross sections are identical and that differences are caused only by the presence on the surface of the primary and secondary electromagnetic fields. Thus the angular dependence of the scattered intensity can be used to determine the assignment of the bands observed, hence the symmetry of the molecule in its adsorption site according to the methods of Section 3.2.3.

It is also interesting to be able to measure the relative values of the diagonal elements of the polarizability tensor in the surface frame of reference. The intensity of the scattered light measured off-normal using high-angle p-polarized radiation is proportional to α'_{zz}. The values of α'_{xx} and α'_{yy} relative to α'_{zz} are measured using normal incidence and normal collection by changing the incident laser polarization. These measurements yield the same kinds of information that any spectroscopic method offers, depending upon what else is known about the system. For example, if the molecule is weakly adsorbed such that its electronic distribution is not greatly altered, the principal values of the polarizability

Table 1. Dependence of Raman Intensities on Excitation and Observation Conditions

θ_{inc}	Inc. pol.	θ_{obs}	α^2_{zz}	α^2_{xz}	α^2_{yz}	α^2_{xy}
70°	p	0°	1.0	0.68	0.68	0
70°	p	55°	4.5	0.38	0.67	0
0°	x	0°	0	0.16	0	0.11
0°	y	0°	0	0	0.16	0.11

tensor will be nearly the same as for the free molecule. This determination of the principal values in the surface frame of reference is equivalent to determining the molecule's orientation on the surface. Alternatively, for the case of strong chemisorption, and if there is independent evidence for the adsorption geometry, these measurements can help us understand how the electronic density has shifted upon chemisorption. We are planning to test these ideas in a new series of experiments.

3.2.3. Selection Rules

The selection rule for infrared absorption and specular EELS states that losses are observed only for those vibrations having a nonzero component of the transition dipole moment along the surface normal. The selection rule is frequently framed in terms of image dipoles; parallel dipoles are screened by their images while perpendicular dipoles are reinforced.

The situation is somewhat more complicated for Raman spectroscopy, since the selection rules depend upon the dipole polarizability rather than the dipole moment directly. Also, the imperfect screening ability of the metal in the visible effects a modification of the image dipole rule. Parallel dipoles are no longer completely screened and the selection rules become more properly "propensity rules" to predict the relative intensities of various vibrational bands in the Raman spectrum. The more exact treatment of the interference of incident and reflected electromagnetic waves at the surface (of which the image dipole formalism is the special case of a perfect conductor) for Raman scattering has been presented by Moskovits[16] and gives some indication of the general trend in surface scattering intensities. Modes that transform as z^2 are expected to be the most intense spectral features, while xz and yz modes should be next in size. x^2, y^2, and xy type vibrations should be significantly weaker than the others. Of course, the actual intensities of the bands will depend on the absolute magnitudes of the polarizability tensor components and the experimental geometry.

In addition to image effects, adsorbate site symmetry may influence the spectral activity.[17] The removal of a symmetry plane due to the presence of an adjacent homogeneous surface will produce an effective molecular symmetry lower than that of the free molecule. If the adsorbate is sensitive to the atomic geometry of the surface, a further decrease in symmetry may occur. These symmetry reductions result in the appearance of bands forbidden in the free molecule. An alternate mechanism for the appearance of forbidden vibrations is the excitation of quadrupole transitions by a large electric field gradient at the surface.[18]

The quadrupole polarizability tensor, A, has components that transform as trinary Cartesian products. An indication of the relative magnitudes of these quadrupole modes was included in the above-mentioned theoretical investigation. Vibrations that transform as z^3 should be strongest, while z^2x and z^2y modes will be less surface-active.

To distinguish which of these several factors dominate the selection rules for Raman scattering by adsorbed molecules we have conducted an experimental investigation involving several different molecules of different symmetries physically adsorbed on two crystal faces of different symmetries. We chose physisorbed molecules to emphasize only the physical interactions mentioned above; chemical effects must await further investigation.

Strong evidence for alteration of the Raman selection rules for an adsorbed molecule is provided by the spectra of a 5-L (uncorrected) submonolayer dose of benzene-d_6 onto the (111) and (110) faces of silver, as shown in Figure 15. The observed spectra are distinctly different from that of the liquid and from each other.

Only two bands appear in the spectrum of benzene absorbed on Ag(111). The vibrations at 945 and 495 cm^{-1} are assigned as the totally

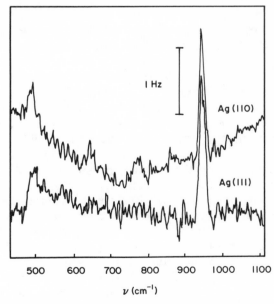

Figure 15. Surface Raman spectra of benzene-d_6 adsorbed on Ag(111) and Ag(110) at 90 K and submonolayer (5 L, uncorrected) coverage.

symmetric ring-breathing mode (a_{1g}) and the symmetric out-of-plane C–H deformation (a_{2u}), respectively.[19] The observation of the latter band, which is normally Raman inactive, immediately suggests either a reduction of the molecular symmetry upon adsorption or the excitation of a quadrupole transition. As benzene is known to adsorb with its ring parallel to the surface (zZ),[15] the simplest reduction symmetry would be C_{6v}, wherein the free molecule (D_{6h}) has lost a reflection plane by imposition of a homogeneous surface. With this symmetry a_{2u} is correlated to a_1 and thereby allowed. Interestingly, this band is the strongest of the infrared active vibrations in the free molecule. The a_{2u} mode belongs to the same representation as z^3, and may therefore also be evidence of a large quadrupole component in the scattering. Distinction between these two alternatives cannot be made on the basis of this spectrum alone. The absence of normally Raman active bands of symmetry e_{1g} and e_{2g} can be explained by the transformation of these modes as (xz, yz) and $(x^2 - y^2, xy)$, respectively, which would produce significantly weaker Raman scattering. EELS investigations show that the benzene spectrum is sensitive to the adsite geometry and have assigned the site symmetry as $C_{3v}(\sigma_d)$.[15] Our results are not inconsistent with this assignment, as the modification to the predicted spectrum is the inclusion of two free molecule forbidden b_{2u} modes at 840 and 1290 cm^{-1} by correlation to a_1, with weaker scattering from all the e modes. The latter b_{2u} band is outside the monitored spectral range, and detection of both these bands may be limited by our sensitivity, since the bands are typically very weak even in the solid state. This system is an experimental demonstration of the presence of surface selection rules, especially that due to surface electric field interference effects. Also, either symmetry reduction or quadrupole interactions are important, but the two cannot be distinguished from this spectrum alone.

The Raman spectrum of the same dose of benzene-d_6 onto the Ag(110) face provides some insight into this problem. This spectrum is significantly different from that of the Ag(111) adsorbed species—further evidence for surface selection rules. Quadrupole polarizability effects should not be greatly altered for various low index faces of the same metal, although small relative changes in intensity of the quadrupolar bands with respect to the dipole modes might be reasonable. The electric field gradient for the various faces is, in fact, expected to vary in proportion to the metal work function,[20] which ranges from 4.52 V for Ag(110) to 4.74 V for Ag(111).[21] In the Ag(110) spectrum, the relative magnitude of the a_{2u} mode is roughly equal to or even greater than the same band in the Ag(111) spectrum, which is opposite the proposed trend. The appearance of at least two other bands of relatively large intensity is not consistent with the quadrupole theory, but can be easily explained by symmetry reduction. The two new bands are assigned as the $650 \text{ cm}^{-1} e_{1g}$ antisymmetric out-of-plane C–H deformation and an e_{2u} out-of-plane C–H deformation at 780 cm^{-1}. The appearance of the e_{2u} mode indicates a symmetry decrease to at least C_{2v}. This low symmetry is a positive indication that the molecule does respond to the adsorption site symmetry. For D_{6h} correlated to C_{2v}, a_{1g}, a_{2u}, e_{2g}, and e_{2u} modes belong to the representation a_1, and are therefore allowed since a_1 has the requisite z^2 transformation property. The e_{2g} modes at 575 and 870 cm^{-1} are not directly observable in this spectrum, which is not surprising considering their relative weakness in the Raman spectrum of the liquid. The appearance of the e_{1g} mode with such intensity is interesting since this mode transforms as (xz, yz), which is allowed, but with decreased magnitude, and the typical free molecule intensity is approximately an order of magnitude less than the 945-cm^{-1} band. This apparent anomaly can be directly ascribed to the C_{2v} symmetry. Disubstituted benzenes with C_{2v} symmetry typically demonstrate a large increase in the intensity of the e_{1g} mode,[19] often to the extent of its being comparable with the a_{1g} band. These two effects compensate to give the observed intensity. Thus, the Ag(110) spectrum provides strong evidence for the importance of symmetry reduction in the surface selection rules for Raman spectroscopy. Quadrupole transitions may indeed contribute to the observed intensity of the a_{2u} mode, but are certainly not the dominant factor.

The comparison of the Raman spectra obtained from a submonolayer dose of benzene-d_6 onto the (111) and (110) faces of silver reveals that two important factors determine the Raman spectral activity of adsorbed molecules. Since the metal is not a perfect reflector in the visible region of the spectrum, rigid image dipole rules are not obeyed. Rather, a set of electric field propensity rules determine the expected relative intensities of different vibrational modes. Our experimental observations confirm the theoretical predictions. As a general rule, z^2 modes will be the most intense spectral features, while bands that transform as xz or yz may appear with reduced intensity. x^2, y^2, and xy modes are very weak scatterers and typically will not appear in the spectra. The adsite geometry will cause a decrease in the molecular symmetry, resulting in the appearance of Raman vibrations that are forbidden in the free species. The new symmetry may also result in increased intensity for some vibrational modes, and to some extent may compensate for the weakened scattering caused by the surface effects on the electric field. Quadrupole polarizability induced dipoles cannot explain the vastly different spectra observed for the two minimally dissimilar surfaces, and therefore must not greatly influence the surface selection rules. This is not to suggest that quadrupole transitions might not play an important role for other surfaces, such as the spherical particle covered silver surfaces used in SERS, which may have a much stronger electric field gradient than that occurring on our smooth, well-ordered surfaces.

For the benzene–silver adsorption system, we find that the best fit with our data indicates that the benzene-d_6 adsorbs at a site with C_{6v} symmetry on the Ag(111) surface and at a C_{2v} site on the Ag(110) surface. With the present detection limits of our surface Raman spectrometer, however, we cannot confirm the $C_{3v}(\sigma_d)$ site symmetry proposed for benzene on Ag(111) on the basis of EELS results, although our data are not inconsistent with such an assignment.

Finally, we have confirmed the general results obtained for the benzene–silver system by obtaining the Raman spectra of pyrazine (D_{2h}) and s-triazine (D_{3h}) on both Ag(111) and Ag(110) surfaces.[22]

3.3. High Surface Area Materials

It is difficult to estimate in general the number of scatterers sampled by the laser on a high surface area material because of multiple scattering of both the incident and Raman shifted radiation. Egerton and Hardin[23] have provided an estimate for both opaque and transparent samples. A typical sample of 2.5 g of a material of $200 \, \mathrm{m^2 \, g^{-1}}$ specific area can be compressed into a disk 25 mm in diameter by 2 mm thick. If a line focused laser is used that illuminates a region $0.5 \times 0.05 \, \mathrm{mm}$ (nearly optimal for filling the entrance slits of a monochromator with an ISPD array), and if the beam penetrates the entire sample, then ca. 10^{17} adsorption sites are sampled (assuming a site density of $10^{14} \, \mathrm{cm^{-2}}$). This is a factor of 10^3 larger than what we routinely observe on single-crystal surfaces and can therefore easily be seen using conventional detection. If we make the conservative assumption that for opaque substrates only 10^{-3} of these sites are sampled, then we have the same number density as for a monolayer of adsorbates on a low surface area material and the same considerations discussed in Section 3.1 apply. Clearly the study of adsorbents and adsorbates by Raman spectroscopy is well within the reach of modern techniques. Two problems are often encountered in the study of high surface area materials—fluorescence and sample heating. Extensive discussions of the fluorescence problem have been given[23-25] and will not be repeated here. The fluorescence has been ascribed to the presence of hydrocarbon and transition metal impurities and can be minimized by careful sample preparation. For long-lived fluorescence, discrimination can be accomplished using pulsed excitation and gated detection since Raman scattering is an instantaneous process.[26]

Sample heating is a bigger problem for high surface area materials because they tend to adsorb most of the laser radiation and are poor thermal conductors. Various schemes of sample spinning or rastering the exciting laser have been used to minimize this problem.[27,28] Multichannel detection should essentially eliminate problems with sample heating by reducing the laser power required to obtain a spectrum, and experiments along these lines are in progress, both in our laboratory and elsewhere.[29]

4. Instrumentation

4.1. Laser Sources

In the early days of Raman spectroscopy, samples were excited by one of the intense lines of a mercury arc. Because the brightness of this source is rather low, acquiring a spectrum was a tedious proposition at best. The discovery of the laser in 1960 revolutionized many areas of spectroscopy, but its impact was felt most strongly in the revitalization of Raman spectroscopy. The extreme brightness of the laser, combined with its narrow bandwidth, coherent properties, short time resolution, and wide range of accessible frequencies have made possible a panoply of new experiments.

Lasers for Raman spectroscopy can be conveniently divided into two categories, continuous wave (cw) and pulsed. For linear or spontaneous Raman spectroscopy and where no time resolution is required, continuous lasers are generally preferred to minimize undesired multiphoton processes that may lead to sample degradation. The most popular cw laser source for spontaneous Raman spectroscopy has been the argon ion laser, which provides a number of lines in the blue-green region of the spectrum (457.9–514.5 nm) as well as in the ultraviolet (336.6–363.8 nm). Krypton ion lasers are also used often when extended wavelength coverage is desired; they provide strong lines in the red (647.1 nm) and the violet (413.1) that complement the spectral coverage of the argon ion laser. Finally, either of these lasers can be used to pump a cw dye laser, if intermediate frequencies are required. The choice of exciting frequency can be very important. Since light scattering intensities scale with the fourth power of the frequency, blue and even ultraviolet excitation would seem to be preferred. This advantage alone is usually offset by the greater experimental difficulties found when working in the ultraviolet, so that the vast majority of Raman studies are conducted using visible excitation. If the molecule of interest has an accessible electronic state, then tuning into resonance can lead to an enormous enhancement of the Raman scattering cross section—some six orders of magnitude. Exploitation of the resonance Raman effect can lead to great increases in both sensitivity and selectivity, focusing only upon chromophoric parts of large molecules, for example. Finally, for fluorescent molecules, the excitation frequency should be chosen to avoid exciting the fluorescent state, since the fluorescence would completely mask the much weaker Raman scattering.

Pulsed lasers have found little application so far in surface Raman spectroscopy. Most pulsed dye lasers have extremely low duty cycles (10^{-6}) so that to achieve the modest average power required for spontaneous Raman scattering requires extremely high peak powers that can result in desorption and decomposition. High repetition rate, injection locked excimer lasers hold great promise for the future, however. Deep ultraviolet excitation has already been shown to lead to very large scattered intensities;[30] the narrow bandwidth of the injection locked laser and much higher repetition rates should lead to great improvements using this approach.

The laser bandwidth and spectral purity are important in Raman spectroscopy. Condensed phase Raman linewidths from molecules are of the order of $5\,cm^{-1}$ so that the ion laser's multimode linewidth $(0.1\,cm^{-1})$ provides sufficient resolution. At least one etalon should be added to a broadband cw dye laser, however, to narrow the line sufficiently. In addition to the lasing emission, ion lasers emit fluorescence from the ionized plasma. Although weak compared to the laser line strength, these plasma lines can overwhelm the Raman scattering if not filtered out before reaching the sample. Unenhanced Raman scattering from molecules on single-crystal surfaces provides perhaps the most stringent test of the adequacy of plasma line rejection, since not only is the signal level low, but the plasma radiation is very effectively elastically scattered by the substrate. A number of approaches have been tried to eliminate this problem. For less critical applications, simple dispersing prisms (either 90° Brewster or 60° equilateral) or gratings can be used with an aperture to transmit only the laser line. This scheme works well if the Raman signals are strong, and if the frequency shifts of interest are relatively large. To improve upon this simple scheme, a lens may be inserted to focus the laser output onto a slit, and a second lens inserted later in the optical train to recollimate the laser beam. This setup is adequate if one laser line is used most of the time; realignment following a change in excitation frequency can be a nuisance. Narrow bandpass (1.0 nm) interference filters provide an alternative solution to the plasma line problem. They are convenient to insert into the optical train, and provide excellent rejection. If many laser lines are used routinely, however, a large number of filters do not handle very much power due to optically absorbing blocking dyes and cements. Interference filters also generally suffer from relatively low throughput. We have been very pleased, however, with the results of a new interference filter, custom designed at our request (Omega Optical). By eliminating the blocking dyes, the transmission of the filter was increased to >70% and its power handling capability dramatically improved. Blocking extends only to include frequencies shifted by $4000 \, \text{cm}^{-1}$ or less, but this is really no limitation for vibrational Raman spectroscopy.

The most versatile and effective solution to the plasma line problem we have found so far is a prism monochromator especially designed for this purpose (Applied Photophysics). While more expensive than the other approaches mentioned, its narrow bandpass (1.0 nm), high through-put (75%), and coaxial input–output have made a real difference in the day-to-day operation of our laboratory. A monochromator of this type is clearly the method of choice for the most demanding applications where frequent changing of laser lines is required.

4.2. Sample Configurations

4.2.1. Single-Crystal Samples

Single-crystal surfaces may be examined under ultrahigh vacuum (UHV), or at elevated pressures *in situ,* using surface Raman spectros-copy. Standard UHV techniques have been used in our laboratory; Raman capabilities were simply added to a commercial UHV chamber. All that is required are two lenses, one to focus the laser onto the sample surface and the second to collect the scattered radiation. Recalling that the optimum angle of incidence is near 60° for metal surfaces in the visible, and that the scattered intensity also peaks near 60°, the simplest experimental geometry is to have the excitation and observation planes orthogonal to one another. This arrangement eliminates any interference from the specularly scattered beam as would result in a coplanar geometry. The geometry of the experiment is outlined in Figure 16. Because unenhanced Raman scattering is so weak, every effort must be made to optimize excitation and detection efficiency. We have found it necessary, therefore to mount all of the optics inside the UHV chamber. The excitation plane is vertical in our chamber; a standard high-angle

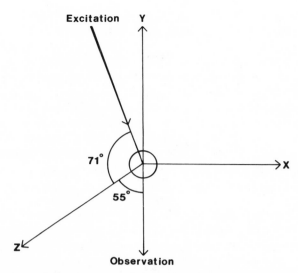

Figure 16. Experimental geometry for unenhanced surface Raman spectroscopy.

(71°) port designed for an electron gun was fitted with a viewport and a stainless steel sleeve that contains a 1-in.-diam, 2-in. focal length planoconvex lens. The convex side of the lens is oriented towards the laser for minimum spherical aberration—the focal length chosen produces a spot diameter of less than 50 μm. Because of the high angle of incidence, the laser beam forms an elliptical spot on the crystal, with an aspect ratio of 3:1.

To collect as much of the scattered light as possible, very low f-number optics are required. Since the working distances of these lenses are very short, it is preferable to mount them inside the vacuum chamber located a focal distance away from the illuminated region of the sample. The lens then produces a collimated beam which can be transmitted efficiently through a viewport to a secondary focusing lens which is chosen to image the laser spot on the entrance slits of a monochromator, under optimal f-matching conditions. We chose a vidicon camera lens (D.O. Optical 25 mm focal length, $f/0.95$) for this application, because of its high speed, high level of correction, and small diameter, the last feature allowing us to use a standard $2\frac{3}{4}$-in. viewpoint to transmit the collimated beam. Special care was taken to make the lens UHV compatible. It was disassembled and its interior dimensions measured carefully so that a stainless steel housing could be made to hold the elements. The focus was set at infinity, and the aperture removed. Thus the only potential problem was the optical cement on one doublet. We have frankly been amazed at how rugged the lens has turned out to be; it has survived well over a hundred bakeouts at 200°C with no signs of damage. Focusing of both the input lens and the collection lens was accomplished by sliding them along a sleeve designed for that purpose, while the chamber was up to air. We have found that, once set, the focus remains stable after many cycles of pumping and baking.

4.2.2. High Surface Area Materials

Studies of supported materials and adsorbates have been conducted using commercial Raman spectrometers, with their standard illumination optics. Discussions have been given about the performance of different excitation geometries, but these probably reflect more the characteristics of the individual instruments used than any fundamental distinctions. Of the common choices 0° (forward scattering) is clearly useful only on thin, transparent samples. Backscattering (180°) is a useful geometry, the trick being to get the laser beam to the sample at normal incidence. Finally, the 90° configuration which is standard on the current generation of spectrometers is extensively used; the only potential problem is specular scattering of the laser beam into the monochromator. If one is not constrained to use standard Raman illuminators, then a simple geometry would be to collect normal to the sample surface, and excite at some angle like 45°, using a cylindrical lens to focus the laser to a line image whose axis is parallel to the spectrometer entrance slit. This arrangement would eliminate problems with specular scatter and lower the power density at the sample reducing the likelihood of degradation.

Very simple cells have been used to obtain Raman spectra of species adsorbed on high surface area materials. The sample cell need only provide support for the pressed or powdered sample, optical access, and control over temperature and pressure of adsorbate. Most cells used in studies to date have been made of glass or quartz, with optical flats cemented in place. Standard glass vacuum line technology has been employed with strong caution presented against the use of vacuum grease (which is highly fluorescent) in connectors or stopcocks.[23] A number of designs have been discussed in the literature including those designed for variable temperature operation.[23-25,31] Low temperatures (77 K) can be very useful in studying the adsorption of molecular gases, while high temperatures are required to study reaction intermediates under typical industrial conditions. It is surprising that UHV hardware has not been more widely used to construct sample cells for Raman studies high surface area materials. These off-the-shelf components can be assembled quickly and easily and are relatively inexpensive. They offer the advantage of easy heating and cooling, and, most importantly are ultraclean. We have just constructed a cell using standard components that is bakable to 250°C, can provide sample temperatures from 77 to 1000 K, and has interchangeable sample holders and windows so that the same cell can be inserted into an FTIR for *in situ* infrared work.[32] Sample heating under laser irradiation can be a severe problem, especially with opaque samples such as supported metals. Sample rotation[27,28] or rotation of the laser beam[33] across the sample face alleviates this problem. Sample heating should also be reduced to negligible levels when multichannel detector technology is applied to Raman spectroscopy of supported materials and adsorbates. The multiplex advantage will allow one to reduce the power required to obtain a spectrum by at least two orders of magnitude. This approach is under development in our laboratory and others.[29]

4.3. Conventional Detection Systems

Multichannel Raman spectrometers have become available with greatly improved reliability over the past five years, so the Raman spectroscopist contemplating the purchase of a new instrument is faced with the choice of a conventional scanning instrument or a multichannel spectrometer. In a scanning double monochromator, the spectrum is obtained by scanning the grating and detecting the signal with a photomultiplier. This is a serial process, each wave number of the spectrum being measured sequentially in time. In a multichannel instrument the entire spectral region of interest is dispersed across an array detector and all wave numbers are detected simultaneously. This is a parallel process (the electronic analog of a photographic plate) and confers a multiplex advantage equal to the number of channels in the detector array, usually of the order of 10^3. All other things being equal, the multichannel spectrometer will record a spectrum at the same signal-to-noise ratio a factor of 10^3 faster, or will produce a spectrum in the same time as the scanning instrument with a signal-to-noise improvement of 32. This is clearly an enormous advantage in virtually any application, and I expect multichannel spectrometers eventually to supplant scanning instruments. Scanning monochromators are capable of higher resolution, and are available at lower cost and therefore may be the instruments of choice if these factors are of primary concern. In the next few paragraphs, I will discuss each of these systems in more detail.

4.3.1. Scanning Monochromators

Raman scattering comprises only a tiny fraction of the total light scattered from a sample, and the task of the monochromator is to separate the weak inelastic scattering from the elastic scattering (Raman spectroscopists usually lump all troublesome forms of elastic and quasi-elastic light scattering under the term "Rayleigh scattering"). For Raman scattering from neat gases and liquids, a single monochromator is usually sufficient, providing five orders of magnitude of stray light rejection for Raman shifts greater than 100 cm^{-1} or so. For studying Raman spectra of adsorbed species, the Rayleigh line is so intense that the stray light overwhelms the Raman signal using a single monochromator. If one is interested in relatively large Raman shifts ($>800 \text{ cm}^{-1}$ or so), then we have found that the addition of a colored glass filter (Schott or Corning) that absorbs the Rayleigh eliminates the strayed light problem. Indeed, for this range of frequencies, this is the preferred configuration for most applications, since the transmission and luminosity (light-gathering power) of a single monochromator is so much greater than one with multiple gratings. The resolution of a small (0.25–0.3 m) instrument, equipped with a moderately densely ruled grating (1800 lines/mm) is ca. 1 cm^{-1}, which is adequate for virtually all surface work. The new generation of blazed holographic gratings are superb. They offer the best of both worlds, the efficiency of blazed conventionally ruled gratings combined with the absence of ghosts and much lower stray light of holographic gratings.

During the past ten years the double monochromator and triple spectrograph have emerged as the dominant instruments for Raman spectroscopy. A scanning double monochromator is essentially a pair of single monochromators arranged in series, the first monochromator serving to prevent the Rayleigh light from entering the second. The stray light level is thus reduced to the square root of that obtainable with a single monochromator. The gratings, whose dispersions add, are driven synchronously as the monochromator scans. These instruments were designed for routine analytical use and tend towards long focal lengths, high resolution, and moderate luminosity. A typical 0.85-m double monochromator can provide 0.2 cm^{-1} resolution.

4.3.2. Photomultipliers

Photomultipliers for Raman spectroscopy must have good red response and relatively low dark currents. Nearly opposite approaches would be taken to optimize each of these characteristics, so practical designs are a compromise. The Hammamatsu R928, and RCA 31034 are currently the most popular choices. The R928 is a rugged, side-on multialkali tube (approximately S-25 response) with a room temperature dark current of ca. 10^3 counts per second. This dark current level is too high for Raman spectroscopy so the tube is cooled to $-20°C$. The 31034 has a GaAs photocathode with an extremely flat response that extends into the near infrared (ca. 900 nm). This photomultiplier must be cooled to less than $-30°C$, since its room temperature dark current is very high (10^5 counts per second). Cost and ruggedness favor the Hammamatsu tube; the extended red response (necessary for krypton ion laser excitation, for example) would make the RCA tube a better choice in more demanding applications.

4.3.3. Counting Electronics

Both analog and digital electronics can be used to measure the output of a photomultiplier. At high light levels the photomultiplier anode pulses overlap strongly and can be integrated into a dc current using a simple RC filter. An electrometer provides a series of attenuators so that a convenient display of the photomultiplier current can be recorded on a strip chart recorder. A bucking current can be applied to suppress any dc background (fluorescence, for example) and thus devote the dynamic range to the signal of interest. In fact this zero suppression feature prompted Hendra to suggest that analog detection is a superior method for surface Raman spectroscopy.[34]

Photon counting has all but supplanted analog detection during the past decade. Since a photon is a discrete entity it should be detected as a single event or count. A photon striking the photocathode of a photomultiplier ejects a photoelectron, which then, through secondary emission at many dynode surfaces, results in an electron cascade. The signal at the anode is a pulse of typically 10^{-8} s duration and variable intensity depending upon the statistics of the electron multiplication processes. This weak (one picocoulomb of charge, typically) signal is amplified by a factor of 10^1–10^2 and then fed to a discriminator. The lower level of the discriminator is set high enough to pass only those signals that originate at the photocathode. The upper level can be used to discriminate against multiphoton events (which are only a problem at high count rates). The pulses are then counted and either converted to an analog signal or stored in a computer. Computer-controlled systems can be used to average multiple scans and have essentially infinite zero suppression which can be very useful for surface Raman spectroscopy. Photon counting is clearly to be preferred over analog detection because photons are inherently discrete, because it eliminates all background other than photocathode dark current, and because it is so easy to interface with a computer for signal processing.

4.4. Multichannel Detection

4.4.1. Spectrographs

As multichannel detectors have become more popular, new spectrographs have been designed to work with them. The triple spectrograph is the current workhorse in this area. The first two monochromators act as a variable bandpass filter that selects the spectral region of interest and passes the radiation onto a third spectrograph for dispersion onto the detector. The gratings in the filter stage are arranged for subtractive dispersion so that the radiation that is incident on the entrance slit of the third monochromator is spatially homogeneous. Since array detectors have a fixed number of elements there is always a tradeoff between spectral coverage and resolution. Several gratings can be mounted on a turret in the spectrograph stage to provide convenient choices of these parameters. Since the spatial resolution of an array detector is somewhat lower than the minimum slitwidth of a scanning monochromator, and since maximum spectral coverage is usually the goal, the triple spectrograph uses a shorter focal length and has a higher luminosity than the scanning instrument. Following the manufacturer's recommendation, the Spex Triplemate provides a bandpass of ca. 1000 cm^{-1} in the visible at a dispersion of ca. 1 cm^{-1} per channel.

4.4.2. Detector Arrays

The availability of commercial multichannel optical detectors has revolutionized many areas of spectroscopy, permitting formerly impossible experiments to be conducted routinely. As mentioned above, the multiplex advantage produces vastly improved spectra in the same time or identical spectra in much less time. Immunity from long-term drifts in source intensity or detector stability is an important secondary benefit.

Two devices dominate the current market for spectroscopic applications, the silicon intensified target vidicon (SIT) and the intensified silicon photodiode array detector (ISPD). Talmi and Busch[35] have written an excellent, up-to-date comparison of the most popular imaging detectors which will serve as a more detailed guide than can be presented here.

The SIT tube was the first multichannel detector adapted (from television cameras) for spectroscopic use. This vacuum tube device is capable of two-dimensional imaging; for spectroscopic application pixels are grouped into vertical columns or channels onto which the image of the spectrometer entrance slit is focused. Thus, the usual display of intensity versus wavelength is produced. The SIT tube has an integral image intensifier to boost the gain so that its sensitivity is comparable to that of a photomultiplier. Photons striking the photocathode (typically S-20) cause the emission of photoelectrons which are accelerated by a potential of ca. 10 kV before striking the silicon target. Electrostatic focusing ensures that the optical image is transformed accurately into an electron image. Each fast electron results in the formation of approximately 2500 electron–hole pairs in the array of silicon photodiodes. If the readout noise of the scanning electron beam can be kept below 2500 electrons per channel per frame scan, then single-photoelectron sensitivity would be achieved, just as in a photomultiplier. In practice, however, approximately two photoelectrons are required to register a count.

There are essentially three sources of noise in the SIT detector. The photocathode emits electrons thermionically; because of the extremely small effective photocathode area, however, this source of noise is insignificant. The readout process generates noise. This has been reduced to essentially one count per frame scan and can be rendered insignificant by extended integration on the target. Far and away the dominant source of noise arises from thermally generated charge carriers in the silicon target. At room temperature dark count rates of the order of 10^3 counts per second are typical, which is too high for the low signal levels encountered in surface Raman spectroscopy. The solution is to cool the tube to reduce the dark count rate. Unfortunately, the resulting improvement in signal-to-noise is not as dramatic as one gets upon cooling a photomultiplier. SIT tubes suffer from a phenomenon known as lag; readout efficiencies are highly nonlinear for small amounts of charge. At room temperature, the high dark current biases the target into the linear regime so that reliable results are obtained. Cooling reduces the dark current by up to four orders of magnitude so that this source of bias is no longer effective and the signal itself provides the bias. Obviously this leads to severe nonlinearities, as well as an overall reduction in sensitivity. We have found that the only solution to this problem is to provide some uniform, low-level illumination of the target superimposed upon the signal. In our experiments, this arises naturally, since the metal single-crystal substrates we study all show a weak continuum inelastic light scattering. The signal plus background are integrated on the target (the reading electron beam is inhibited during this process) until a signal level approaching saturation of the A/D converter is reached (2^{14}), and then the target is read by multiple scans (25) to recover the charge completely. Extended integration essentially eliminates the readout noise and one is left only with the shot noise of the signal plus background. Background scans are taken of the substrate prior to adsorption, to normalize channel-to-channel gain variations by subtraction. We are confident of our ability to detect spectra where peak count rates are as low as one count per second in the presence of backgrounds as high as ten counts per second. Spectra can be obtained at lower count rates, or with lower backgrounds, but we always check very carefully for linearity in these cases by running multiple experiments with various scan parameters.

The ISPD is a second generation detector, especially designed for spectroscopic applications. A microchannel plate image intensifier is used in this device for amplification. Microchannel plates consist of a very large number of continuous dynodes formed in small channels in a glass substrate; application of a potential of a few hundred volts provides a gain of 10^4. The electrons impinge on a phosphor, whose output is optically coupled to a silicon photodiode array as the readout device. The array employs photodiodes to detect the electrons, as does the vidicon, but the readout is effected by on-chip shift registers.

Each of these detectors has certain advantages. Because the technology is more nature, the SIT is now less expensive than the ISPD. The vidicon is capable of two-dimensional imaging; the ISPD is not. The SIT is a much more rugged detector; it has been indestructible in my experience. The ISPD requires great care. Both detectors can be gated for time-resolved experiments; the ISPD is somewhat faster and the SIT cannot be gated when it is cooled, a severe disadvantage. Since the ISPD is assembled from the diode array and an image intensifier, there is somewhat greater flexibility in the choice of cathode materials; consequently improved response can be obtained, particularly in the red. For ultraviolet response, the ISPD is far superior, since there are no glass elements in the optical path before the photocathode as are found in the SIT. The ISPD provides better spectral linearity, since it is determined by the geometry of the silicon array and not by the precision with which a scanning electron beam can be positioned. The ISPD is also lag free, resulting in linear behavior over a wide range of light fluxes. Finally, the noise is certainly no worse that the SIT, though not markedly better in our experience. A greater number of channels is available (700–1024, depending upon the manufacturer) for the ISPD than the SIT (500).

The resolution obtainable with both of these detectors is not quite as good as that of a scanning monochromator, since the former have pixels spaced on 25-μm centers, while monochromator slits can easily be closed to 10 μm. The situation is even worse than this, however, because of interchannel cross talk. Even for a perfectly focused image of less than 25 μm width, the signal from the detector will occupy at least three channels. With the SIT tube, the intensity of the side channels is usually less than 50% of that of the center channel; slightly worse results are found for the ISPD because of the microchannel plate image intensifier. In either case, if the Rayleigh criterion is adopted as measure of resolution (a valley, however slight, between overlapping peaks) it is clear that, at best, resolution of 2 channels (50 μm) times the linear dispersion of the spectrograph is to be expected. The manufacturers should be consulted for detailed recommendations for any particular application. As a general guide, if two-dimensional imaging, ruggedness, or cost are important considerations, then the SIT would serve well. The ISPD is the detector of choice if other considerations are paramount.

Finally, I should mention that third generation detectors are just around the corner. These will be based on charge-coupled devices (CCD), and should offer the best of both worlds. The noise of the CCD can apparently be reduced to the point where image intensification is no longer required. This will represent a quantum leap (no pun!) in multichannel detection capability; a wide spectral range (silicon is sensitive to 1000 nm), vastly improved quantum efficiency (60%–70%), and two-dimensional imaging. Several commercial versions are already on the market, and I am sure that we will see other manufacturers providing new products soon.

5. Examples and Applications

5.1. Adsorbates on Single-Crystal Surfaces

Raman spectroscopy, as a probe of adsorption on low-area single-crystal surfaces, is in its infancy. Only a few studies have been reported, but I hope that the technique will gain in popularity so that we will see new results appearing with greater frequency.

Our research group provided the first example of a Raman spectrum of a molecule adsorbed on a well-characterized single-crystal surface without enhancement.[6] Our objective was to demonstrate the feasibility of the technique, emphasizing the important advantages offered by multichannel optical detection. As a bonus we discovered that we could identify the products of a surface decomposition reaction using Raman spectroscopy.

For this investigation we chose the adsorption of nitrobenzene on Ni(111). Nitrobenzene was picked as the adsorbate because it has a relatively large and accurately known Raman scattering cross section. In addition, since it is a relatively high boiling liquid, it can easily be condensed at low temperatures to form a multilayer film, which was important for optimizing the experimental conditions. Ni(111) was chosen as the substrate for several reasons. First, it is a typical transition metal of some catalytic importance. Second, its optical properties (dielectric function) are such that it will not support the conduction electron resonances responsible for SERS, with visible excitation. To make absolutely certain that there was no possibility of SERS we worked only with atomically flat, smooth surfaces. The combination of the flat geometry with the nickel dielectric function guaranteed that there was no electromagnetic SERS mechanism operative.

The sample surface was cleaned and characterized under ultrahigh vacuum by argon ion bombardment and annealing cycles and by Auger electron spectroscopy and low-energy electron diffraction (LEED). Surface impurities were typically less than 1% of a monolayer and the surface was well ordered, as determined by the sharp LEED spots and low background intensity. The Raman spectrometer was designed according to the considerations outlined in Section 3, and was described in Section 4. A standard ultrahigh vacuum chamber was equipped with focusing and collection optics, oriented at 70° and 55° to the surface normal, respectively. 150 mW of 514.5 nm argon ion laser power was used for excitation. A single monochromator and a cooled SIT detector were used to record the spectrum. The dispersion of the instrument was ca. 3 cm^{-1} per channel and a glass cutoff filter was used to discriminate against the scattered Rayleigh limit. Detailed descriptions of the instrument are provided elsewhere.[6]

A multilayer film some 5 nm thick was provided by a 100-L exposure (uncorrected) of nitrobenzene on a Ni(111) crystal held at 100 K. Figure 17a shows the spectrum that resulted. Every band in the spectrum except the small feature near 1215 cm^{-1} corresponds to a band in liquid nitrobenzene. The frequencies are only slightly shifted, as expected for physisorption, but the intensities are altered somewhat, perhaps indicating partial ordering of the adlayer. Prominent features include the C≡C stretches at 1580 and 1470 cm^{-1}, the NO$_2$ antisymmetric and symmetric stretches at 1515 and 1340 cm^{-1}, respectively, the C–H in-plane bend at 1160 cm^{-1}, the C–N stretch at 1100 cm^{-1}, and the trigonal ring breathing mode at 990 cm^{-1}. The spectrum, which required 500 s of signal averaging, shows an excellent signal-to-noise ratio and illustrates the power of Raman spectroscopy in studying thin films.

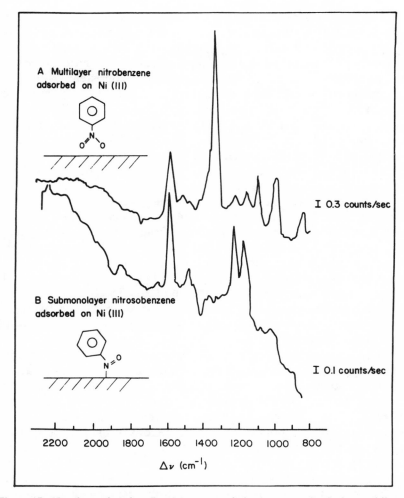

Figure 17. Unenhanced surface Raman spectra of nitrobenzene adsorbed at multilayer (A) and submonolayer (B) coverage on Ni(111) at 100 K. The multilayer film resulted from a 100 L (uncorrected) dose while the submonolayer film was produced by a 3 L (uncorrected) dose. 150 mW of 514.5 nm excitation was used. Spectral resolution was set at 20 cm^{-1} and the signals were acquired in 500 s (A) and 1000 s (B), respectively.

Reducing the dose to 3 L (uncorrected) produced the spectrum in Figure 17b. This dose corresponds to approximately half monolayer coverage. The gross differences in appearance indicate either a drastic increase in symmetry, which could render some bands forbidden, or the occurrence of some chemistry. Note in particular the complete absence of the NO$_2$ symmetric stretch, and the increased intensity of the band at 1215 cm^{-1} relative to its intensity in the 100-L spectrum. A literature search revealed that it had been suggested that nitrobenzene adsorbs dissociatively on nickel films at low temperatures to give nitrosobenzene and chemisorbed oxygen.[36] Our spectrum is consistent with this suggestion, the 1215-cm^{-1} band being the N=O stretch of chemisorbed nitrosobenzene, which is shifted from its gas phase value of 1523 cm^{-1}. This shift presumably reflects some backbonding into the π^* orbitals of the N=O group reducing its bond order, hence its vibrational frequency. Our simple picture of the geometry of chemisorbed nitrosobenzene is shown in Figure 17b. To confirm the assignment we adsorbed nitrosobenzene on the surface at room temperature. A spectrum identical to that of Figure 17b resulted, giving us confidence in the assignment.

An important question is whether the nitrosobenzene spectrum we observed resulted from some enhancement of the scattering cross section due to a chemical SERS mechanism (see Section 6.4). We ruled out this possibility by careful calibration of our detection efficiency.[6]

Thus the adsorption of nitrobenzene on Ni(111) provided the first example of an unenhanced Raman spectrum of a molecule adsorbed on a well-characterized surface. The first layer chemisorbs dissociatively to form nitrosobenzene; subsequent layers of nitrobenzene can be condensed without reaction.

As a second example, we present the spectrum of ethylene chemisorbed on Ag(110). Silver catalysts are used commercially for the partial oxidation of ethylene and there is great interest in the reaction mechanism. We are in the process of studying this reaction over single-crystal model catalysts at elevated pressures, using Raman spectroscopy to identify and monitor the concentrations of the surface species *in situ*. In order to assess the feasibility of this approach we first wanted to obtain the Raman spectrum of the reactants adsorbed on silver. Ethylene is a very weak Raman scatterer, its cross section being more than a factor of 10 smaller than that of nitrobenzene, so it provides a stringent test of our sensitivity. It is known from EELS studies that ethylene adsorbs only to a limited extent on clean Ag(110) at low temperatures, but that the extent of adsorption can be greatly increased if the surface is first predosed with oxygen at room temperature.[37] The oxygen dissociates in the troughs of this crystal to form given oxygen atoms which decrease the electron density on the neighboring silver atoms. Ethylene, being nucleophilic, can then adsorb onto these partially positive silver atoms.

Figure 18 shows the Raman spectrum of ethylene adsorbed at submonolayer coverage on a Ag(110) surface that had been predosed with oxygen to half-monolayer coverage. The oxygen layer formed a

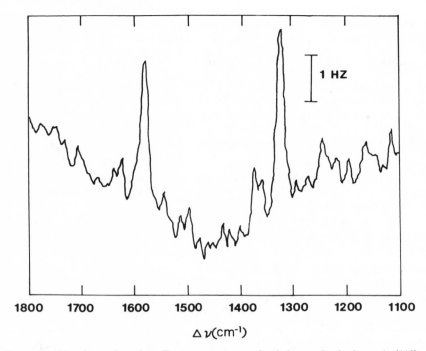

Figure 18. Unenhanced surface Raman spectrum of ethylene adsorbed on Ag(110) at 100 K, at submonolayer coverage. The surface was predosed with oxygen at 300 K until a 2 × 1 LEED pattern was observed. It was then exposed to 100 L (uncorrected) of ethylene at 100 K. Excitation conditions were the same as for the spectrum of Figure 17.

distinct 2 × 1 LEED pattern. The spectral region monitored includes two bands, the C=C stretch at ca. $1580 \, \text{cm}^{-1}$ and the symmetric C–H in-plane deformation at ca. $1320 \, \text{cm}^{-1}$. These frequencies are ca. $20 \, \text{cm}^{-1}$ lower than in the gas phase and suggest a weak interaction (π-bonding) with the surface. The EELS results show that the frequency of the Ag–O stretch is unperturbed, suggesting that the ethylene absorbs on a silver atom. The Raman spectrum of ethylene on silver corroborates the EELS results and is further evidence of the sensitivity of unenhanced Raman spectroscopy.

5.2. High Surface Area Materials

5.2.1. Supports, Catalyst Structure

Raman scattering from oxides tends to be very weak. While this is an advantage for studying molecules adsorbed on the surfaces of these kinds of supports, it is clearly a disadvantage for studying the structures of the support materials themselves. Egerton and Hardin[23] have reviewed the literature on Raman spectroscopy of common oxide adsorbents. For the most part, it has been an unsuccessful enterprise. Silicas, aluminas, silica-aluminas, zeolites, a few other metal oxides, and supported metals and metal oxides have been examined. With the silicas it is difficult to eliminate Raman scattering from sample windows, which of course should be very similar. Aluminas tend to fluoresce so strongly that the intrinsic Raman spectra are obscured. No unambiguous Raman scattering was observed from metals supported on oxides. The zeolites have been much more successfully investigated. Angell,[38] Egerton et al.,[39] and Buechler and Turkevich[40] have reported the Raman spectra of a large number of both natural and synthetic zeolites.

Of the various catalysts that have been studied by Raman spectroscopy, the most successful results have been obtained for the molybdate catalysts that are widely used industrially for hydrodesulfurization (HDS) and hydrodenitrification (HDN). The catalyst is typically a molybdate supported on alumina, and it may be promoted by various additives such as nickel and cobalt. There is a rich surface chemistry to be explored for this system, trying to understand how different preparation conditions (e.g., impregnation, drying, and calcining) affect the structure of the catalyst and how the active catalyst differs from the initially prepared material through the extrinsic sulfiding and coking that occurs.

Oganowski et al.[41] obtained the infrared and Raman spectra of several mixed oxides and concluded that the active catalyst was MoO_3 supported on M_9MoO_4. Brown and Makovsky,[27] using a rotating sample, obtained the Raman spectra of a commercial cobalt-molybdate catalyst supported on silica-alumina. They were able to identify both bridging ($\Delta v = 877 \text{ cm}^{-1}$) and terminal ($\Delta v = 948 \text{ cm}^{-1}$) M–O stretching vibrations. They found no evidence for bulk MoO_3 ($\Delta v \approx 1000 \text{ cm}^{-1}$). The addition of cobalt produced no change in the Raman spectrum. Sulfiding led to the formation of MoS_2. A used catalyst showed extensive coking. (The Raman spectrum of graphitic carbon is characteristic and very intense.) In subsequent studies this group[42] probed the coverage dependence of the state of the surface and found evidence for MoO_4^{2-}, $Al_2(MoO_4)_3$, and MoO_3. They proposed that the commonly held belief that the active catalyst was a monolayer of MoO_3 with bulk properties was not correct; at 15% or less Mo by weight no bulk MoO_3 bands were observed. Finally, regeneration of the commercial catalyst by exposure to either dry nitrogen or oxygen did not return all of the spectral features.

Medema et al.[43] identified a similar sequence of compound formation in the Co–Mo/γ-Al$_2$O$_3$ system. For pure molybdenum or alumina MoO$_4^{2-}$ is initially present on the surface and is converted to a polymolybdate phase in which the molybdenum is octahedrally coordinated. This is followed by the formation of a "bulk" Al$_2$(MoO$_4$)$_3$ compound and then finally by MoO$_3$. They identified Co$_3$O$_4^-$ and CoAl$_2$O$_4^-$ on the surface of γ-Al$_2$O$_3$ following cobalt impregnation, not by the Raman spectra but by the color of these anions. When Co was added to the molybdenum system it apparently encouraged the formation of the polymolybdate phase, but the polymer structure was not changed significantly. MoO$_3$ and Al$_2$(MoO$_4$)$_3$ were converted to CoMoO$_4$.

Jeziorowski and Knozinger[44] studied both γ- and η-aluminas during impregnation, drying, and calcination to form molybdate catalysts. They impregnated the support at both low and high pH and determined that the initial form of the molybdenum oxide had little effect on the species finally produced. MoO$_4^{2-}$ aggregates in solution at low pH to form polymolybdates, chiefly Mo$_7$O$_{24}^{6-}$. They suggested that the isopolyanions in solution degrade near the support surface so that the primary adsorbate is MoO$_4^{2-}$. These authors also identified a polymolybdate surface phase but pointed out that unambiguous identification of the molybdenum coordination geometry is difficult since the vibrational frequencies spanned by both octahedral and tetrahedral complexes overlap considerably.

Cheng and Schrader[45] have studied the formation of molybdate catalysts, supported on γ- and α-alumina and silica. At less than 5% Mo loading the spectra resemble those of polymolybdates. Addition of Co greatly increases the complexity of the spectra which do not match any known Mo compounds but most closely resemble those of precipitated molybdates. They suggest that heteropolymolybdates are suitable model compounds, having two marker bands in the 800–1000 cm^{-1} region of the system. The sharper, more intense high-frequency band moves to higher frequencies in larger aggregates. The broader, weaker band is an indication of distortion of the Mo environment. These authors point out that the spectral assignments are not unique; bands from both tetrahedral and octahedral Mo structures are found in the same region of the spectrum, and since distorted geometries are common for these systems ambiguities can result. As a rough guide, the tetrahedral M–O stretch is found near 900 cm^{-1}; for polymolybdates it is near 950 cm^{-1}, while for octahedral MoO$_3$ it is closer to 1000 cm^{-1}. Their results support the earlier conclusion that the surface species formed are independent of pH of the impregnation solution. The Raman bands are very broad, indicating heterogeneity of the local environment on the surface.

The oxide form of these molybdate catalysts is almost certainly not the active form for HDS. During activation and use, they become sulfided. Schrader and Cheng[46] have investigated the sulfiding of Mo/γ-Al_2O_3 catalysts using Raman spectroscopy as an *in situ* diagnostic technique. Using a controlled atmosphere rotating sample cell they investigated two different sulfiding procedures. Sulfiding at 400°C using 10% H_2S in H_2 produces MoS_2. Stepwise sulfiding using the same reagents at 150°, 250°, and 350°C produced molybdenum oxysulfide, reduced molybdate, and MoS_2 surface layers. These authors presented a model for the partially sulfided and reduced catalyst. The low-temperature polymolybdate phase is partially reduced under low-temperature sulfiding conditions yielding an oxysulfide phase that contains both oxygen vacancies and Mo–Mo bonds. Sulfiding at higher temperatures results in further replacement of oxygen leading to sulfur-enriched oxysulfides and MoS_2 layers. Exposure of both of the oxysulfide phases to air partially reoxidizes the surface, removing some oxygen vacancies. Finally, extensive oxidation of either MoS_2 or the oxide form results in MoO_3.

Schrader and Cheng[7] have carried this approach even further by using pyridine as a probe of surface acid sites. As mentioned below in more detail, the frequencies of the pyridine ring breathing modes near $1000\ cm^{-1}$ are sensitive measures of the local pyridine environment. For example, the symmetric ring-breathing mode which is found at $992\ cm^{-1}$ in the liquid occurs at $998\ cm^{-1}$ for pyridine hydrogen-bonded to a surface hydroxyl on alumina, at $1004\ cm^{-1}$ to a hydroxyl group that bridges molybdenum and aluminum and at $1017\ cm^{-1}$ when bound to a Lewis acid site (an aluminum atom with an open sextet of electrons) on alumina. In their study of Co-Mo/γ-Al_2O_3 HDS catalysts Schrader and Cheng found that Bronsted sites were formed upon the addition of Mo to γ-Al_2O_3. At high loadings, where MoO_3 dominated the surface, pyridine adsorption was suppressed. The addition of cobalt produced a new adsorption site whose vibrational frequency was intermediate between those found for pyridine adsorbed on Bronsted and Lewis acid sites. Schrader and Cheng[48] very recently reported a study of the sulfiding of Co-Mo/γ-Al_2O_3 catalysts. High-temperature sulfiding resulted in stable MoS_2 structures. Stepwise sulfides revealed a competition between reduction and sulfur incorporation for $CoMoO_4$, Co_3O_4, MoO_3, and polymolybdate phases. The cobalt appears to increase the extent of catalyst reduction. They extended their earlier model to include the role of the cobalt.

5.2.2. Adsorbates

A large number of adsorbates has been studied on high surface area materials. The motivation for this work has varied. It has included, for example, the use of Raman spectroscopy to measure adsorption isotherms, to study the distribution of acidic sites, to determine the influence of the substrate on the Raman selection rules, and to follow the source of chemical reactions catalyzed by the surface. In this section I review each of these topics briefly.

Hendra and Loader[34] were the first to use laser Raman spectroscopy to examine adsorption on a number of oxide supports including various silica, aluminas, and silica-aluminas. They found physical adsorption for most of the molecules studied, as deduced from the minor shifts in the adsorbate vibrational frequencies and linewidths. Adsorption isotherms were determined by the BET method and the coverage was adjusted by varying the ambient pressure. Several centrosymmetric molecules were studied to see if the surface interaction was sufficiently strong to break the inversion symmetry and allow normally forbidden lines to be observed. For CS_2 and *trans*-dichloroethylene, spectra essentially identical to the liquid spectra were obtained. For the former no Raman forbidden bands were observed in the adsorbed state, while for the latter the same intensity ratios between forbidden bonds (which are weakly allowed in the liquid spectra) and allowed bands were seen, indicating a very minor perturbation.

Pyridine has been widely used to probe surface acidity, since it can act as either a Bronsted or Lewis base, and since it displays characteristic frequency shifts and intensity alterations in its two ring-breathing modes near 1000 cm^{-1}. Table 2 shows the frequencies of both the symmetric and trigonal ring-breathing vibrations for pyridine in various environments. The symmetric ring-breathing mode shifts to higher frequencies as pyridine is first hydrogen-bonded, then protonated, and finally fully coordinated. Although the frequency of the trigonal ring-breathing mode remains more or less constant, its intensity relative to that of the

Table 2. Frequencies and Relative Intensities of Pyridine Ring-Breathing Vibrations in Various Environments[a]

Environment	Bonding	$v_{20}(\text{cm}^{-1})$	$v_1(\text{cm}^{-1})$	$v_{12}(\text{cm}^{-1})$	$I_{v_{12}}/I_{v_1}$
Liquid	—	3057	991	1031	0.75
10% in CCl_4	—	3060	991	1030	0.91
10% in $CHCl_3$	Hydrogen bonded	—	998	1038	0.80
10% in H_2O	Hydrogen bonded	3072	1002	1035	0.45
10% in HCl	Pyridinium ion	3109	1010	1029	0.34
$PyH^+ BF_4^-$	Pyridinium ion	—	1011	1032	0.15
Solid pyridine oxide	Coordinated	—	1016	1043	0.15
Py : $ZnCl_2$	Coordinated	3080	1025	1050	0.04

[a] Data from Refs. 49–51.

symmetric mode varies considerably depending upon the strength of the interactions.[51] Oxide surfaces have both Bronsted and Lewis acid sites, the former being surface hydroxyls, the latter typically aluminum atoms having an incomplete octet of electrons. On silica, for example, pyridine adsorbs at low coverages to give bands near $1010 \, \text{cm}^{-1}$ and $1032 \, \text{cm}^{-1}$ with an intensity ratio of approximately $2:1$. These frequencies and intensities are very similar to those observed for a solution of pyridine in water and are therefore assigned to pyridine hydrogen bonded to surface hydroxyls. As the surface coverage is increased, bands at $992 \, \text{cm}^{-1}$ and $1032 \, \text{cm}^{-1}$ grow with an intensity ratio closer to $1:1$. These frequencies and intensity ratios are similar to the liquid and thus are assigned to physisorbed pyridine. On alumina, there is evidence for a Lewis acid site, as indicated by a single band in the region of $1010–1018 \, \text{cm}^{-1}$, which is very similar to the results obtained for pyridine coordinated to other Lewis acids.

Pyridine adsorption can be used to determine the site density for chemisorption, as contrasted to the physisorption site density probed by conventional BET methods. For example, Cooney and Tam[52] monitored the intensity of the $1008 \, \text{cm}^{-1}$ band of pyridine hydrogen bonded to silica as a fraction of coverage and found a Langmuir adsorption isotherm. The surface area calculated on the basis of Bronsted acid sites was $80 \, \text{m}^2 \, \text{g}^{-1}$, which is far less than the $325 \, \text{m}^2 \, \text{g}^{-1}$ quoted by the manufacturer on the basis of N_2 adsorption and analysis of the resulting BET isotherm. Clearly the Raman technique is promising for measuring the density of specific chemical sites, a result not available by other methods.

There have been several interesting studies of surface-catalyzed reactions using Raman spectroscopy. Hendra and Loader[53] found that acetaldehyde polymerizes on silica to form the cyclic trimer paraldehyde. No evidence of the parent molecule was found. In another study, Hendra and co-workers[54] followed the isomerization of 1-hexene on alumina. If the alumina was activated by heating only to room temperature, no reaction occurred. If, on the other hand, the substrate was activated by heating to 1200 K, both *cis*- and *trans*-2-hexenes were observed. The bands observed were only slightly shifted from their liquid values, indicating that only physisorbed reactants and products were being monitored.

As a final example, consider the adsorption of acetone on γ-alumina.[55] Above $200°C$, the spectrum of the adsorbed acetone (which is strongly perturbed relative to the liquid) changes dramatically. It is postulated that two surface acetone molecules react to form adsorbed mesityl oxide. Similar results were found for butanone on γ-alumina.[55]

There has been a series of interesting papers published recently by Krasser and his group using Raman spectroscopy to examine the adsorption of small molecules on supported transition metal surfaces.[56–58] They have studied benzene on nickel and platinum supported on silica, and hydrogen and carbon monoxide either singly or coadsorbed on supported nickel catalysts. Although the results were not always reproducible they observed enhancement factors of 10^3 or so for these systems and were able to identify molecularly adsorbed species. If the sample preparation conditions can be brought under control so that a constant enhancement factor is achieved from sample to sample (which could even be as low as unity and still result in detectable spectra), the method promises to be complementary to the established infrared technique for studying catalytic reactions over supported metal catalysts.

6. Surface-Enhanced Raman Scattering

6.1. Introduction

In the previous sections I have discussed Raman scattering by molecules adsorbed on low-area substrates, and on high-area materials. In the former case, extreme care is required to obtain spectra, and it is difficult to discriminate against Raman scattering from molecules in an ambient phase. In the latter case sensitivity and selectivity are both improved by the much higher surface area to volume ratio. Since selectivity and sensitivity are always important goals it is no wonder that the discovery that the Raman scattering by molecules adsorbed on certain surfaces was over a millionfold more intense than expected was greeted with great enthusiasm. Surface-enhanced Raman scattering (SERS), as it was to become called, would fulfil both of these goals simultaneously. Given the potential power of this method in surface and interface analysis, and the (later to be discovered) complexity of the mechanism, it is not surprising that SERS became a dynamic field of study. In this section I introduce the phenomenon, outlining the salient experimental observations. I then discuss briefly the two classes of mechanisms that contribute to the overall enhancement, giving only the essential physics necessary for a general understanding. Finally, I propose some experiments designed to provide a critical test of the several enhancement theories available and will conclude with some applications that I found particularly interesting.

SERS was discovered nearly ten years ago by Fleischman et al.,[59] who observed anomalougly intense Raman scattering from pyridine adsorbed on a silver electrode surface which had been roughened electrochemically to increase its surface area. These investigators did not find the scattered intensity particularly remarkable, attributing it to the large number of adsorption sites available on a highly roughened surface and the possibility of multilayer adsorption. Van Duyne and Jeanmaire,[60] and independently Albrecht and Creighton,[61] repeated this experiment using electrodes that were roughened less extensively and observed even higher intensities. These groups proposed that an enormous increase—a factor of 10^5–10^6—in the Raman scattering cross section must have occurred upon adsorption. Enhancement was subsequently found for a variety of molecules adsorbed on several different kinds of surfaces. The quest for the mechanism of the enhancement has been difficult and exciting, because there appear to be at least two very different kinds of mechanisms that contribute in various proportions depending upon the details of the particular system. Also, as in any rapidly developing field, the experimental results have sometimes been conflicting and irreproducible. There is now general agreement, however, about the following experimental observations:

1. SERS occurs for a variety of molecules adsorbed on the surfaces of relatively few metals. The effect has been confirmed for silver, copper, gold, lithium, potassium, and sodium, metals that are highly reflective in the visible.
2. Surface roughness appears to be required. The major contribution can be either submicroscopic roughness (features of dimension 10–100 nm) or atomic scale roughness (steps, kinks, adatoms, or vacancies, for example) depending upon the system.
3. The enhancement may be remarkably long ranged, depending upon the surface topography. Appreciably enhanced scattering has been observed for molecules separated by tens of nanometers from the surface.
4. The excitation profile (dependence of the scattered intensity upon exciting frequency) deviates markedly from ω^4, displaying broad resonances.
5. The intensities of the Raman bands generally fall off with increasing vibrational frequency.
6. The Raman bands are completely depolarized.
7. Molecules adsorbed in the first layer are often distinguishable from those absorbed in adjacent layers by shifted vibrational frequencies, by the appearance of new bands, and by larger enhancement factors.
8. Selection rules for Raman scattering are relaxed, resulting in the

appearance of modes that are normally Raman inactive and even infrared inactive in the gas phase.

9. Vibrational frequencies and excitation profiles are both functions of applied potential in electrochemical experiments. The excitation profiles of different vibrational modes are different.

10. Surfaces that support SERS invariably display a weak continuum inelastic scattering, even in the absence of an adsorbed species.

Many mechanisms have been proposed to explain the experimental observations listed above. In the early days of SERS, a number of interesting proposals were made that suggested that mere proximity to a surface would lead to enhanced Raman scattering.[62] We showed recently that this is not the case by demonstrating that the Raman scattering from pyridine adsorbed on an atomically smooth silver surface was unenhanced. This experiment, along with other experiments of a suggestive nature, allow us to conclude that there are only two broad classes of mechanisms that contribute to SERS: electromagnetic and chemical. The electromagnetic theories ascribe the enhancement to the presence of large electromagnetic fields produced by excitation of conduction electron resonances in the metal substrate. The chemical theories assert that enhanced cross sections arise from the participation of adsorption-induced resonances as intermediate states in the scattering process, by analogy to resonance Raman scattering. There is good evidence that both mechanisms operate to various extents, depending upon the specific system. In the following sections, I outline the physical basis for each of these enhancement mechanisms and review the supporting experimental evidence.

6.2. Electromagnetic Enhancement

The intensity of Raman scattering by adsorbed molecules is given by

$$I_R = \sigma' I_L (\partial \sigma / \partial \Omega) \Omega \tag{30}$$

where σ' is the adsorbate surface density, $(\partial \sigma / \partial \Omega)$ the differential Raman scattering cross section, I_L the laser intensity and Ω the solid angle over which the scattered photons are collected. Unexpectedly large Raman signals can result if the surface area (hence the adsorbate density) is larger than the geometrical area, if the Raman cross section is larger in the adsorbed state than for the free molecule, or if the laser flux is higher at the surface than expected. Using radioactive labels to count molecules, it has been shown for SERS active systems that the increase in surface area contributes very little to the overall enhancement.[63] It is certainly reasonable to expect that the Raman cross section for adsorbed molecules will, in general, be different than for gas phase molecules. Even in the case of physisorption, where perturbations of the molecular electronic states are minor, preferred orientations can lead to enhanced scattering. More dramatic effects are expected if the molecule forms a strong chemical bond to the surface; the extreme case occurs when a new electronic state is created that happens to be resonant with the exciting laser. Such resonances are responsible for increases in the Raman cross section of many orders of magnitude for molecules in the gas phase or in solution; similar changes could be expected for adsorbed molecules. Changes in the Raman cross section upon adsorption form the basis for the chemical enhancement theories, a simple model for which is discussed later. Finally, the electromagnetic field of the exciting laser can be effectively amplified by the presence of a polarizable solid. The calculation of these enhanced fields is a problem in classical electrodynamics.

It has been known for a long time that special geometries could produce electromagnetic fields at interfaces that are much larger than the applied fields. These enhanced fields result from the polarization of the metal and their magnitudes can easily be calculated from Maxwell's equations. Physically, the enhanced local fields can be ascribed to the excitation of either propagating or localized surface plasmons, fundamental excitations of the two-dimensional electron gas. In fact, surface plasmon excitation was used to increase the sensitivity of surface Raman spectroscopy before the generality of SERS was recognized.[64] When a thin (50-nm) silver film is deposited on a hemispherical prism made from a glass of high refractive index, propagating surface plasmons can be excited as the angle of incidence is chosen to effect momentum conservation. The enhanced evanescent electromagnetic fields at the silver surface lead to about a factor of 10^2 increase in the Raman scattering from a monolayer of adsorbed molecules. A grating can also be used to excite surface plasmons, the grating wave vector providing the necessary momentum. Again, enhancement factors of 10^2 or so are observed.[65] Far more widespread than the use of propagating surface plasmons, however, has been the use of plasmons localized on the surfaces of small particles.

The surfaces of SERS-active electrodes are very rough on a submicroscopic scale. The dissolution and redeposition of many layers of silver produce a surface that is covered with nearly spherical bumps of diameter 20–100 nm. In addition to roughened electrodes, SERS has been observed for molecules adsorbed on colloids, on silver island films, and on silver ellipsoids prepared by evaporation on silicon posts. These small structures all support electromagnetic resonances which are localized surface plasmons. On resonance, the electromagnetic field in the interior of the particle can be significantly larger than the applied field, owing to the polarization of the solid. By the boundary conditions on Maxwell's equations, the field at the particle surface must be similarly increased.

The essential physics is contained in the expression for the dipole field induced in a small spherical particle. As long as the particle radius is less than 2% of the wavelength of the light, the electrostatic Rayleigh approximation is valid. The dipole moment is given by

$$\mu = \frac{\varepsilon_1(\omega_L) - \varepsilon_2}{\varepsilon_1(\omega_L) + 2\varepsilon_2} r^3 \mathbf{E}_L \tag{31}$$

where ε_1 is the complex dielectric function of the sphere, ε_2 the dielectric function of the surrounding medium, r the sphere radius, and $\mathbf{E}_L$ the incident electric field amplitude. The induced dipole is driven resonantly for frequencies where Re $\varepsilon_1 = -2\varepsilon_2$, leading to an enormous increase in the electric field amplitude at the particle surface. The field is uniform inside the particle, and decays with the dipole decay law outside the particle. Thus the small particle acts to localize the electromagnetic field of the light, effectively increasing the laser power at the molecule.

The incident field drives the molecular oscillator, which radiates at the Stokes' shifted Raman frequency. The Raman radiation can also excite the dipolar resonance of the sphere; in effect the sphere acts as an antenna for the outgoing radiation. Thus, very crudely, the sphere amplifies the Raman effect by the fourth power of the enhanced surface electromagnetic field.

Since the incident laser field and the Raman scattered field are at different frequencies, the last statement is not quite correct. More precisely, the total classical enhancement factor for a sphere (electrostatic limit) is given by

$$G = \left[\frac{\varepsilon_1(\omega_L) - \varepsilon_2}{\varepsilon_1(\omega_L) + 2\varepsilon_2} \right]^2 \left[\frac{\varepsilon_1(\omega_S) - \varepsilon_2}{\varepsilon_1(\omega_S) + 2\varepsilon_2} \right]^2 (r/r + d)^{12} \tag{32}$$

where L and S refer to the laser and scattered fields, respectively. This simple model provides qualitative understanding of many of the experimental results. Resonance occurs when Re $\varepsilon_2 = -2\varepsilon_1$ at either the laser frequency or the scattered frequency. Both excitation and emission can be resonant simultaneously if the vibrational frequency shift is not too large. At resonance, the enhancement is proportional to $(\text{Im } \varepsilon_2)^{-4}$; this relation shows why metals of high reflectivity (low loss) are better enhancers. The long-range nature of the enhancement is described by the dipole decay law—molecules 5 nm away from a 25-nm radius particle still show 10% of the enhancement of first layer molecules. Apparent discrepancies between the range dependence reported by different laboratories can be rationalized on the basis of different surface preparations resulting in different size roughness features. For example, the enhancement is effectively confined to the first monolayer for small particles (<50 nm). Excitation profiles, which deviate from normal ω^4 behavior, are dominated by the small particle conduction electron resonances. The falloff in intensity of high-frequency vibrations is also explained: the driving field and scattered field cannot simultaneously excite the particle resonance if they are of very different frequencies. That both fields are not resonant simultaneously also explains the different excitation profiles for different bands: maxima for higher-frequency vibrations occur at shorter wavelengths, as the scattered field is brought closer to resonance. The distribution of molecular orientations and the variation of the electric field of the dipole over the surface of the sphere result in large depolarization ratios.

Within the Rayleigh approximation, the enhancement is independent of the particle size. For particles much larger than a few percent of the wavelength of the light, higher multipoles are excited. The resonance shifts to the red, broadens, and the maximum enhancement decreases. For example, a full electrodynamic calculation predicts an enhancement factor of 10^6, for a 5-nm radius silver sphere, excited at 382 nm. For a 50-nm sphere, the peak enhancement is only 10^4, occurring for 500 nm excitation.[66]

Although many aspects of the electromagnetic theory have been verified qualitatively, there has not yet been an experiment in quantitative agreement with the theory. Isolated colloids appear to be the system of choice with which to test the electromagnetic theory; unfortunately it is difficult to keep colloids from aggregating. The interactions between colloid particles are difficult to treat theoretically, and only simple models have been tried. There are several possible reasons why this important experimental test has not yet been successful. First, the predicted maximum in the excitation profile occurs in a difficult region where continuous laser sources are not available. In an early experiment, a monotonic increase in the scattered intensity was observed as the exciting frequency was swept towards the ultraviolet; lack of available laser lines precluded the extension of the scan past the predicted peak, however. This experiment is being repeated using pulsed lasers, and the results are eagerly awaited.[67] Second, there have been many suggestions that the dielectric function of a small particle may be much different than that of the bulk metal. Electron scattering at the interface, and quantum size effects,[68-70] will increase the imaginary part of the dielectric function, reducing the quality of the resonance. Thus, depending upon the size of the colloidal particles, the enhancement may be sufficiently reduced to create serious detection problems.

A great deal of effort has been devoted towards increasing the sophistication of the theory and creating new experimental tests. For example, isolated spheres have been treated in both the electrostatic and electrodynamic limits as described above. Isolated spheroids, for which the enhancement may be much larger, have also been modeled in both limits. The important problems of interacting particles (aggregated colloids, for example), or bumps on a conducting plane (for which both interparticle coupling and coupling to the substrate are important) have only been treated via simple models.[71] These situations await the development of more sophisticated electrodynamic treatments. Since this discussion is not intended to be a critical review of SERS, the reader is referred to other articles for more detailed discussions of this point.[3,4]

6.3. Chemical Enhancement

There are several lines of evidence that suggest that the electromagnetic mechanism is not the only one that contributes to the enhancement. In all cases, the enhancement calculated from electromagnetic theories is an overestimate of the actual contribution from that source. The theories all leave out important interactions that would reduce the calculated electrodynamic enhancement. These include, for example, interparticle interactions and corrections to the dielectric function of small particles. That the calculated and measured enhancement factors have often been in close agreement has been taken as good evidence for the exclusivity of the electromagnetic mechanism. Clearly, as the electromagnetic theories get better, there will become more room for an additional enhancement mechanism.

Evidence for an additional mechanism beyond electromagnetic includes the first layer effect, quenching of SERS by the presence of submonolayer amounts of foreign atoms, and the potential dependence of SERS excitation profiles. In experiments where very thin films are prepared on surfaces, molecules in the monolayer immediately adjacent to the surface often show enhancements that are one to two orders of magnitude greater than that expected from extrapolation of the intensities in the adjacent multilayers.[72] The vibrational bands for these species are often shifted in frequency from those in the condensed layers, indicating stronger interactions with the surface. Using a technique known as underpotential deposition, very small amounts of foreign metal atoms can be deposited on an electrode surface, with no danger of forming more than a monolayer. SERS activity is irreversibly lost when as little as 3% of the surface is covered by metals such as thallium and lead, indicating that only a small fraction of the surface is SERS active.[73,74] Finally, the intensities of the SERS signals depend upon potential in a way that is more complicated than can be explained by concentration changes alone.

Although the first two experiments mentioned above suggest the importance of chemisorption and SERS activity, they are not inconsistent with a purely electromagnetic mechanism. Suppose, for example, that the surface is covered not only with bumps that can be seen in the electron microscope, but with a small number of smaller bumps. Suppose further that these small bumps have a greater proportion of atomic scale defects (which are necessary to form a curved surface). Since chemisorption at defects is often stronger than at atomically smooth terraces, the defects simply glue the molecules to the smaller bumps, which may be minority species on the surface, but which have enormous enhancement factors. Thus the SERS signal could arise solely from adsorption on the small bump, and the first two observations above could also be taken as support of an electromagnetic mechanism. I mention this possibility to illustrate how careful one must be in designing experiments that are conclusive.

The potential-dependent excitation profiles provide the best evidence to date in support of a chemical enhancement mechanism. Several groups have reported that the Raman intensities are strong functions of the applied electrode potential—factors of 10 or so variation have been observed.[75,76] These variations have been interpreted as arising from charge transfer excitations which serve as new resonant intermediate states in resonant Raman scattering. Recall that the Raman polarizability can be expressed by the Kramers–Heisenberg dispersion equation,

$$\alpha = \sum_i \frac{\langle f| H |i \rangle \langle i| H |g \rangle}{(E_i - E_g) - h\nu + i\Gamma_i} + \sum_i \frac{\langle f| H |i \rangle \langle i| H |g \rangle}{(E_i - E_g) + h\nu} \qquad (33)$$

where i, j, f are the initial, intermediate, and final states of the molecule, E_i and E_j are the energies of the states i and j, $h\nu$ is the photon energy, and Γ is the width of the intermediate state. Resonance enhancement of the Raman cross section occurs when $h\nu = E_i - E_g$. Most molecules studied by SERS do not have electronic transitions in the visible portion of the spectrum; consequently new resonant intermediate states must have been created by adsorption in order for this mechanism to be operative. Charge transfer states have been proposed to be the resonant intermediate states. The frequencies of these charge transfer excitations depend upon the electrode potential. In the simplest picture, an adsorbate energy level is shifted and broadened into a resonance by tunneling of the conduction electrons between the metal and molecule. The charge transfer excitation can then take place either from the Fermi level to an unoccupied portion of the adsorbate resonance or the optical electron can originate in the occupied portion of the adsorbate orbital and terminate in empty metal states above the Fermi level. Confirmation of this general picture has been provided by the potential dependence of the excitation profiles. For metal–molecule charge transfer, more negative potentials push the Fermi level up and the charge transfer excitation red shifts. The opposite, of course, happens for molecule–metal charge transfer. Both kinds of charge transfer excitations have been reported for molecules adsorbed at electrode surfaces.

Persson has presented a simple theory with which to estimate the charge transfer contribution to the enhancement.[77] He used a modified Newns–Anderson Hamiltonian from chemisorption theory, with the electron–photon interaction given by the dipole approximation and the vibronic interaction given using the leading term in the Herzberg–Teller expansion. Assuming a Lorentzian functional form for the adsorbate density of states, he was able to generate the enhancement factor numerically with the position (referenced to the Fermi level) and width of the adsorbate affinity level and the laser frequency as parameters.

Direct evidence for the existence of charge transfer states in SERS active systems was provided by electron energy loss (EELS) experiments. In addition to the valence electronic excitations, Demuth and co-workers observed new bands at low energies for a number of molecules adsorbed on both single-crystal surfaces and low-temperature evaporated films.[78,79] The new charge transfer absorptions occur in the visible, where almost all SERS excitations have been carried out. On the single-crystal surfaces, the line shape shows an onset indicating the participation of the bulk density of states, and the spectrum is rather weak. On low-temperature evaporated films, a much more intense nearly symmetric band is observed, which has been assigned to adsorption at defects. The optical electron originates in this case from a very narrow density of states, essentially an atomic orbital. Using parameters taken from Demuth's experiment, Persson calculated an enhancement factor of 30, in good agreement with a number of experimental observations. Since the width of the donor density of states is an important parameter in the theory (this has also been pointed out by Ueba,[80]) and since the charge transfer excitations are much more intense on low-temperature evaporated films, it has been proposed that defects of atomic scale lead to the greatest enhancement.

In an effort to determine the role of defects in the chemical mechanism, we have developed a novel approach. Rather than work with surfaces with the largest enhancements and study small contributions to the total, we have chosen to work with surfaces on which the enhancement is minor (a factor of at most 4) and then create defects of known structure. The advantage of this approach is greater dynamic range— factors of 10 enhancement are easily seen on top of a very low background—and the important ability to characterize the atomic scale nature of the defect using various electron spectroscopies. In addition, the electronic spectra of molecules adsorbed on these surfaces can be obtained by high-resolution electron energy loss spectroscopy, so that detailed comparisons with theory can be made.

We have shown previously that the Raman scattering from a monolayer of pyridine adsorbed on the low-index faces of silver is unenhanced.[81] Thus the Ag(111) surface serves as a reference for frequencies and intensities. Defects can be created on single-crystal surfaces by cutting them along high index directions to make stepped and kinked surfaces. Adatoms can be deposited by thermal evaporation or by underpotential deposition. We have so far examined three stepped and kinked silver surfaces, chosen because of previous reports of enhancement.[82] On Ag(521) and Ag(987), which are stepped and kinked surfaces with (111) terraces, pyridine is physically adsorbed, and the Raman spectrum is unenhanced. On Ag(540), however, pyridine is chemisorbed, but the Raman spectrum is still unenhanced. This experiment shows that chemisorption alone is not sufficient to produce SERS. It will be very interesting to obtain the electronic spectrum of this system so that the position of the charge transfer state (if any) can be determined. Our hope is that experiments like these will shed some light on the nature of SERS active sites and the relationship between chemisorption at defects, charge transfer, and Raman scattering intensities.

6.4. Applications of SERS

Now that the factors that contribute SERS intensity have been sorted out, at least qualitatively, the focus of SERS research has shifted towards applications. While quantitative work clearly requires a complete understanding of the mechanism, there is a great deal to be learned about interfacial processes that can be gleaned from the spectral changes themselves.

SERS is beginning to find important applications in many areas of chemistry, including chemical analysis, corrosion, lubrication, and heterogeneous catalysis.

Thin films are used as protective coatings in many areas of technology and any microscopic understanding of the structures, formation mechanisms, and mode of action of these coatings would enable us to design materials with desired properties. Sandroff et al.[83] used SERS to study the conformation of a simple amphiphile (1-hexadecane thiol) adsorbed on a silver island film. The position of the C–S stretching frequency showed that the molecule was bound to the surface through the sulfur. The C–C stretching region has bands that are markers for the trans or gauche conformational isomers of the chain. The adsorbed molecule has a solidlike structure as deduced from a comparison of its spectrum with that of the liquid and solid phases. Exposure to water had no effect on the conformation—strong hydrophobic interactions and closest packing of the surfactant excluded the water. CCl_4 exposure, on the other hand, makes the molecule more liquidlike. Thus CCl_4 facilitates disorder perhaps by impregnation. Wetting was suggested to be associated with conformational changes in amphiphiles.

SERS has also been used to study charge transfer upon adsorption. Tetrathiofulvalene (TTF), a strong electron acceptor, was fully oxidized when adsorbed on silver island films.[84] The frequency of the C=C stretching modes has been shown to be a quantitative measure of the amount of charge transferred in this compound. Impurities on the surface were important in the charge transfer process; they raised the work function of the surface so that is was greater than the ionization potential of the molecule, a necessary condition. The oxidation process on gold colloids was slow and could be followed in real time. Initially TTF was oxidized to $TTF^{0.3+}$, which slowly converted to the $+1$ state. The rate of the reaction was thought to be determined by the rate of impurity adsorption on the sol.

Lubrication is another area in which SERS has been successfully applied. Organic sulfides like diphenyl or dibenzyl sulfide are extreme wear additives. Current thought is that these compounds decompose to form metal sulfides which are then easily sheared. Sandroff et al.[85] showed that these compounds cleaved to form mercaptides, probably via monosulfide intermediates. The phenyl derivative lies flat on the surface while the benzyl derivative stands up. These results are at odds with the accepted mechanism of lubrication because they suggest that Ag_2S is formed before the mercaptide. The combination of SERS as an in situ diagnostic with classic wear tests would be a powerful approach to the study of lubrication.

Hester et al.[86] have used SERS to study the electropolymerization of thionene, which is important in modifying electrodes for certain applications. Jeziorowski et al.[87] have looked at the role of oxalic acid as a corrosion inhibitor on copper. They found that copper oxalate forms a protective film.

There have been several applications of SERS in heterogenous catalysis. Otto et al.[88] have observed several dioxygen species adsorbed on the surfaces of coldly evaporated silver films. They have assigned the spectra to superoxide and peroxide species. These intermediates have long been proposed to be relevant to the ethylene epoxidation reaction, in which silver is used as the catalyst. It would be very interesting to be able to run the epoxidation reaction over a silver catalyst, using SERS to monitor reaction intermediates in situ. Wokaun et al. have also applied SERS to the study of catalytic reactions.[89] They have been able to prepare copper surfaces that show enhancement at elevated temperatures. Copper foil was roughened by sandblasting and etching; this treatment resulted in an enhancement of 10^6 for nonresonant molecules. They have used this surface to study the adsorption and bonding of m-toluidine to copper surfaces.

Finally, SERS has been used to study decomposition reactions on surfaces. Moskovits and DiLella[90] have followed the thermal decomposition of 1,3,5-trifluorobenzene on silver. The spectral lines of the parent molecule disappeared as the sample was warmed to 150 K and a number of new bands grew in, including a broad band in the C–H stretching region characteristic of aliphatic or olefinic species. DiLella has pursued this problem further, showing that monofluorobenzene decomposes to form several products tentatively assigned as fluoroactylene and an aliphatic hydrocarbon.[91]

7. Outlook

I am confident that the application of Raman spectroscopy to interesting problems in surface physics and chemistry will become more important and will continue to contribute to our understanding of interfaces and adsorbed molecules. I expect general improvements in technology over the next few years, which should result in much greater sensitivity. We are exploring the use of large aperture ellipsoidal reflectors to replace the collection lenses we currently use. I anticipate a factor of 5 improvement in collection efficiency and much greater convenience in performing experiments involving tangential surface electromagnetic fields. Another important improvement will be the availability of charge-coupled device (CCD) detectors. When optimized for spectroscopy these detection arrays should offer at least a factor of 5 improvement in quantum efficiency compared with current SIT and ISPD detectors. I am also very enthusiastic about the application of ultraviolet excitation to surface Raman spectroscopy. The ω^4 dependence of the scattered intensity above would yield another factor of 5 improvement in sensitivity if the excitation wavelength was decreased from the green 514.5-nm argon line to the ultraviolet 363.8-nm line, for example. More dramatic improvements could be realized using deep ultraviolet excitation, since most molecules will show either preresonant or resonant enhancement of the Raman cross sections in that frequency region. Thus a factor of 10^2–10^3 improvement in sensitivity can be realistically expected over the next few years.

Apart from purely technical advances, new experiments are being attempted which promise to increase our understanding further. Among the most exciting of these is surface-enhanced hyper-Raman scattering. Since the surface optical fields on the rough surfaces and in SERS experiments are so intense, hyper-Raman spectroscopy (two-photon excited Raman spectroscopy) is feasible. Since hyper-Raman is a three-photon process it is characterized by a parity selection rule opposite to that of normal Raman spectroscopy, and in fact yields the same spectrum as does infrared reflection or EELS.[92] The combination of SERS and SEHRS then yields the complete vibrational spectrum of an adsorbed species.

Acknowledgments

I am deeply indebted to my first students, Vickie Hallmark, Keenan Brown, Camille Howard, and Dave Mullins, who built the lab with me and did all of the experiments reported here. Without their hard work and confidence, I would never have been asked to write this article. Horia Metiu of the University of California at Santa Barbara provided the initial spark that aroused my curiosity in this field, and he has been a valuable friend and advisor since. I am also indebted to Phaedon Avouris at IBM and Mike White of this department for their continued valuable advice. Finally I am grateful to The Robert A. Welch Foundation, The National Science Foundation, The Research Corporation, The University of Texas Research Institute, The Alfred P. Sloan Foundation, and The Camille and Henry Dreyfus Foundation for their support.

References

1. R. P. Van Duyne, in: *Chemical and Biochemical Application of Lasers* (C. B. Moore, ed.), Vol. IV, pp. 101–185, Academic, New York (1978).
2. *Surface Enhanced Raman Scattering* (R. K. Chang and T. E. Furtak, eds.), Plenum Press, New York (1982).
3. A. Otto, in: *Light Scattering in Solids* (M. Cardona and G. Güntherodt, eds.), Vol. IV, pp. 289–418, Springer-Verlag, Berlin (1984).
4. H. Metiu, Surface enhanced spectroscopy, *Prog. Surf. Sci.* **17**, 153 (1984).
5. A. Campion, Surface enhanced Raman spectroscopy, *Comments Solid State Phys.* **11**, 107–123 (1984).
6. A. Campion, J. K. Brown, and V. M. Grizzle, Surface Raman spectroscopy without enhancement: nitrobenzene on Ni(111), *Surf. Sci.* **115**, L153–L158 (1982).
7. D. A. Long, *Raman Spectroscopy*, McGraw-Hill, New York (1977).
8. R. S. Tobias, Raman spectroscopy in inorganic chemistry, *J. Chem. Ed.* **44**, 1–8 (1967).
9. R. G. Greenler and T. C. Slager, Method for obtaining the Raman spectrum of a thin film on a metal surface, *Spectrochim. Acta* **29A**, 193–201 (1973).
10. J. D. Jackson, *Classical Electrodynamics*, Wiley, New York (1962).
11. *American Institute of Physics Handbook*, 3rd ed., pp. 6.124–6.155, McGraw-Hill, New York (1972).
12. J. D. E. McIntyre and D. E. Aspnes, Differential reflection spectroscopy of very thin surface films, *Surf. Sci.* **24**, 417–434 (1971).
13. P. J. Feibelman, Surface electromagnetic EELS, *Prog. Surf. Sci.* **12**, 287–406 (1982).
14. D. R. Mullins, *Surface Raman Spectroscopy: Investigations on Smooth, Stepped and Kinked Crystal Surfaces and Angular Resolution Studies*, Ph.D. dissertation, The University of Texas at Austin (1984).
15. Ph. Avouris and J.E. Demuth, Electronic excitations of benzene, pyridine and pyrazine adsorbed on Ag(111), *J. Chem. Phys.* **75**, 4783–4794 (1981).
16. M. Moskovits, Surface selection rules, *J. Chem. Phys.* **77**, 4408–4416 (1982).
17. H. Nichols and R. M. Hexter, Site symmetry of surface adsorbed molecules, *J. Chem. Phys.* **75**, 3126–3136 (1981).
18. J. K. Sass, H. Neff, M. Moskovits, and S. Holloway, Electric field gradient effects on the spectroscopy of adsorbed molecules, *J. Phys. Chem.* **85**, 621–623 (1981).
19. F. R. Dollish, W. G. Fateley, and F. F. Bentley, *Characteristic Raman Frequencies of Organic Compounds*, Wiley, New York (1974).
20. P. J. Feibelman, private communication.
21. A. W. Dweydari and C. H. B. Mee, Work function measurements on (100) and (110) surfaces of silver, *Phys. Status Solidi A* **27**, 223–230 (1975).
22. V. M. Hallmark and A. Campion, Selection rules for surface Raman spectroscopy, *J. Chem. Phys.* **84**, 2933 (1986).
23. T. A. Egerton and A. H. Hardin, The application of Raman spectroscopy to surface chemical studies, *Catal. Rev. Sci. Eng.* **11**, 71–116 (1975).
24. R. P. Cooney, G. Curthoys, and N. T. Tam, Laser Raman spectroscopy and its application to the study of adsorbed species, *Adv. Catal.* **24**, 293–342 (1975).
25. W. N. Delgas, G. Haller, R. Kellerman, and J. H. Lunsford, *Spectroscopy in Heterogeneous Catalysis*, pp. 58–85, Academic, New York (1979).
26. R. P. Van Duyne, D. L. Jeanmaire, and D. F. Shriver, Mode-locked laser Raman spectroscopy—A new technique for the rejection of interfering background luminescence signals, *Anal. Chem.* **46**, 213–222 (1974).

27. F. R. Brown and L. E. Makovsky, Raman spectra of a cobalt oxide-molybdenum oxide supported catalyst, *Appl. Spectrosc.* **31**, 44–46 (1977).

28. C. P. Cheng, J. D. Ludowise, and G. L. Schrader, Controlled-atmosphere rotating cell for *in situ* studies of catalysts using laser Raman spectroscopy, *Appl. Spectrosc.* **34**, 146–150 (1980).

29. G. L. Schrader, private communication.

30. P. Hargis, private communication.

31. R. Hester, in: *Raman Spectroscopy, Theory and Practice* (H. A. Szymanski, ed.) Vol. 2, pp. 141–173, Plenum Press, New York (1970).

32. A. Campion and N. Somers, unpublished results.

33. N. Zimmerer and W. Kiefer, Rotating surface scanning technique for Raman spectroscopy, *Appl. Spectrosc.* **28**, 279–281 (1974).

34. P. J. Hendra and E. J. Loader, Laser Raman spectra of adsorbed species, *Trans. Faraday Soc.* **67**, 828–840 (1971).

35. Y. Talmi and K. W. Busch, in: *Multichannel Image Detectors* (Y. Talmi, ed.), Vol. 2, American Chemical Society, Washington, D.C. (1983).

36. K. Nishi, K. Chinomi, Y. Inoue, and S. Ikeda, X-ray photoelectron spectroscopic study of the absorption of benzene, pyridine and nitrobenzene on evaporated nickel and iron, *J. Catal.* **60**, 228–240 (1979).

37. C. Backyx, C. P. M. DeGroot, and P. Biloen, Electron energy loss spectroscopy and its applications, *Appl. Surf. Sci.* **6**, 256–272 (1980).

38. C. L. Angell, Raman spectroscopic investigation of zeolites and absorbed molecules, *J. Phys. Chem.* **77**, 222–227 (1973).

39. T. A. Egerton and A. H. Hardin, Raman spectra near 1000 cm^{-1} of pyridine adsorbed on a series of partially exchanged Y-zeolites, *47th National Colloid Symposium of the American Chemical Society,* Ottowa (1973).

40. E. Buechler and J. Turkevich, Laser Raman spectroscopy of surfaces, *J. Phys. Chem.* **76**, 2325–2332 (1972).

41. W. Oganowski, J. Hanuza, B. Jezowska-Trzebiatowski, and J. Wrzyszcz, Physicochemical properties and structure of $MgMoO_4$-MoO_3 catalysts, *J. Catal.* **39**, 161–172 (1975).

42. F. R. Brown, L. E. Makovsky, and K. H. Rhee, Raman spectra of supported molybdena catalysts, *J. Catal.* **50**, 162–171 (1977).

43. J. Medema, C. Van Stam, V. H. J. deBeer, A. J. A. Konigs, and D. C. Koninsberger, Raman spectroscopic study of Co-Mo/γ-Al_2O_3 catalysts, *J. Catal.* **53**, 386–400 (1978).

44. H. Jeziorowski and H. Knözinger, Raman and ultraviolet spectroscopic characterization of molybdena on alumina catalysts, *J. Phys. Chem.* **83**, 1166–1173 (1979).

45. C. P. Cheng and G. L. Schrader, Characterization of supported molybdate catalysts during preparation using laser Raman spectroscopy, *J. Catal.* **60**, 267–294 (1979).

46. G. L. Schrader and C. P. Cheng, *In situ* laser Raman spectroscopy of the sulfiding of Mo/γ-Al_2O_3 catalysts, *J. Catal.* **80**, 365–385 (1983).

47. G. L. Schrader and C. P. Cheng, Laser Raman spectroscopy of Co-Mo/γ-Al_2O_3 catalysts, characterization using pyridine adsorption, *J. Phys. Chem.* **87**, 3675–3681 (1983).

48. G. L. Schrader and C. P. Cheng, Sulfiding of cobalt molybdate catalysts: Characterization by Raman spectroscopy, *J. Catal.* **85**, 488–498 (1984).

49. P. J. Hendra, J. R. Horder, and E. J. Loader, The Raman spectrum of pyridine adsorbed on oxide surfaces, *J. Chem. Soc. A,* 1766–1770 (1971).

50. T. A. Egerton, A. H. Hardin, Y. Kozirovski, and N. Sheppard, Reduction of fluorescences from high-area oxides of the silica, γ-alumina, silica-alumina and

Y-zeolite types of Raman spectra for a series of molecules adsorbed on these surfaces, *J. Catal.* **32,** 343–361 (1974).

51. R. O. Kagel, Raman spectra of pyridine and 2-chloropyridine adsorbed on silica gel, *J. Phys. Chem.* **74,** 4518–4519 (1970).

52. R. P. Cooney and N. T. Tam, The determination of the surface area of silica by laser Raman spectroscopy, *Aust. J. Chem.* **29,** 507–513 (1976).

53. P. J. Hendra and E. J. Loader, Laser Raman spectra of sorbed species: Physical adsorption on silica gel, *Nature* **216,** 789–790 (1967).

54. I. D. M. Turner, S. O. Paul, E. Reid, and P. J. Hendra, Laser-Raman study of the isomerization of olefins over alumina, *J. Chem. Soc. Faraday Trans. I* **72,** 2829–2835 (1976).

55. H. Winde, Zur ramanspektroskopischen untersuchung sorbierter molekule, *Z. Chem.* **10,** 64–67 (1970).

56. W. Krasser, H. Ervens, A. Fadini, and A. J. Renouprez, Raman scattering of benzene and deuterated benzene chemisorbed on silica supported nickel, *J. Raman Spectrosc.* **9,** 80–84 (1980).

57. W. Krasser, Enhancement of Raman scattering by molecules adsorbed on small nickel particles, *Proceeding VIIth International Conference on Raman Spectroscopy,* pp. 420–421 (1980).

58. W. Krasser, A. Fadini, and A. J. Renouprez, The Raman spectrum of carbon monoxide chemisorbed on silica-supported nickel, *J. Catal.* **62,** 94–98 (1968).

59. M. Fleischman, P. J. Hendra, and A. J. McQuillan, Raman spectra of pyridine adsorbed at a silver electrode, *Chem. Phys. Lett.* **26,** 163–166 (1974).

60. D. J. Jeanmaire and R. P. Van Duyne, Surface Raman spectroelectrochemistry, *J. Electroanal. Chem.* **84,** 1–20 (1977).

61. M. G. Albrecht and J. A. Creighton, Anomalougly intense Raman spectra of pyridine at a silver electrode, *J. Am. Chem. Soc.* **99,** 5215–5217 (1977).

62. T. E. Furtak and J. Reyes, A critical review of theoretical models of surface enhanced Raman scattering, *Surf. Sci.* **93,** 351–382 (1980).

63. J. G. Bergman, J. P. Heritage, A. Pinazuk, J. M. Worlock, and J. H. McFee, Cyanide coverage on silver in conjunction with surface-enhanced Raman scattering, *Chem. Phys. Lett.* **68,** 412–415 (1979).

64. Y. J. Chen, W. P. Chen, and E. Burstein, Surface-electromagnetic wave-enhanced Raman scattering by overlayers on metals, *Phys. Rev. Lett.* **36,** 1207–1210 (1976).

65. P. N. Sanda, J. M. Warlaumont, J. E. Demuth, J. C. Tsang, K. Christmann, and J. A. Bradley, Surface-enhanced Raman scattering from pyridine on Ag(111), *Phys. Rev. Lett.* **45,** 1519–1523 (1980).

66. B. J. Messinger, K. V. VonRaben, R. K. Chang, and P. W. Barber, Local fields at the surface of noble metal minispheres, *Phys. Rev. B* **24,** 649–657 (1981).

67. M. Moskovits, private communication.

68. M. Meier and A. Wokaun, Enhanced fields on large metal particles: dynamic depolarization, *Opt. Lett.* **8,** 581–583 (1983).

69. P. Apell and D. R. Penn, Optical properties of small metal spheres: Surface effects, *Phys. Rev. Lett.* **50,** 1316–1319 (1983).

70. W. A. Kraus and G. C. Schatz, Plasmon resonance broadening in small metal particles, *J. Chem. Phys.* **79,** 6130–6139 (1983).

71. V. Laor and G. C. Schatz, The role of surface roughness in surface enhanced Raman spectroscopy (SERS): The importance of multiple plasmon resonances, *Chem. Phys. Lett.* **82,** 566–570 (1981).

72. M. Moskovits and D. P. DiLelle, in: *Surface Enhanced Raman Scattering* (R. K. Chang and T. E. Furtak, eds.) pp. 243–274, Plenum Press, New York (1982).

73. T. Watanabe, N. Yanagihara, K. Honda, B. Pettinger, and L. Moerl, *Chem. Phys. Lett.* **96,** 649–655 (1983).

74. T. E. Furtak, abstract, American Chemical Society 185th National Meeting, Seattle, Washington, March, 1983.

75. J. Billmann and A. Otto, Electronic surface state contribution to surface enhanced Raman scattering, *Sol. State. Commun.* **44,** 105–108 (1982).

76. J. J. McMahon, T. P. Dougherty, J. Riley, G. T. Babcock, and R. L. Carter, Surface enhanced Raman scattering and photodimerization of pyridyl substituted ethylenes at a silver electrode surface, *Surf. Sci.* **158,** 381 (1985).

77. B. N. J. Persson, On the theory of surface-enhanced Raman scattering, *Chem. Phys. Lett.* **82,** 561–565 (1981).

78. J. E. Demuth and P. N. Sanda, Observations of charge-transfer states for pyridine chemisorbed on Ag(111), *Phys. Rev. Lett.* **47,** 57–60 (1981).

79. D. Schmeisser, J. E. Demuth, and Ph. Avouris, Metal-molecule charge transfer excitation on silver film, *Chem. Phys. Lett.* **87,** 324–326 (1982).

80. H. Ueba, Role of defect-induced charge transfer excitation in SERS, *Surf. Sci.* **129,** L267–L270 (1983).

81. A. Campion and D. R. Mullins, Normal Raman scattering from pyridine adsorbed on the low-index faces of silver, *Chem. Phys. Lett.* **54,** 576–579 (1983).

82. A. Campion and D. R. Mullins, Unenhanced Raman scattering from pyridine absorbed on stepped and kinked silver surfaces under ultrahigh vacuum, *Surface Sci.* **158,** 263 (1985).

83. C. J. Sandroff, S. Garoff, and K. P. Leung, Surface-enhanced Raman study of the solid/liquid interface, *Chem. Phys. Lett.* **96,** 547–551 (1983).

84. C. J. Sandroff, D. A. Weitz, J. C. Chung, and D. R. Herschbach, Charge transfer from tetrathiafulvalene to silver and gold surfaces studied by surface-enhanced Raman scattering, *J. Phys. Chem.* **87,** 2127–2133 (1983).

85. C. J. Sandroff and D. R. Herschbach, Surface enhanced Raman study of organic sulfides adsorbed on silver, *J. Phys. Chem.* **86,** 3277–3279 (1982).

86. R. E. Hester, K. Hutchinson, W. J. Albery, and A. R. Hillman, Raman spectroscopic studies of a thionene modified electrode, *Proc. 9th International Conference on Raman Spectroscopy,* p. 724 (1984).

87. H. Jeziorowski and B. Moser, Raman spectroscopic studies of the chemisorption of oxalic acid on copper oxide surfaces in aqueous media, *Proc. 9th International Conference on Raman Spectroscopy,* p. 726 (1984).

88. C. Pettenkofer, I. Pockrand, and A. Otto, Surface enhanced Raman spectra from oxygen on silver, *J. Phys. (Paris) C* **10,** 463 (1983).

89. S. K. Miller, A. Baker, M. Meier, and A. Wokaun, Surface enhanced Raman scattering and the preparation of copper substrates for catalytic studies, *J. Chem. Soc. Faraday Trans. I* **80,** 1305–1312 (1984).

90. M. Moskovits and D. P. DiLella, in: *Surface Enhanced Raman Scattering* (R. K. Cheng and T. E. Furtak, eds.), pp. 243–274, Plenum Press, New York (1982).

91. D. P. DiLella, R. R. Smardzewski, S. Goha, and P. A. Lund, Surface-enhanced Raman study of the catalytic decomposition of fluorobenzene on silver, *Surf. Sci.* **158,** 295 (1985).

92. R. P. Van Duyne, private communication.

9

Infrared Spectroscopy of Adsorbates on Metals: Direct Absorption and Emission

P. L. Richards and R. G. Tobin

1. Introduction

Measurements of the vibrational spectra of monolayers or submonolayers of adsorbates on surfaces present a severe challenge to the
infrared spectroscopist. In many cases, multiple surfaces cannot be used
and small signals must be measured that are superimposed on background radiation that can be many orders of magnitude stronger. To be
successful, an experiment must be both well conceived and well executed.
The conventional approaches are transmittance and reflection–absorption
spectroscopy. Hoffmann[1] and Ryberg[2] have reviewed the reflection–
absorption technique. Many approaches have been used to enhance the
size of the surface signal relative to the backgrounds. These include
multiple reflection,[3,4] attenuated total internal reflection,[5] surface
electromagnetic waves,[6] and Stark modulation,[7] as well as direct
measurements of absorption and emission. In this chapter we review
some aspects of infrared technology that are relevant to this measurement problem and then describe direct absorption and emission techniques as applied to the vibrational spectroscopy of adsorbed atoms and

P. L. Richards and R. G. Tobin • Department of Physics, University of California, and
Materials and Molecular Research Division, Lawrence Berkeley Laboratory, Berkeley,
California 94720. *Present address for R. G. T.:* AT & T Bell Laboratories, Murray Hill,
New Jersey 07974-2070.

molecules on metal surfaces. The techniques described will have applications to other surface spectroscopies such as atoms or molecules adsorbed on transparent insulators and to chemiluminescence, but the discussion will focus on the important, and especially difficult, problem of the vibrational spectroscopy of adsorbates on metals.

The high conductivity of metals ensures that the infrared electric vector at the position of the adsorbate is very nearly perpendicular to the surface. The infrared beam must have a large angle of incidence to produce this perpendicular field. Under these conditions, 10%–20% of the incident beam is absorbed in the metal and 80%–90% is reflected. A very much smaller amount is absorbed in the adsorbed layer. The large metallic absorption arises from the fact that the penetration depth for photons in the metal is much larger than atomic or molecular dimensions. Measurement of the reflected beam appears to be the most straightforward way to obtain vibrational information for the adsorbate. This information, however, comes superimposed on a background that is 80%–90% of the incident infrared beam. Variations in this background, whether due to source fluctuations, spectrometer instability, or photon statistics, appear as noise.

The background encountered in a reflection–absorption experiment is not fundamental. Techniques that measure the power absorbed in the sample, or the power emitted by the sample, have lower backgrounds by a factor of order 10 than reflection experiments. Since most carefully designed experiments are limited by the ability to cancel these backgrounds, this can be a worthwhile improvement. One quantity of interest is the ratio of the strength of the molecular signal to the signal absorbed by, or reflected by, the metal. In Figure 1b we show this ratio as a function of incident angle for CO chemisorbed on Ni. We see that the fractional molecular signal peaks at large angles of incidence for both absorption and reflection, but that the peak is both wider and higher for absorption than for reflection. The advantage of direct absorption measurements is even greater for metals with higher conductivity such as Cu, Ag, Au, or Pt. Since interesting surface species are being studied which have peak absorptivities less than 10^{-4}, however, a background signal corresponding to an absorptivity of 10^{-1} or even 10^{-2} is far from negligible. For this reason it is in principle easier to measure a given small adsorbate signal on a lossless dielectric substrate than on a metal.

The large background encountered in any infrared experiment on a metal substrate requires careful attention to the modulation scheme used. Ideally, the modulation in an ir surface experiment serves two functions: to distinguish the surface from the substrate background, and to avoid

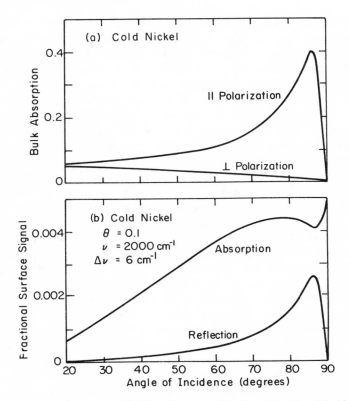

Figure 1. (a) Absorption of a cold Ni surface as a function of angle of incidence and direction of polarization. (b) Ratio of the peak absorption of the C–O stretch vibration of chemisorbed CO to the absorption and the reflectance of a cold Ni surface.

low-frequency noise and drifts. In general, a single modulation technique may not accomplish both objectives.

The surface signal can be isolated by comparing spectra measured with and without an adsorbate present. The drawback of this approach is that it represents a modulation on the time scale of minutes to hours; it is therefore very susceptible to low-frequency drifts. A second form of faster modulation is generally needed in addition. Sensitivity to drifts is minimized if the fast modulation is also partially surface selective. Two modulation schemes that also distinguish surface from substrate contributions are polarization modulation[4,8,9] and wavelength modulation[2,10,11]

Polarization modulation exploits the fact that only the p-polarized ($\parallel$ polarized) component of the infrared beam contains the surface signal. A rotating polarizer[4,8] or photoelastic modulator[9] switches the polarization of the beam. After demodulation, the detector signal is proportional to the difference in intensity between the two polarizations. When the spectrometer efficiency is strongly polarization dependent as is typically the case with a grating instrument, a second polarizer, fixed at an angle between the s- and p-orientations, should be used after the switched polarizer. In principle, the fixed polarizer keeps the polarization state of the light entering the spectrometer fixed, regardless of the orientation of the switched polarizer. In practice, imperfections in available infrared polarizers limit the usefulness of this approach. Nonetheless, over selected frequency ranges,[8] or with a less polarized spectrometer,[9,12] polarization modulation can be used very effectively.

In the case of wavelength modulation,[10,11] the wavelength incident on the detector is modulated, giving a signal after demodulation proportional to the derivative of the intensity with wavelength. This approach enhances sharp spectral features over broad ones. It therefore emphasizes the adsorbate signal if it is sharper than other spectral features in the background. The infrared emissivity of metals is essentially featureless. The spectrometer efficiency, however, must have no sharp structure in the frequency range of interest. For such a system, wavelength modulation can be very useful.[10,11]

In the preceding discussion, it is tacitly assumed that background variations are the primary limit to the sensitive surface measurements. We believe that this is true for well-designed experiments, but it is far from obvious. Traditionally, infrared spectroscopy has been source brightness or, nearly equivalently, detector noise limited. Developments in infrared technology have occurred, however, which make it possible to avoid detector noise in the types of experiments under discussion.

A difficult spectroscopic problem, such as the measurement of the vibrational properties of monolayers of adsorbates on metal surfaces, can benefit from the most careful experimental analysis. In order to make this analysis comprehensible, we present in Section 2 a summary of the present status of some relevant techniques of infrared spectroscopy. In Section 3 we describe an experiment that measures the heat deposited in the metal substrate by the infrared flux from a blackbody source. This experiment is based on the technology of low-temperature bolometric detectors developed for astronomy at submillimeter wavelengths. It provides high sensitivity and broad wavelength coverage, but is not easily compatible with the techniques used to produce clean single-crystal surfaces. In this respect its limitations are similar to some of the acoustic detection techniques that are used with laser sources.[13]

In Section 4 we describe an infrared emission experiment in a cooled environment that has been used to measure the vibrational frequencies of CO on clean, well-characterized single-crystal surfaces of Ni and Pt. This experiment is compared and contrasted with other emission experiments designed to measure monolayer adsorbates on metals. Again, infrared technologies recently developed for astronomy can be used to optimize such measurements.

2. Status of Relevant Technology

2.1. Overview

For many years infrared spectra were measured in the vibrational frequency range from 400 to 4,000 cm^{-1} using thermal sources, diffraction grating spectrometers, and room temperature detectors. These techniques are convenient, and sensitive enough to do many important measurements. Spectroscopists who needed high resolution, or longer wavelengths, increasingly used cooled photon detectors and/or cooled bolometric detectors. Since ratios of signal-to-noise remained a problem, the prejudice arose that infrared spectroscopy is detector noise (or source brightness) limited.

Fourier transform infrared spectroscopy (FTIR) is being used increasingly. In addition to the multiplex advantage over dispersion spectrometers that is available when the spectroscopy is detector noise limited, Fourier Michelson spectrometers can have high resolution with relatively high throughput and can cover very wide spectral bands without changes in optical components. The computer necessary for Fourier analysis has proved very useful for coadding spectra and accurate data manipulation, especially for applications such as surface spectroscopy in which large backgrounds must be subtracted. The factors that influence the choice of an infrared spectrometer will be discussed in more detail below.

Tunable laser sources suitable for vibrational spectroscopy have become increasingly available for near and middle infrared wavelengths. These sources have been extraordinarily useful for very high resolution measurements, for nonlinear effects, and for short pulse measurements which require large power density in a narrow spectral band. They have also greatly expanded the usefulness of a variety of modulation and detection techniques for measuring weak absorptions.[13,14]

Lasers are of unquestioned value for surface Raman spectroscopy[15] and surface nonlinear spectroscopies,[16] which may prove very useful in the future. The role of laser sources in linear infrared spectroscopy of monolayer coverages of molecules on metal surfaces is not obvious. Because of the large background signals, sources are required that have very high amplitude stability. The spectral range of interesting molecular vibrations is wider than that covered by the more generally available tunable infrared lasers. Infrared free electron lasers now under development, however, promise a wider tuning range. If high stability can be maintained, then these new lasers may prove very useful for linear surface infrared spectroscopy. Since only moderate resolution is required, the narrow linewidth of infrared lasers is not a critical advantage. Useful reviews of the performance of tunable lasers are available.[17,18]

The power available from laser sources permits the use of relatively inefficient modulation or detection schemes which can in principle enhance the adsorbate signal relative to the background. A microphone[14] or a thermometer attached to the sample can measure the absorbed power, rather than the reflected power. As is discussed in Section 3 below, however, the thermometric technique can also be used effectively with thermal sources. In either case, they are not easy to make compatible with the cleaning procedures used to produce well-characterized single-crystal surfaces. This difficulty is avoided by a technique that detects the thermal distortion of the sample surface due to absorbed laser power.[13] Another technique uses an electric field to Stark shift the signal. This enhances the adsorbate signal relative to the metallic absorption.[7] The achievable frequency shifts are small compared to typical linewidths, but the approach is very promising when stable tunable lasers are available.

The synchrotron radiation from electron storage rings is widely used for uv and x-ray spectroscopy. There has been some interest in using the infrared output from such facilities. Calculations of the infrared radiation from storage ring sources show that they are considerably brighter than conventional laboratory thermal infrared sources. Unfortunately, as we discuss below, the throughput available at most infrared wavelengths is considerably less than is used for most laboratory infrared spectroscopy. Consequently there is only a moderate improvement in useful infrared power, except for experiments that can only make use of very small throughput.

In order to characterize the image that can be formed from a storage ring source (or from any partially coherent source) it is useful to define a normalized throughput $A\Omega v^2$, where A is the area of a focal spot, Ω is the solid angle of convergence, and v is the wave number. This quantity is variously called the phase space volume or the number of modes. The latter name arises from the fact that a diffraction-limited beam (with a single transverse mode) has $A\Omega = v^{-2}$.[19] For a thermal source the power per mode in the wave number interval dv is

$$p_v \, dv = hc^2 v[\exp(hcv/kT) - 1]^{-1} \, dv \qquad (1)$$

for one polarization. In the Rayleigh–Jeans limit (1) reduces to the Johnson noise expression

$$p_v \, dv = kTc \, dv \qquad (2)$$

Incoherent radiation, such as that typically obtained from a thermal source, can be described as consisting of a large number of these modes. The Planck spectral distribution, giving the spectral power into a throughput $A\Omega$, can be obtained from equation (1) by multiplying by the number of polarizations times the number of modes $P_v = 2A\Omega v^2 p_v$. It is convenient to describe the output of any source (thermal or nonthermal) in terms of distribution of modes and an equivalent Rayleigh–Jeans temperature $T_{eq} = p_v/kc$, which is a measure of the power per mode. The effective temperature T_{eq} for a thermal source is equal to the physical temperature for $hcv \ll kT$ and falls exponentially for $hcv \gg kT$ as is shown in Figure 2.

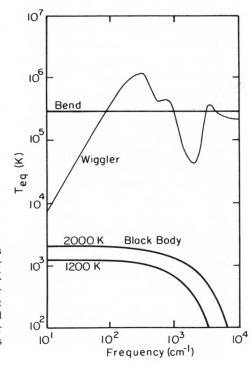

Figure 2. Infrared power per mode as a function of photon frequency for thermal sources at 1200 and 2000 K compared with the time averaged power computed for a bending magnet and a wiggler magnet on a proposed synchrotron light source. The power per mode is given in temperature units defined in the text.

The equivalent temperature for a storage ring source with a tightly focused electron beam is proportional to the beam current and is essentially independent of frequency in the infrared. The value of T_{eq} calculated[20] for synchrotron radiation from a bending magnet of a proposed new storage ring, the advanced light source (ALS) is compared with typical blackbody sources in Figure 2. Calculations for existing storage rings such as the National Synchrotron Light Source give a slightly higher T_{eq} at low infrared frequencies due to a larger beam current, but with some falloff at high frequencies due to the spread of the electron beam.[21] Also shown in Figure 2 is the output expected from one type of magnetic wiggler introduced into a straight section of the ALS. Although such wigglers are very useful in increasing T_{eq} at higher frequencies, the type shown here is of limited value in the infrared. Interest in storage ring infrared sources arises from the fact that T_{eq} is 10^3–10^4 times higher than for thermal sources at infrared frequencies.

A complete comparison between thermal and storage ring sources, however, must include consideration of the available distribution of modes, which determines the available throughput. A typical Fourier spectrometer designed for vibrational spectroscopy using a thermal source operates with an $f/1$ focus on a detector area of $1\,mm^2$. This corresponds to $A\Omega\nu^2 = 10^4$ modes at $\nu = 1000\,cm^{-1}$ and 10^6 modes at $\nu = 10,000\,cm^{-1}$. The infrared beam from a bending magnet on a storage ring by contrast is fan shaped. It is diffraction limited (one mode high) in the vertical direction and has the full range of bend angles in the horizontal direction. If the same detection system is used, the number of available modes is only 10^2 at $1000\,cm^{-1}$ and 10^3 at $10,000\,cm^{-1}$. Thus the total power from the storage ring in this application is about equal to that from a thermal source in the vibrational frequency range. It is substantially larger, however, at far infrared wavelengths.

The comparison is somewhat more favorable for the storage ring in the reflection absorption surface spectroscopy experiment. The narrow range of angles shown in Figure 1 over which a large fractional surface signal is obtained is well adapted to the fan-shaped beam from the storage ring. About one order of magnitude more signal in the vibrational frequency range can be expected for such a source than for a thermal source.

In this review we will argue that when modern detectors are properly used, infrared spectroscopy of molecules on metal surfaces with thermal sources need not be source power or detector noise limited. Therefore only modest advantages can be expected from using a more powerful source, and then only if the source is stable and conveniently tunable.

This argument for the usefulness of thermal sources is valid only if experiments are carefully optimized to minimize ambient photon noise.

The infrared spectral range coincides very closely with the spectral range over which the thermal emission from room temperature objects is important. The presence of this background thermal radiation has a profound influence on the design of many infrared experiments. This ambient photon flux can saturate sensitive detectors. Fluctuations in this flux very often dominate system noise. Sensitive experiments with visible light are usually done in a dark room so that the effects of ambient photons can be neglected. At longer wavelengths the brightness of the ambient thermal photons increases rapidly. In the far infrared, thermal sources are only a factor of 3 brighter than room temperature. Conditions comparable to those achieved at visible wavelengths in a dark room can only be achieved by cooling the "room." One major advantage of bright sources such as lasers or even storage rings is that large ratios of signal to noise can be obtained with less effort required to control the ambient thermal flux.

In this paper we will describe two infrared surface experiments that use thermal sources. The design of these experiments is dominated by the need to reduce ambient photon noise. A quantitative understanding of photon noise is essential for this task. Although the literature contains many descriptions of photon noise,[22] the equations are difficult to use correctly without some knowledge of how they arise. In the next section we present an overview of the subject of photon noise at infrared wavelengths.

2.2. Photon Noise

For our purposes we can distinguish three types of incoherent square law detectors. In photovoltaic detectors a fraction η of the incident photons above a cutoff wave number v_c generates electron–hole pairs that migrate to the electrodes. The photocurrent that arises from a photon rate $\dot{N}$ (which is assumed to lie in the wave number interval between v and $v + dv$) is $I = \eta \dot{N} e$. The noise in this photocurrent, in the ideal case, arises from fluctuations in the rate of arrival of infrared photons.

In extrinsic photoconductive detectors, a fraction η of the incident photons creates either mobile electrons or mobile holes, which move under the influence of an externally applied field. These carriers produce a photocurrent $I = G\eta \dot{N} e$. Here G is the photoconductive gain, which is equal to the ratio of the carrier lifetime to the transit time. It expresses how many times a mobile carrier can cross the detector before it is lost through recombination or trapping. The noise in this photocurrent, in the ideal case, arises both from fluctuations in the rate of arrival of incident photons (which controls the rate of generation of mobile carriers) and from the statistically independent rate of recombination which eliminates these carriers.

In thermal detectors a fraction η of the incident photons dissipate their energy as heat. The resulting temperature change is read out by a thermometer whose output is often a voltage. The output voltage is given by $\eta \dot{N} hc\nu S$, where S is the voltage responsivity in [V/W]. As is discussed in Section 3.2 below, the fundamental noise in thermal detectors comes both from fluctuations in the rate of arriving photons, and from internal thermal fluctuations.

For photon detectors the absorptive quantum efficiency η is essentially zero for $\nu < \nu_c$ and is a slowly varying function of frequency with typical values from 0.1 to 0.5 for $\nu > \nu_c$. For thermal detectors η is called the absorptivity. It is typically a weak function of frequency and >0.5 for all important frequencies.

In order to understand photon noise we focus our attention on the number $\eta \dot{N} t$ of charge carriers generated in a photovoltaic detector in some arbitrary time interval t. Since these photons arrive at random under the conditions of interest for this article, Poisson statistics can be used to compute the root-mean-square fluctuation in this number of charges $\Delta(\eta \dot{N} t)_{\text{RMS}} = (\eta \dot{N} t)^{1/2}$. The RMS output noise current averaged over time t is thus $\Delta(\eta \dot{N} t)_{\text{RMS}} e/t = e(\eta \dot{N}/t)^{1/2}$, where e is the electronic charge. Since the postdetection noise bandwidth associated with averaging over a time $t = 0.5\,\text{s}$ is 1 Hz, the noise current per unit postdetection bandwidth is $\Delta I_{\text{RMS}} = e(2\eta \dot{N})^{1/2}$ [A Hz$^{-1/2}$] for a photovoltaic detector, and $\Delta I_{\text{RMS}} = Ge(4\eta \dot{N})^{1/2}$ [A Hz$^{-1/2}$] for an extrinsic photoconductive detector. By considering the energy associated with photon fluctuations, photon noise in the output of a thermal detector can be written $\Delta V_{\text{RMS}} = hc\nu S(2\eta \dot{N})^{1/2}$ [V Hz$^{-1/2}$] if the detector is cold enough that its own radiation can be neglected.

It is traditional to define the noise-equivalent photon rate NEṄ as the photon rate that must be incident on a detector for the output current to be equal to the root mean square fluctuation per unit bandwidth in the output current. For a photovoltaic detector this condition is $\eta e\text{NEṄ} = e(2\eta \dot{N})^{1/2}$. The resulting noise equivalent photon rate is

$$\text{NEṄ} = \left(\frac{2\gamma \dot{N}}{\eta}\right)^{1/2} \text{[photons s}^{-1}\,\text{Hz}^{-1/2}] \tag{3}$$

where $\gamma = 1$ for a photovoltaic and $\gamma = 2$ for a photoconductive detector. Similarly, we can define a noise equivalent power NEP as the incident signal power required to obtain an output signal equal to the RMS output noise. The result is

$$\text{NEP} = hcv\left(\frac{2\gamma\dot{N}}{\eta}\right)^{1/2} \text{ [W Hz}^{-1/2}] \tag{4}$$

where v is the wave number of the photons. For a cold thermal detector $\gamma = 1$.

The photons that cause the fluctuations in detector output can be the signal from a thermal source with temperature T and emissivity ε that reach the detector through a spectrometer with throughput $A\Omega$ and transmittance τ. Alternatively they may come from the room, or the spectrometer itself. The Planck theory gives the power spectrum $P_v \, dv$ (or photon rate $\dot{N}_v \, dv$) that reaches the detector in the throughput $A\Omega$ and the wave number range from v to $v + dv$,

$$\dot{N}_v \, dv = \varepsilon\tau A\Omega_{\text{eff}}B(v, T) \, dv/hcv \tag{5}$$

where $B(v, T) = 2hc^2v^3[\exp(hcv/kT) - 1]^{-1}$. To be precise,

$$\Omega_{\text{eff}} = \int_\Omega \cos\theta \, d\Omega$$

where θ is the angle between the propagation direction and the normal to the radiating surface.

If photons are incident on the detector over a significant fractional bandwidth, then the mean square fluctuations arising from photons at different frequencies must be added in the appropriate way. For a photon detector, the fraction η of the incident photons at any frequency above the cutoff creates essentially equivalent mobile carriers. A relatively simple expression can be obtained if we assume that η is independent of frequency over the band of interest. For a photovoltaic detector,

$$\Delta I_{\text{RMS}} = e\left[2\eta \int_{v_c}^\infty \dot{N}_v \, dv\right]^{1/2} \text{ [A Hz}^{-1/2}] \tag{6}$$

If we use $\dot{N}_v$ from equation (5) then for either type of photon detector

$$\text{NE}\dot{N} = \left|\frac{4\gamma A\Omega_{\text{eff}}k^3T^3}{\eta c^2h^3} \int_{x_c}^\infty \frac{\varepsilon\tau x^2 \, dx}{e^x - 1}\right|^{1/2} \tag{7}$$

where $x = hcv/kT$. We assume that the infrared band is defined by the frequency dependence of the product $\varepsilon\tau$ of emissivity times transmittance which includes the effects of filters and spectrometers. This noise equivalent photon rate is the most convenient quantity for evaluating the performance of photon detectors. Unfortunately the practice has arisen of converting $N\dot{E}N$ to NEP = $N\dot{E}Nhc\langle v\rangle$, where $\langle v\rangle$ is a weighted average infrared frequency,

$$\langle v\rangle = \int_{v_c}^{\infty} \varepsilon\tau\dot{N}_v v\, dv \Big/ \int_{v_c}^{\infty} \varepsilon\tau\dot{N}_v\, dv \tag{8}$$

This tradition is harmless for narrow infrared bands, but introduces unnecessary complexity when broad bands are involved.

For thermal detectors such as bolometers, the absorptivity η is approximately constant in frequency, but the energy per photon is proportional to frequency. Thus the power fluctuations that arise from photons at different frequencies are weighted properly by summing the mean square voltage fluctuations in the detector output

$$\Delta V_{RMS} = \left[2\eta \int_0^{\infty} \dot{N}_v v^2\, dv\right]^{1/2} hcS \tag{9}$$

so that

$$NEP = \left|\frac{4A\Omega_{eff}k^5 T^5}{\eta c^2 h^3} \int_0^{\infty} \frac{\varepsilon\tau x^4\, dx}{e^x - 1}\right|^{1/2} [W\,Hz^{-1/2}] \tag{10}$$

We now have a complete prescription for calculating the minimum noise to be expected from a detector that is subjected to a known infrared flux. Under certain practical situations, where the detector responsivity and quantum efficiency (or absorptivity) are known from the manufacturer, the values of $\dot{N}$ or p_v can be deduced from the electrical output of the detector. The expected photon noise can then be computed from equations (3) and (4) [or (7) and (10)] and compared with the measured noise. This is one practical way in which the question of whether the detector is photon noise limited under its operating conditions can be answered.

Another approach to estimating photon noise is to compute the incident power from the temperature and the geometry of the infrared sources. As an example consider a source temperature $T = 10^3$ K and a spectrometer throughput $A\Omega = 10^{-2}$ sr cm^2, efficiency (or transmittance) $\tau = 0.1$, and wave number bandwidth Δv, which can be broad for an FTS or narrow for a dispersion spectrometer. Signal photon fluctuations are very important at visible, or higher frequencies, where there are few energetic photons, but do not usually limit infrared measurements. Using the above numbers for $\Delta v = 1$ cm^{-1} at $v = 10^3$ cm^{-1} gives a signal power $p_v\, dv = 3.7 \times 10^{-7}$ W and a signal photon limited NEP = 6×10^{-14} [WHz$^{-1/2}$] for a ratio of signal-to-noise $S/N = 6 \times 10^6$ [Hz$^{1/2}$].

As an example of background photon noise consider 300 K background radiation incident through a 2π solid angle ($\Omega_{\text{eff}} = \pi$) at all frequencies above the detector cutoff on an ideal photon detector with $\eta = 1$. This ideal photon detector limit[22] depends on the detector cutoff frequency. It varies from NEP $= 8 \times 10^{-12}(A)^{1/2}$ [W Hz$^{-1/2}$] for a detector which cuts off at $v_c = 2000 \text{ cm}^{-1}$, has a maximum value of $3.8 \times 10^{-11}(A)^{1/2}$ for $v_c = 800 \text{ cm}^{-1}$ and falls to $1 \times 10^{-11}(A)^{1/2}$ at 100 cm^{-1}. It is known as the background-limited infrared photoconductor (BLIP) limit. It is important to note that this so-called limit can be avoided by reducing the background temperature or by reducing the solid angle or bandwidth with cooled optical components.

When 300 K radiation is incident at all frequencies on an ideal cooled thermal detector with absorptivity $\eta = 1$, the NEP obtained from equation (10) is $4 \times 10^{-11}(A)^{1/2}$ [W Hz$^{-1/2}$], where A is the detector area in [cm^2]. The minimum noise in a room temperature thermal detector is larger than this value because fluctuations in the emitted photons must also be included. Since photon noise and several other sources of noise vary as $A^{1/2}$, detector performance is sometimes specified in terms of the specific detectivity $D^* = A^{1/2}/\text{NEP}[\text{W}^{-1} \text{ Hz}^{1/2} \text{ cm}]$.[22]

2.3. Detectors

Modern photoconductive detectors operated at low temperatures have noise very close to that given by the photon noise limit in equation (4) with $\eta \gtrsim 0.1$ for a wide range of values of the incident photon rate $\dot{N}$. An example of this performance is given[23] in Figure 3. When used with conventional amplifiers[24] these detectors become amplifier noise limited at photon rates $\lesssim 10^7$ photons/s, corresponding to NEP $= 2 \times 10^{-17}$ W Hz$^{-1/2}$ at 10 μm. Specialized amplifiers exist, however, with noise levels which become important only below ~ 100 photons/s.[25] Some of these detectors are listed in Table 1.

These high performance detectors have been developed for the exacting requirements of space infrared astronomy with cooled optics.[26] For nearly all laboratory experiments the background photon rate is so large that a much wider variety of cooled detectors, can approach the background photon noise limit, including nearly all of the well-known photovoltaic detectors such as InSb, HgCdTe, and PbSnTe. These photovoltaic detectors typically have higher operating temperatures and are thus often more convenient to use than the extrinsic photoconductive detectors listed in Table 1. Since most infrared detectors are sold for applications in which they view a large amount of 300 K background, it is

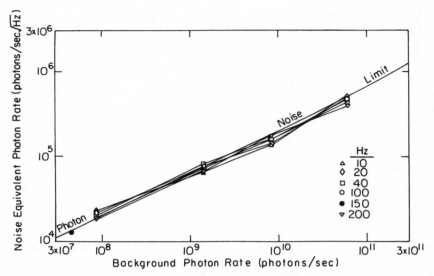

Figure 3. Measurements of noise equivalent photon rate NEṄ versus photon rate $\dot{N}$ for a Ge:Ga detector at 100 cm^{-1} (100 μm). The data agree with the photon noise computed from equation (1) with $\eta = 1$ to the experimental uncertainty, which is a factor ~3 in $\dot{N}$.

sometimes difficult to obtain specifications or test results relevant to low background applications.

Several types of photoconductive and photovoltaic infrared detectors are becoming available in integrated arrays of tens to thousands of detectors. The readout is accomplished by methods related to those used with the Si CCD arrays which are sensitive to visible frequencies.

The most sensitive available thermal detectors are semiconductor bolometers.[27] In low backgrounds such bolometers have NEPs ap-

Table 1. Properties of Some High Performance Photo-conductive Detectors

Detector material	Cutoff wavenumber (cm^{-1})	Cutoff wavelength (μm)
Si:In	1400	7
Si:Bi	625	16
Si:As	370	27
Ge:Be	175	57
Ge:Ga	83	120
Stressed Ge:Ga	50	200

proaching 10^{-15} [W Hz$^{-1/2}$] when operated at liquid ^{4}He temperatures, and at least one order of magnitude better at lower temperatures.[28] Photon noise limited operation can be achieved in almost any imaginable surface experiment. A detailed discussion of bolometric detectors is given in Section 3.

An infrared experiment can be optimized by cooling apertures and filters (or spectrometers) to minimize the background photon rate and then selecting a detector and amplifier system good enough to be limited by the fluctuations in the detected photons coming both from the background and from the source.

2.4. Spectrometers

The most frequently used infrared spectrometers are the Fourier Michelson spectrometer, often referred to as the Fourier transform spectrometer or FTS, the grating spectrometer, and the Fabry–Perot spectrometer. The FTS has the well-known multiplex advantage. If the noise in the system remains constant as the bandwidth over which the signal power falls on the detector is changed, as is the case for inherent detector noise, then the FTS can measure the entire spectrum of n spectral elements in essentially the same time that a grating or Fabry–Perot spectrometer with similar efficiency and throughput takes to obtain a single spectral element with the same signal-to-noise ratio.

The multiplex advantage of Fourier spectroscopy is often used successfully to reduce the effects of the noise from a relatively insensitive, but convenient, room temperature detector such as the pyroelectric detector. It is a common experience that the substitution of a more sensitive cooled detector does not achieve the anticipated degree of improvement. This can occur because the noise is no longer independent of the bandwidth over which signal power falls on the detector.

As we have seen, detectors exist for which detector noise is negligible for nearly all laboratory experiments. When a photon noise limited detector is used, noise varies as the square root of the signal power as the bandwidth is changed, and there is no multiplex advantage. The performance of a single frequency spectrometer is nearly equivalent to that of a multiplexed spectrometer for such optimized experiments. A third case exists in which the noise is proportional to the signal power falling on the detector. Such noise could arise from fluctuations in the source, microphonics in the optics, chopper noise, etc. In this case the Fourier spectrometer has a multiplex disadvantage and takes n times longer to measure the spectrum than the narrow band spectrometer. As with other types of noise, only that portion of the source noise spectrum that is close to the modulation frequency is important. A rapid scan FTS does not have a multiplex disadvantage for slow drifts in source intensity. Such slow drifts are important, however, in any surface experiment that requires the subtraction of spectra taken many minutes apart.

Diffraction grating spectrometers can be used with a linear array of n detectors in the dispersed output. This gives a multichannel advantage of a factor n in time, whatever the source of noise. As high-quality integrated arrays become available with $n \geq 10^2$, the grating spectrometer will often be the device of choice for optimized experiments at moderate resolution, particularly since it is much easier to cool than the Fourier spectrometer.

The FTS and the Fabry–Perot spectrometer can give higher throughput at high resolution that the grating spectrometer.[29] For the relatively low resolution of $\sim 1\,\mathrm{cm}^{-1}$ needed for vibrational spectroscopy of chemisorbed molecules and the limited throughput available with samples at grazing incidence, this difference is often not of great importance.

The FTS is very convenient to use, compared with grating or Fabry–Perot spectrometers, when a wide spectral band is to be covered. It is thus the practical choice for many spectroscopic problems that do not have to be carefully optimized. For difficult problems such as measurement of the vibrational spectra of molecules chemisorbed on metal surfaces, alternative spectrometers should be considered.

2.5. Conclusions for Surface Spectroscopy

From this discussion of infrared techniques, it should be clear that there are a number of alternative approaches to the problem of infrared surface spectroscopy. The required throughput, resolution, spectral range, and ratio of signal to detector or photon noise can be obtained in several ways. The most critical aspect of a measurement of vibrational spectra on metals, however, arises because small signals must be observed that are superimposed on large backgrounds. This aspect of the measurement imposes additional requirements that are very demanding. First, the experiment should be designed to minimize the backgrounds. Second, a background subtraction scheme must be implemented, and third, all parts of the experiment should be extremely stable on the time scale of the background subtraction.

In the following sections of this paper we describe two experiments in which the background is minimized by measuring the sample absorptivity and emissivity, rather than the reflectivity. This is done with some sacrifice in signal level. In both cases, however, the experiments have been optimized so that detector and photon noise do not play an important role. The surface sensitivity achieved is limited by the precision with which the remaining background has been subtracted.

3. Direct Absorption Spectroscopy

3.1. Techniques for Measurement of Absorbed Power

Direct absorption spectroscopy requires some method of detecting a change in the sample caused by the absorption of infrared power by the adsorbed molecules. The most straightforward method, and the one used in the experiments described here, is simply to measure the rise in sample temperature due to the absorbed power using a sensitive thermometer thermally coupled to the sample. In this method the sample itself is used as a bolometric infrared detector.

There are other techniques used to detect absorbed power which involve the generation of sound in the sample, or the thermal expansion of the surface. Typical sensitivities of such techniques are on the order of 10^{-8} W Hz$^{-1/2}$ or worse,[13] which is many orders of magnitude worse than bolometric detection. This sensitivity is not sufficient for use with thermal or storage ring sources. As stable tunable lasers become available over the spectral band of interest for vibrational spectroscopy, these techniques will become more important.

3.2. Sensitivity of Thermal Detection

In the direct absorption measurement described here, a thermometer attached to the metal sample serves as the detector of absorbed infrared power. The structure used is very similar to that of the composite far infrared bolometer,[27,28] which has been brought to a high state of development for use in infrared astronomy. The infrared flux in the surface experiment is very large compared with that encountered in the astronomy experiments, so many of the stringent design requirements for those detectors can be relaxed. Before discussing optimization of the surface experiment in detail, it is useful to review the simple theories of bolometer response and noise.

We consider a bolometer whose electrical resistance $R(T)$ is a known function of its temperature T. The bolometer is assumed to have heat capacity C and to be connected to a heat sink at temperature T_S by a thermal conductance G. Since the absorbed infrared power is partly chopped or interference modulated, it is written $P = P_0 + P_1 e^{i\omega t}$. The bolometer temperature then varies as $T = T_0 + T_1 e^{i\omega t}$. By equating the infrared and electrical power dissipated in the bolometer with the power conducted to the heat sink and stored in the heat capacity we have the linearized equation

$$P_0 + P_1 e^{i\omega t} + I^2 R(T_0) + I^2 \frac{dR}{dT} T_1 e^{i\omega t} = G(T_0 + T_1 e^{i\omega t} - T_S) + i\omega C T_1 e^{i\omega t}$$
(11)

The steady state part of equation (11)

$$P_0 + I^2 R(T_0) = G(T_0 - T_S)$$
(12)

gives the value of G required to keep the bolometer cold. The time varying part of equation (11) can be solved for the temperature responsivity T_1/P_1. A more useful quantity is the voltage responsivity $S = V_1/P_1 = IT_1(dR/dT)/P_1$, which gives the voltage response of a bolometer that is biased with a constant current I,

$$S = \frac{I(dR/dT)}{[G - I^2(dR/dT)] + i\omega C} = \frac{I\alpha R}{G_{\text{eff}}(1 + i\omega\tau_{\text{eff}})}$$
(13)

We have followed the conventional practice by defining an effective thermal conductance $G_{\text{eff}} = G - I^2(dR/dT)$ and an effective time constant $\tau_{\text{eff}} = C/G_{\text{eff}}$. The temperature dependence $R(T)$ is characterized by the parameter $\alpha = (dR/dT)/R$ evaluated at $T = T_0$. The most useful thermometers for composite bolometers or surface absorption experiments are made from heavily doped and compensated semiconductors, e.g., Si or Ge, which conduct by hopping of carriers between impurity sites. This process[30] gives $R(T) = R_H \exp[(T_H/T)^{1/2}]$ so that $\alpha = -1/2(T_H/T^3)^{1/2}$. The proper choice of thermometer parameters is desirable to minimize certain noise mechanisms. If the bolometer bias current is obtained from a constant voltage source and a load resistance $R_L > R$, which is cooled to T_S, then noise from the bias circuit can be made negligible.[31] The most important sources of noise are then the photon noise discussed in Section 2, the Johnson noise from the bolometer resistance R, and the thermal fluctuation noise from the thermal conductance G.

In order to compare the magnitudes of these noise contributions, it is conventional to express each one as an NEP referred to the bolometer input. The photon noise is given in this form in equation (10). The mean square voltage spectral density for Johnson noise is $4kTR$ per unit bandwidth. The resulting contribution to the NEP for unit bandwidth is

$$\text{NEP}_J = \left(\frac{4kTR}{|S|^2}\right)^{1/2} [\text{W Hz}^{-1/2}] \tag{14}$$

Energy fluctuations, which occur whenever an object, such as a bolometer, with heat capacity C is connected to a heat sink through a thermal conductance G, can be computed by conventional thermodynamic arguments.[32] The thermometer attached to the bolometer will read a mean square temperature fluctuation $\langle(\Delta T)^2\rangle = kT^2/C$. This total fluctuation can be written as an integral over a temperature spectral density of the form $\langle(\Delta T)^2\rangle_\omega = 4kT^2/G_{\text{eff}}(1 + \omega^2\tau_{\text{eff}}^2)$. The frequency dependence comes from that of $|S|^2$ in equation (10). Then if $\omega\tau_{\text{eff}} \ll 1$, the temperature spectral density is $4kT^2/G_{\text{eff}}$ per unit bandwidth. Converting to a power fluctuation at the detector input we obtain

$$\text{NEP}_T = (4kT^2 G_{\text{eff}})^{1/2} [\text{W Hz}^{-1/2}] \tag{15}$$

Since NEP is a measure of a signal-to-noise ratio and since the frequency dependence of S enters in both the signal and the noise, equation (15) is valid for all frequencies.

In direct absorption measurements on surfaces it may be inconvenient to reduce C until $\omega\tau_{\text{eff}} \ll 1$. If experiments are done with $\omega\tau_{\text{eff}} > 1$ then the responsivity S is reduced. This in turn increases the NEP due to Johnson noise and amplifier noise. There is no penalty in sensitivity until one of these noise sources becomes comparable with the photon noise.

Since photon, Johnson, and thermal contributions to the noise are statistically independent, the observed NEP[31] is the square root of the sum-of-the-squares of the contributions in equations (10), (14), and (15).

For a surface absorption experiment using a thermal source and a rapid scan FTS the throughput is limited by the sample to $A\Omega \simeq 10^{-2}\,\text{sr cm}^2$ and the optical bandwidth can be thousands of cm^{-1}. For ideal performance, the steady state infrared power P_0 should not be much larger than the modulated power P_1, and the bolometer should be photon noise limited. In practice it is difficult to reduce the ambient background to this value in a UHV system. Excess ambient background power P_0 increases the photon noise and requires a larger G to keep the bolometer cold. This in turn increases NEP$_T$. It also reduces the responsivity S which increases NEP$_J$. It is thus of highest importance to use cooled baffles, filters, and apertures to keep the throughput and spectral bandwidth for P_0 nearly equal to those for P_1.

The arguments sketched above show that even if there is no source noise, the noise in a system that has been properly optimized for a given spectral bandwidth will generally increase as the spectral bandwidth is increased. Under these conditions a multiplex advantage cannot be expected from the FTS. A multiplex disadvantage might even arise. Cooled filters can then be used with profit to limit the spectral bandwidth. There is often a tradeoff between the convenience of a wide spectral range and the possibility of a higher signal-to-noise ratio with a narrow range.

3.3. The Direct Absorption Instrument

In this section, we will describe an instrument for direct absorption spectroscopy developed at Berkeley by Bailey et al.[33] This description will include the spectrometer, the optical system, the UHV chamber, and the sample configuration. We will then compare the observed performance of the system with that estimated from the detector parameters. Even though the detector achieves photon noise limited performance, the surface experiment is limited—as in most surface infrared experiments— by other instabilities that affect the background subtraction. Nevertheless, the instrument is capable of detecting as little as 0.002 monolayers of CO on an evaporated silver film.

The spectrometer used in this experiment is an EOCOM 7001P rapid scan FTS. It has a spectral range of $400–4000 \, \text{cm}^{-1}$, with a maximum resolution of $0.065 \, \text{cm}^{-1}$. Although the instrument is purged continuously with dry air, small amounts of water vapor always remain. The change in the water absorption with time can easily dominate the surface signal when spectra taken hours apart must be compared. Thus the system has so far been used primarily at frequencies greater than $1800 \, \text{cm}^{-1}$, where water vapor absorption is small. This limitation can be overcome with the use of an evacuated spectrometer, and such a modification is planned.

The infrared beam from the spectrometer enters the vacuum chamber through an indium-sealed KRS-5 window. It is then focused onto the sample by a KRS-5 lens located inside the vacuum chamber. At the sample, the beam has a half-angle of $9.1°$, and is incident at an angle of $83°$ to the surface normal. The system throughput, for a $4 \times 8 \, \text{mm}$ sample, is then $A\Omega = 3.1 \times 10^{-3} \, \text{sr} \, \text{cm}^2$. The power absorbed by the sample from the spectrometer beam is typically $16 \, \mu\text{W}$.

The minimum mirror speed of the spectrometer is $0.07 \, \text{cm} \, \text{s}^{-1}$ corresponding to a modulation frequency of $280 \, \text{Hz}$ for an infrared frequency of $2000 \, \text{cm}^{-1}$. This property of the spectrometer is an important constraint on the detector design, giving $\omega\tau_{eff} = 1$ at $\tau_{eff} = 0.6$ ms.

As we have already discussed, the FTS offers little fundamental advantage over a dispersion instrument, since neither the multiplex advantage nor the throughput advantage is significant. It is, however, a stable and convenient instrument with a rapid modulation frequency.

In this direct absorption experiment, the sample is part of a bolometric infrared detector. The requirements of low heat capacity and sensitive thermometry impose severe constraints on the operating temperature and the type of sample that can be measured. In order to obtain adequate response time and detection sensitivity a heat sink temperature $T_S \sim 1.2$ K was used. The samples were evaporated metal films deposited on a dielectric substrate (sapphire) which has a high Debye temperature and high thermal conductivity at low temperature. From the point of view of heat capacity, most crystalline dielectric or semiconducting substrates could be used. It would be very desirable to be able to use single-crystal metal substrates. Because of the large heat capacity of metals at liquid helium temperatures, however, metal thicknesses would have to be $\lesssim 0.1$ mm. Such samples are difficult, but not impossible, to produce by grinding[34] or by epitaxial growth.[35,36] A second obstacle to the use of single-crystal samples is the high annealing temperatures traditionally used while cleaning the sample. The thermometer technology used at present involves epoxy and/or indium solder, so it is limited to temperatures below 150°C. A technique for heat sinking the thermometer during annealing has been developed which was successful but inconvenient.[33] Other techniques emphasizing higher temperature materials seem possible, but have not been demonstrated.

The decision to use evaporated film substrates greatly simplifies the vacuum system, though at some cost in sample characterization. Since a film can be deposited at the beginning of each experiment, elaborate facilities for cleaning and characterizing the surface are not required. Moreover, true ultrahigh vacuum conditions are needed only while the experiment is actually in progress. At these times a cold finger at ~ 1.2 K must extend into the chamber, so cryopumping by the liquid helium temperature surfaces can be used to reduce the pressure in the vicinity of the sample. At other times, a base pressure of 10^{-9} torr or even higher is adequate.

Based on these considerations, the design shown in Figure 4 was chosen. The primary vacuum chamber is pumped by a 15-cm cryotrapped diffusion pump. A separate vacuum chamber surrounds the containers for liquid nitrogen and liquid helium. A cold finger, which is an extension of the liquid helium tank, reaches into the UHV chamber and terminates in

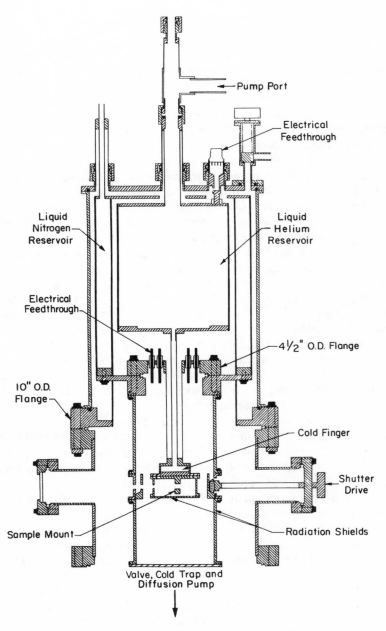

Figure 4. Vertical cross section of cryostat and sample chamber for direct absorption experiment.

a copper plate, to which the sample assembly is attached. The sample is surrounded with a radiation shield, also at liquid helium temperature. There are several holes in the radiation shield, providing access to the sample from the outside. A liquid-nitrogen-cooled radiation shield, with holes aligned with those in the helium-cooled shield, surrounds the cold finger. Between the two shields is a rotatable shutter at liquid nitrogen temperature, which selects only certain directions of access to the sample, depending upon its position.

This elaborate system of shielding is necessary in order to minimize the amount of room temperature radiation reaching the sample during the infrared measurement, yet provide access to the sample as needed for film deposition and dosing with adsorbate molecules. With the shields fully closed to minimize incident radiation, only 1.8 μW of the background loading is absorbed in the sample, much less than the power from the spectrometer. Because of the effective shielding, it is not practical to dose the sample by filling the chamber with gas. Instead, an effusive beam doser with direct line of sight to the sample is used.

It should be noted that the need for such cold shielding makes the use of standard surface analysis tools such as LEED and Auger spectroscopy difficult. A two-level system, with an unshielded level for surface analysis and preparation at relatively high temperature, and a shielded level for the infrared measurement would be useful for measurements of well-characterized samples.

The detector configuration chosen is a composite design developed for infrared astronomy,[27,28] which is easily adapted for a wide range of infrared power levels. A drawing of the bolometer is shown in Figure 5. The front surface of the sapphire substrate is coated with 50 Å of Cr followed by 1000 Å of Au; this metallic coating ensures good adhesion of the sample metal films that are subsequently deposited. The thermometer is a small (~1 mm) cube of neutron transmutation doped Ge with ion-implanted ohmic contacts.[28,30] It is attached to the substrate with a small quantity of thermally conductive epoxy.[37] Copper leads, 0.005 cm in diameter, are attached to the chip with indium solder. A thermocouple consisting of 0.005 cm diameter copper and constantan wires is also glued to the back of the substrate. The thermocouple is used to monitor the temperature when the sample is deliberately heated above a few degrees kelvin. At low temperatures, the thermal conductivity of alloys such as constantan is negligible compared to that of pure, nonsuperconducting metals such as copper, so the three copper leads provide the dominant thermal conductance, G, between the bolometer and its heat sink. Tungsten wires are used both to support and to heat the sample.

The measured properties of the bolometer, under operating condi-

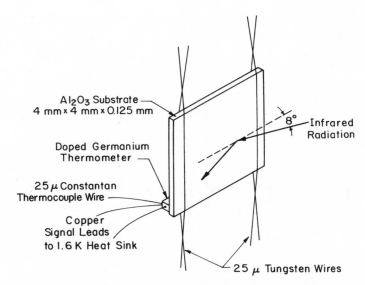

Al₂O₃ Substrate
4 mm × 4 mm × 0.125 mm

Doped Germanium
Thermometer

25 μ Constantan
Thermocouple Wire

Copper
Signal Leads
to 1.6 K Heat Sink

8°
Infrared
Radiation

25 μ Tungsten Wires

Figure 5. Sample assembly for direct absorption experiment on evaporated films.

tions are given in Table 2. It can be seen that the calculated and measured dc responsivities are in close agreement. The time constant gives $\omega\tau_{\text{eff}} \approx 15$ at 300 Hz, so that $|S|$ is reduced by a factor of 15 from its dc value. Since Johnson noise and amplifier noise are still negligible, this roll-off in responsivity does not reduce the sensitivity.

The measured noise spectrum of the detector circuit is dominated by large peaks at the harmonics of 60 Hz. The mirror speed of the interferometer is chosen so that the resulting peaks in the infrared spectra do not interfere with the spectral range of interest. Between the peaks, a white noise level of 75 nV Hz$^{-1/2}$ is observed at a modulation frequency of 300 Hz. This level corresponds to an NEP of 1.9×10^{-12} W Hz$^{-1/2}$, in excellent agreement with the value estimated from the known properties of the bolometer and the infrared loading. Evidently the detector system approaches photon noise limited performance.

We will see, however, that the ultimate sensitivity of the surface spectroscopy experiment is limited at a higher level by other effects. It is, of course, for precisely this reason that the direct absorption and emission techniques are attractive, since the substrate signal to be subtracted is much lower than in reflection–absorption spectroscopy.

We can use the known NEP of the detector, and the spectrum of the absorbed power, to estimate the sensitivity of the system to a small spectral feature on the smooth background of the substrate absorption.

Table 2. Typical Operating Conditions for the Sample in the Direct Absorption Experiment, Including Both Calculated and Directly Measured Values for Responsivity and Noise from Equations (4) and (10)–(12)

Sink temperature	$T_S = 1.2\,\text{K}$
Bolometer temperature	$T_0 = 1.8\,\text{K}$
Electrical resistance	$R = 3.3\,\text{M}\Omega$
Bias current	$I = 0.5\,\mu\text{A}$
Temperature coefficient	$\alpha = 2.2\,\text{K}^{-1}$
Thermal conductance	$G = 3 \times 10^{-5}\,\text{W}\,\text{K}^{-1}$
Effective thermal conductance	$G_{\text{eff}} = 3.2 \times 10^{-5}\,\text{W}\,\text{K}^{-1}$
Time constant	$\tau_{\text{eff}} = 7.9\,\text{ms}$
Heat capacity	$C = 2.2 \times 10^{-8}\,\text{J}\,\text{K}^{-1}$
Calculated dc responsivity	$S_C = 7.5 \times 10^5\,\text{V}\,\text{W}^{-1}$
Measured dc responsivity	$S_M = 5.8 \times 10^5\,\text{V}\,\text{W}^{-1}$
Absorbed photon power	$P_0 = 16\,\mu\text{W}$
Average frequency	$v = 10^3\,\text{cm}^{-1}$
Contributions to the NEP at 300 Hz:	
Calculated photon noise	$\text{NEP}_P = 8.0 \times 10^{-13}\,\text{W}\,\text{Hz}^{-1/2}$
Calculated Johnson noise	$\text{NEP}_J = 4.7 \times 10^{-13}\,\text{W}\,\text{Hz}^{-1/2}$
Calculated thermal noise	$\text{NEP}_T = 7.6 \times 10^{-14}\,\text{W}\,\text{Hz}^{-1/2}$
Calculated total noise	$\text{NEP} = 9.3 \times 10^{-13}\,\text{W}\,\text{Hz}^{-1/2}$
Measured noise voltage	$\Delta V_{\text{RMS}} = 75\,\text{nV}\,\text{Hz}^{-1/2}$
Measured NEP = $\Delta V_{\text{RMS}}/S_M(300\,\text{Hz}) = 1.9 \times 10^{-12}\,\text{W}\,\text{Hz}^{-1/2}$	

The power absorbed in a silver substrate in a 4-cm^{-1} bandwidth at $v = 2000\,\text{cm}^{-1}$ has been measured to be approximately $2.3 \times 10^{-8}\,\text{W}$. Based on the known NEP, we calculate that the system should have a sensitivity to fractional changes in the absorptance of $8.3 \times 10^{-5}\,\text{Hz}^{-1/2}$. That is, in a 1-Hz bandwidth, the system should be able to detect a surface feature with an absorptance less than 10^{-4} of the substrate absorptance, with a signal-to-noise ratio of unity.

In Figure 6, we show the spectrum of an evaporated metal film that has been exposed to $10^{-3}\,\text{L}$ CO at a temperature of 2 K. A spectrum of the clean film has been subtracted, and the result divided by the reference. An additive constant is included for clarity of display. The total integration time for the spectrum was approximately 2400 s. Taking into account the adsorption of background CO during the measurement of the spectra, we estimate that the CO coverage was less than 0.002 monolayers. This low level of background adsorption is due to the low-temperature shields that surround the sample. The integrated intensity of the CO signal, expressed as a fraction of the bulk absorptance, is 4.2×10^{-4}, and the noise level, in the same terms, is 8.5×10^{-5},

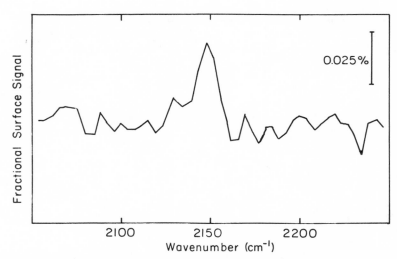

Figure 6. Infrared spectrum of 10^{-3} L CO chemisorbed at 2 K on a silver film deposited at $T_D = 4$ K, showing a single band at 2148 cm^{-1}. The integration time was ~40 min. The spectrometer resolution was 8 cm^{-1}.

corresponding to a sensitivity of 5.9×10^{-3} Hz$^{-1/2}$. This sensitivity is some 70 times worse than that calculated from Table 2. Clearly, as is usually the case with surface infrared experiments, it is not the detector sensitivity, but errors in the cancellation of the background that limit the ultimate sensitivity of the measurement. Nevertheless, Figure 6 clearly illustrates that the direct absorption technique is capable of detecting extremely small quantities of adsorbates, and very weak vibrational signals.

3.4. Experimental Results: CO on Ag

As an example of the use of infrared absorption spectroscopy, we summarize here some results for CO adsorption on evaporated silver films. These data were obtained by Dumas *et al.*,[38] using the apparatus described in the preceding section. They form part of an extensive investigation of CO adsorption on noble metal films. The complete results have been published elsewhere.[38,39]

We will discuss three aspects of CO adsorption on evaporated silver: the dependence of the adsorption behavior on the temperature T_D of the sapphire substrate during the deposition of the silver film, the shift of the CO stretch frequency with coverage, and the integrated intensity of the CO stretch band.

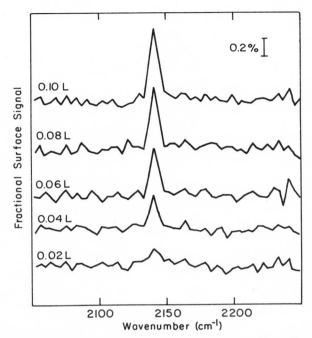

Figure 7. Infrared spectra of CO physisorbed at 2 K on a silver film deposited at $T_D = 300$ K, as a function of exposure in Langmuirs showing a single narrow band at 2143 cm^{-1}. The spectrometer resolution was 4 cm^{-1}.

It is found that the nature of the adsorption of CO depends strongly on T_D. For $T_D < 150$ K, only physisorbed CO can be detected when the sample is exposed to CO at 2 K. Figure 7 shows a sequence of infrared spectra, as a function of CO exposure, in Langmuir (L) for a film deposited at $T_D = 300$ K, which is well above the threshold temperature. The frequency of the single band, 2143 cm^{-1}, is equal to that for gas phase CO. This frequency, and the lack of any frequency shift with increasing coverage, are characteristic of physisorption. This result is consistent with the known behavior of CO on single-crystal silver surfaces, which also do not support CO chemisorption.[40]

For a silver deposition temperature $T_D = 4$ K, which is well below the threshold temperature, the spectra, shown in Figure 8, are quite different. At low coverages, a single band due to chemisorbed CO is observed. The band appears initially at 2148 cm^{-1}, and shifts to lower frequency as the coverage increases. The band reaches its full intensity at an exposure of 0.4 L. Evidently the film deposited at low temperature

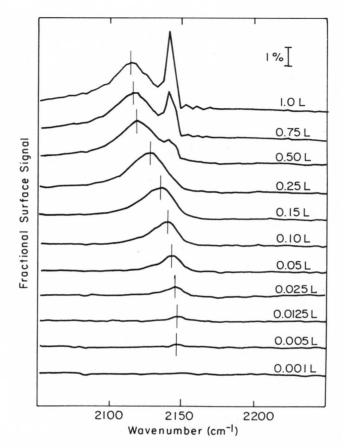

Figure 8. Infrared spectra of CO chemisorbed at 2 K on a silver film deposited at $T_D = 4\,K$. A broad band due to chemisorbed CO appears at low coverage at $2148\,cm^{-1}$, and shifts to lower frequency with increasing exposure. At 0.4 L, a sharp band due to physisorbed CO appears at $2143\,cm^{-1}$, and the band due to chemisorbed CO shifts further to low frequency. The spectrometer resolution was $4\,cm^{-1}$.

contains special active sites at which CO chemisorbs, in a concentration corresponding to ~40% of a monolayer on a single crystal. For exposures greater than 0.4 L, a sharp peak at $2143\,cm^{-1}$, due to physisorbed CO, appears in the spectrum. This physisorbed peak can be removed by heating to 25 K, while the band due to chemisorbed CO persists up to 80 K.

The shift of the vibrational band with exposure is also of con-siderable interest. Shifts of the band to higher frequency with increasing coverage are commonly observed for CO on transition metals.[41,42] This shift is attributed to the combined effects of a dynamic dipole–dipole interaction and a static, chemical interaction between neighboring molecules, mediated by the metal. For CO on Cu single crystals, the latter effect is known to cause a shift to lower frequency, with the result that the two effects very nearly cancel, resulting in little or no net shift.[10]

In the present case, for CO chemisorbed on a silver film deposited at low temperature, a large shift (26 cm^{-1}) to lower frequency is observed. The static and dynamic contributions to the shift can be separated by coadsorbing different isotopes. Because the isotopes have different resonant frequencies, they have negligible dynamic dipolar interaction.[43] However, the static chemical shift should be unaffected by the different mass of the neighboring coadsorbed molecules. Thus the observed shift is attributed entirely to the static effect, and the difference between the shifts observed in the two experiments can be assigned to the dynamic interaction. Figure 9 summarizes the results of such a measurement. At low coverage, the chemical contribution fully explains the shift. At higher

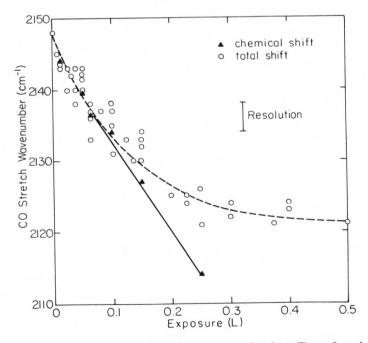

Figure 9. Total frequency shift of the CO stretch vibration from Figure 8, and chemical shift deduced from isotopic substitution as a function of exposure. The lines are drawn as a guide for the eye.

coverage, there is a difference between the total shift and the chemical shift, which is attributed to the dynamic interaction.

By means of coadsorption experiments with physisorbed CO and argon, it is found that a large part of the chemical shift occurs even when the coadsorbing molecules do not interact chemically with the surface at all. This effect is attributed to the influence of the local work function on the bonding of the chemisorbed molecule. A detailed discussion of this effect appears in the complete paper.[38]

The dynamic coupling is very well modeled by dipole–dipole coupling, including image effects as described by Persson and Ryberg,[44] with no need to invoke any additional dynamic coupling through the metal.[45] Figure 10 shows the dynamic shift, as deduced from Figure 9, and the integrated intensity of the chemisorption band, as a function of coverage. The lines are a fit to the data using the dipole coupling theory, with the vibrational polarizability $\alpha_v = 0.27 \pm 0.015$ Å^3 and interaction potential $\bar{U}(0) = 0.04 \pm 0.002$ Å^{-3}. Again, we refer the reader to the full paper for a detailed discussion.[38]

One of the reasons for investigating adsorption on noble metals deposited at low substrate temperature is that such systems are known to exhibit large enhancements in Raman cross section over the values observed either in the gas phase, or for adsorption on smooth single crystals.[15] This phenomenon is known as surface-enhanced Raman scattering (SERS). A large part of the enhancement associated specifically with rough evaporated films has been attributed to electromagnetic resonances at visible frequencies.[15] It has also been suggested, however, that such films contain special Raman-active sites, not present on annealed films or single crystals, at which dynamic charge transfer greatly increases the Raman cross section.[15] It would be expected that molecules adsorbed at such sites, if they exist, would also exhibit much greater infrared absorption than commonly observed for adsorbed molecules. It has already been shown that rough silver films contain special chemically active sites, not present in annealed films, at which CO chemisorbs. It is interesting to inquire whether these might also be the postulated Raman-active sites, by comparing the infrared absorption by chemisorbed CO on rough silver films with that observed for CO on single crystals.

The polarizability α_v, used in the dynamic coupling model, allows such a comparison to be made. For a given fractional coverage of CO, the integrated intensity of the infrared absorption band is simply proportional to α_v. The value 0.27 Å^3 found for CO on evaporated silver is identical to that previously measured for CO on Cu(100).[10] There is thus no unusual enhancement of the infrared cross section associated with

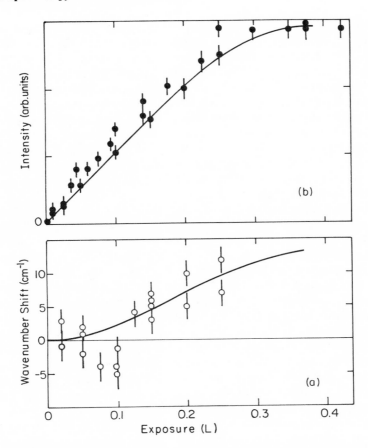

Figure 10. (a) Dynamic contribution to the frequency shift, derived from data in Figure 9, for chemisorbed CO on a silver film, $T_D = 4\,K$. (b) Integrated band intensity, from Figure 4. The lines are theoretical fits to the data, using the Persson–Ryberg dipole coupling model with $\alpha_v = 0.27\,\text{Å}^3$ and $\bar{U}(0) = 0.04\,\text{Å}^{-3}$.

the special chemisorption sites found on the rough silver film, and no evidence for the hypothetical Raman-active sites.

3.5. Surface Calorimetry

The sample design described here is optimized for the detection of small temperature changes caused by the dissipation of extremely small quantities of power in the sample. In the experiments discussed so far, the source of this power has been infrared radiation, and the purpose has been to measure the vibrational spectrum of the adsorbate layer. Such a sample configuration, however, is also ideally suited for the measurement of surface thermodynamic quantities at low temperature—that is, a bolometer is also an excellent differential calorimeter. In this section, we will discuss how the detector technology described here can be used for the measurement of various surface thermodynamic quantities.

Perhaps the easiest measurement of an important thermodynamic quantity that can be made with such an instrument is the direct determination of the binding energy of an adsorbate as a function of coverage. Normally, this quantity is measured indirectly, by thermal desorption spectroscopy, which generally requires additional assumptions about the kinetics of desorption.

When a molecule chemisorbs, virtually all of its binding energy is dissipated in the substrate as heat. If a flux of N molecules per second is incident on the sample from an effusive beam doser, or a molecular beam, and each molecule has a binding energy $E_b \sim 1\,\mathrm{eV}$, then the power deposited in the sample is $P = NE_b$. The kinetic energy of molecules from a 300 K source is $\sim 10^{-4}\,\mathrm{eV}$. For the detector we have described, the NEP at low frequencies and in the absence of infrared loading from the spectrometer is of order $8 \times 10^{-14}\,\mathrm{W\,Hz^{-1/2}}$. For a particle rate of $N = 10^{10}\,\mathrm{s^{-1}}$, the sensitivity to changes in E_b is then $5 \times 10^{-5}\,\mathrm{eV\,Hz^{-1/2}}$. In practice the accuracy of the measurement would certainly be limited by other uncertainties, predominantly in the determination of the molecular flux N. Geraghty et al.[46] have used a configuration similar to the one described here to measure the heat of adsorption of pyridine on Ni at 8 K.

Another thermodynamic measurement that can be performed with a bolometric detector is that of the heat capacity due to an adsorbate. Such measurements are particularly important in the study of surface phase transitions. In such an experiment, the infrared loading on the sample would be minimized, and a calibrated ac power applied by Joule heating of an auxiliary heater attached to the sample. The frequency chosen should be above $1/\tau_{\mathrm{eff}}$, so that the response of the bolometer is dominated by the heat capacity C, rather than by the thermal conductance, G. The change in the sample heat capacity due to the addition of an adsorbate will appear as a change in the amplitude of the temperature oscillation induced by the applied power.

This experiment can be approximately analyzed with the model used in Section 3.2, which includes an applied power of the form $P_0 + P_1(1 + e^{i\omega t})$. At a sufficiently high modulation frequency, temperature fluctuations become negligible, and the noise is limited by Johnson noise in the thermometer. In this limit, the minimum detectable change δC in the heat capacity is given by

$$\left(\frac{\delta C}{C}\right)^2 = 4k\left(\frac{T_0}{T_1^2}\right)\frac{1}{I^2 R\alpha^2} \tag{16}$$

where the parameters are as defined in Section 3.2.

For the bolometer described above, the noise equivalent δC is $1.1 \times 10^{-14}\,\mathrm{J\,K^{-1}\,Hz^{-1/2}}$ at 1.8 K, and $2.2 \times 10^{-12}\,\mathrm{J\,K^{-1}\,Hz^{-1/2}}$ at 4 K, for $T_1 = 0.01$ K. The rapid decrease in sensitivity with increasing temperature is due primarily to the T^3 dependence of C and the exponential temperature variation of R. For comparison, the heat capacity of a layer of 10^{14} molecules, in a two-dimensional Debye model with a surface Debye temperature of 100 K, is $2.7 \times 10^{-11}\,\mathrm{J\,K^{-1}}$ at 1.8 K, and $1.3 \times 10^{-10}\,\mathrm{J\,K^{-1}}$ at 4 °K, varying as T^2. From these estimates, it is clear that the bolometer should be able to detect the adsorbate contribution to the heat capacity rather easily. By carefully optimizing the detector and the operating parameters for heat capacity measurements in a particular temperature range, it should be possible to achieve even greater sensitivity.

4. Infrared Emission Spectroscopy

4.1. The Emission Spectrometer

In this section we will describe a specific instrument for emission spectroscopy, indicating how the concepts and technologies discussed in the preceding sections influenced the design. A less tutorial description has been published elsewhere.[47]

The idea of using infrared emission spectroscopy for the study of surface vibrational modes is not new, but experiments using room temperature spectrometers have been plagued by low sensitivity. Griffiths[48] has reviewed work prior to 1975 on multilayer films on flat metal surfaces. A similar application has been the study of oxide growth on copper[49] and molybdenum[50] surfaces. Another approach has been to investigate monolayer adsorption on high surface area dispersed catalysts.[51–54] Blanke and Overend[4] proposed enhancing the sensitivity of the technique for flat surfaces by using a multiple reflection geometry, and demonstrated improved signal-to-noise ratio in the spectrum of a nine-layer Langmuir–Blodgett film. All of these experiments used room temperature spectrometers, with sample temperatures ~400–500 K. Allara et al.[55] used a liquid nitrogen cooled FTS to obtain monolayer sensitivity on a flat metal surface at 300 K.

The goal of the instrument described here was to extend greatly the capabilities of infrared emission spectroscopy as a tool for surface science. It is the first system capable of measuring emission signals from submonolayer coverages of adsorbates on clean, well-characterized, single-crystal metal substrates, over the frequency range from 400 to 3000 cm^{-1}, with moderate ($<10 \text{ cm}^{-1}$) resolution. The useful sample temperatures depend on the frequency chosen but are typically $\gtrsim 250$ K. The most important factors affecting the design of the instrument were the small ratio (10^{-4}–10^{-2}) of the adsorbate emission to the substrate emission, and the availability of photoconductive detectors for the frequencies of interest that can approach the signal photon noise limit for the emission signal from the substrate alone. The first consideration required careful attention to modulation techniques and to system stability. The second motivated a major effort to minimize the amount of background radiation incident on the detector.

At 400 cm^{-1}, the photon fluxes from blackbodies at 300, 100, and 5 K are in the ratio 10^{45} to 10^{43} to 1. At higher frequencies the ratios increase exponentially. From these numbers it is clear that if the contribution to the radiation reaching the detector from the spectrometer is to be kept small compared to that from the sample at ~300 K (which is also the source), then all optical elements should be cooled at least to liquid nitrogen temperature. Since the Si:Sb photoconductive detector chosen required liquid helium cooling in any case, it was decided to cool the entire spectrometer to ~5 K. In this way, drifts due to spectrometer emission were kept negligible compared with the small adsorbate signals. It is possible, however, to do surface emission work, especially at higher infrared frequencies, with a liquid nitrogen cooled instrument.[55]

A diffraction grating spectrometer was chosen over an FTS system for several reasons. Since the detection is photon noise limited, no multiplex advantage was expected for an FTS. A multiplex disadvantage was possible. Since the throughput is sample limited, little throughput advantage could be realized from the FTS. The grating spectrometer was easier to cool. As will be discussed later, the possibility exists of using arrays of detectors in the dispersed output of the grating spectrometer to achieve a multichannel advantage. The optical layout of the emission system is shown in Figure 11.

The sample is maintained in an ultrahigh vacuum environment. An indium-sealed KRS-5 lens acts both as a vacuum window and as a focusing element, collecting the emitted radiation from the sample and focusing it on the entrance slit of the spectrometer. The lens is not cooled, but its emissivity is very low in the frequency range of interest. Nevertheless, variations in the small amount of radiation it emits can be a

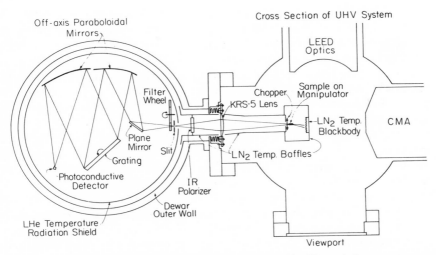

Figure 11. Optical layout of infrared emission apparatus, with LHe-cooled spectrometer on the left and ultrahigh vacuum system on the right.

significant source of drift. Since the reflectivity of KRS-5 is ~40%, careful LN_2-cooled shielding has been used to minimize the reflection of radiation from warm objects into the beam. Even so, residual reflected radiation can cause slow drifts in the detector output. The effects of such drifts can be minimized by proper modulation of the signal.

The spectrometer is a compact Czerny–Turner design consisting of a planar steering mirror, off-axis paraboloidal collimator and camera mirrors, and a rotatable diffraction grating on an aluminum substrate. Six different gratings are used to cover the full spectral range with adequate resolution and efficiency, but the spectrometer must be opened to change gratings. All of the optical elements are bolted to a 30-cm-diam aluminum plate, which is screwed to the cold plate of a modified commercial liquid helium cryostat. Radiation shields at helium and nitrogen temperatures surround the spectrometer. Because of the large mass that must be cooled, it requires about two hours and eight liters of liquid helium to cool the spectrometer initially; thereafter, two liters of helium every twelve hours is sufficient to keep the system cold.

A filter wheel that can be rotated from outside the cryostat is located behind the entrance slit. Because of the low sample temperature, high orders of diffraction are eliminated by the exponential rolloff of the blackbody spectrum, so the usual low pass filters are not required. High pass filters are necessary, however, for measurements at frequencies well above the peak of the blackbody emission from the sample to avoid scattered low frequency radiation. A 2 mm thickness of LiF or MgF_2 is used for measurements above $1500\ cm^{-1}$. Another aperture of the filter wheel carries a polystyrene film for frequency calibration. Frequency calibration can also be checked by observing the high-order diffraction lines of He:Ne laser light directed on the sample through a window in the UHV sample chamber. The frequencies of many of the absorption bands in polystyrene do not appear to shift significantly with temperature.

An infrared polarizer is located on the LN_2 cooled shield of the spectrometer in front of the entrance slit. It is used to reject the s-polarized ($\perp$-polarized) light from the sample which contains no adsorbate contribution.

The heart of the spectroscopic system is the detector. It is an extrinsic Si:Sb photoconductor sensitive to frequencies greater than $330\ cm^{-1}$, mounted in an integrating cavity to maximize quantum efficiency and provided with two apertures to minimize stray radiation. The cavity aperture gives a slit width of 1 mm and a system throughput of $2.5 \times 10^{-3}\ sr\ cm^2$. This should be compared with the maximum useful throughput of $2.5 \times 10^{-2}\ sr\ cm^2$ available from our 15×5 mm samples in the angular range from 65° to 85°.

The photon rate at the detector is in the range $\dot{N} = 10^9–10^{10}$ photons s^{-1}. We use a conventional transimpedance amplifier[24] with a feedback resistor in the range $10^9–10^{10}\ \Omega$. This resistor is cooled to 5 K to reduce Johnson noise. The first stage of the amplifier is a commercial[56] dual JFET package heat sunk to the 5 K cold plate. The noise in this detection system is dominated by photon fluctuations for $\dot{N} > 10^8$ photons s^{-1}, at modulation frequencies above 10 Hz. At lower frequencies, system instabilities introduce additional noise and drift.

In order to perform any emission experiment, it is necessary to observe the sample against a background at a different temperature. To reduce background loading on the detector and to minimize the effect of instabilities, that background should be much colder than the sample. In our experiment this is achieved by including a liquid-nitrogen-cooled stage in the UHV chamber. Attached to the stage are shields to prevent stray ambient radiation from entering the beam, and a cold blackbody behind the sample, which prevents ambient radiation from reflecting off the sample into the spectrometer. The shields located in the UHV system are coated with gold black,[57,58] the reflectivity of which has been measured to be <1% for $v > 500\ cm^{-1}$. The emission spectrum of the gold black shows that it is well heat sunk to the underlying metal.

The emission system is sufficiently stable, even on the long time scale involved in the comparison of spectra with and without adsorbed molecules, that vibrational spectra can be obtained in a dc mode,[47] without the additional fast modulation discussed in Section 1. More consistent base-line subtraction is achieved, however, when the emitted radiation from the sample is chopped. The chopper consists of a metal vane coated with gold black attached to a torsion bar, and driven to oscillate at its resonant frequency of 20 Hz by magnetic coils, giving a 40-Hz optical modulation. This chopper is essentially a UHV-compatible version of a commercial oscillating vane chopper.[59] It is mounted on the cold stage, so that the vane is cooled by conduction to ~100 K. When the vane passes in front of the sample, the spectrometer looks at a cold, black surface.

Leakage signals past shields and emission signals from inadequately cooled objects in the optical train can contribute to the photon noise and, more important, can give signals that drift with time. Locating the chopper as close as possible to the sample ensures that many such signals are not chopped, and so do not contribute slow drifts to the measurement. The chopper cannot discriminate against drifts in the temperatures of the sample and the cold stage, or of ambient radiation leaking around the input shield. The sample temperature is regulated to within 0.04 K using a thermocouple sensor spot-welded to the back of the sample. Better regulation would be desirable and could be obtained by using an infrared sensor. The cold stage temperature is maintained at ~100 K with a continuous flow of LN_2. Cooling it below 50 K with a mechanical refrigerator would eliminate this source of drift. It is difficult to completely prevent ambient temperature radiation inside the UHV system from entering the optical path. The cooled shield must have openings so that the sample can be moved into position for LEED and other surface characterization experiments.

Since the temperatures of the sample and the cold stage reference are very different, the measurement is sensitive to drifts in system gain. Two sources of gain drift have required attention. The amplitude of the chopped signal depends on the chopper drive amplitude, which must be adjusted to minimize sensitivity to amplitude fluctuations. Also, the amplifier gain depends on the temperature of the load resistor[24] through its temperature coefficient of resistance. This effect can be minimized by selecting load resistors. An alternative solution to the gain drift problem is to use a reference source at or near the sample temperature. It should be black to avoid structure on its emission spectrum, and small to give a signal comparable to that of the sample. This approach has not been attempted.

If we assume that photon noise limits the sensitivity of the instrument, we can calculate the detection threshold for small molecu-

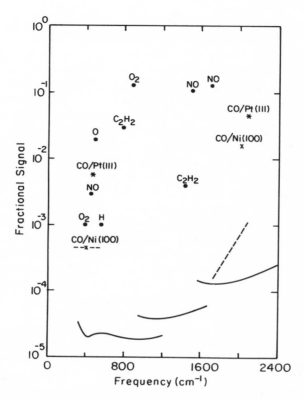

Figure 12. The solid curves show the calculated detection threshold for our emission apparatus if it were limited by photon noise. The dashed lines show the experimental sensitivity without modulation. The ×'s show the measured surface signal from CO on Ni(100). The dots and chemical symbols indicate the signal levels to be expected from monolayer coverages of various adsorbates on Pt(111).

lar signals on a large background from the substrate. Figure 12[47] shows the result of such a calculation, assuming a substrate of emissivity 0.1 and a temperature of 300 K and using measured values of the spectrometer efficiency and detector quantum efficiency. The three solid lines, corresponding to three different gratings, indicate the minimum adsorbate signal (expressed as a fraction of the substrate emission) that could be measured with a signal-to-noise ratio of unity in a 1-Hz bandwidth. These lines represent the theoretical limit to the performance of the instrument if all extraneous noise sources were eliminated.

The dots shown in the figure represent the expected fractional signal due to a saturation coverage of various adsorbates on Pt(111). The values were estimated from published electron energy loss (EELS) spectra, using Ibach's[60] comparison of ir and EELS cross sections. The ×'s show our measured signal levels for the two observable modes of CO on Ni(100); these measurements will be discussed in more detail below. The asterisks show the measured signal levels for CO on Pt(111). The strength of the technique, especially at the lower frequencies, where the vibrational modes tend to be weak, is immediately apparent.

In practice the photon noise limit has not been reached. The dashed lines show the measured sensitivity of the apparatus. The excess noise is limited by fluctuations in the stray radiation that enters the spectrometer despite the cold baffles. Since this source is modulated along with the sample emission, it is not greatly reduced by chopping. The chopper does, however, make it possible to achieve this level of performance quite consistently; in the dc mode, extra drifts often make the signal-to-noise ratio much worse. If the noise due to stray radiation were eliminated, the next limiting factor would probably be the stability of the source temperature.

4.2. Experimental Results: CO on Ni(100)

The infrared emission apparatus has been used to measure both the intramolecular carbon–oxygen stretching vibration of CO on Ni(100),[61] (Figure 13), and the low-frequency molecule–substrate vibration for the same system[62] (Figure 14). Measurements of the molecule–substrate vibration of CO on Pt have also been made.[63] The observed linewidth of the C–Ni mode, 15 cm^{-1}, is in excellent agreement with the theoretical value of 13.9 cm^{-1}, which was subsequently calculated by Ariyasu et al.,[64] from the assumption that vibrational energy decay into substrate phonons dominates the linewidth. If this explanation of the linewidth is correct, then measurements of such adsorbate–substrate mode linewidths may be a very useful probe of energy transfer processes at surfaces. An attempt to verify the line-broadening mechanism by measuring the temperature dependence of the linewidth is still in progress. To date, only the infrared emission technique has been successful in measuring the linewidths of such weak, low-frequency vibrational modes as the C–Ni stretch. Other surface infrared spectroscopies have been limited to frequencies above ~1000 cm^{-1}.

In the case of the C≡O stretch vibration data of Figure 13, the observed linewidth is certainly not lifetime dominated. In fact, the narrowest lines, at very low coverage and at saturation, are not resolved at an instrumental resolution of 18 cm^{-1}. (A factor of 2 better resolution, combined with higher efficiency, is now available with a new grating for

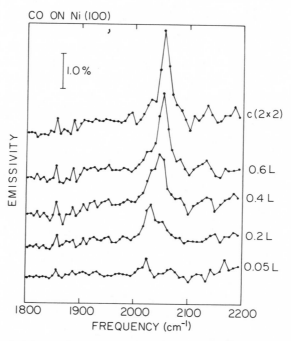

Figure 13. IR emission spectra of CO on a clean Ni(100) surface at 310 K. The instrumental resolution was ~18 cm^{-1}.

this frequency range.) The line shape is nonetheless interesting, for two reasons: First, the linewidth at saturation is a factor of 2 smaller than previously reported for this system.[65] This fact, combined with the absence of any band due to bridge-bonded CO—in agreement with LEED data[66] and recent electron energy loss spectroscopy results,[67] but in contrast to other published vibrational data[68,69]—leads us to believe that our surface was cleaner than in some other experiments. We found that even minute amounts of contamination (a few percent of a monolayer of carbon) gave rise to line-broadening and the appearance of bridging CO.

The second interesting feature is the asymmetric shape of the broadened bands at intermediate coverages. The bands appear to consist of two unresolved lines. Neither the resolution nor the sensitivity of the experiment was adequate for a conclusive interpretation, but we have suggested an explanation in terms of inhomogeneous broadening related to island formation at low coverages.[61] An alternative explanation of the line shape in terms of vibrational dephasing may also be possible. This

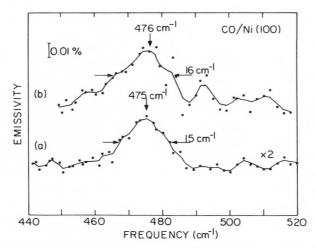

Figure 14. Infrared emission spectra from a saturation coverage of CO on Ni(100) at 310 K. The instrumental resolution was 2.5 cm^{-1}, and a linear base line has been subtracted from the curves. The solid lines are obtained by computer smoothing of the data. (a) Spectrum of a disordered CO layer on a partially contaminated surface. (b) Spectrum of an ordered $c(2 \times 2)$ CO overlayer on a clean surface.

mechanism has been found to be important for bridge-bonded CO on Ni(111)[70,71] and on Pt(111).[72] Measurements at higher resolution and over a range of temperatures will be needed to clarify this issue, and such experiments are planned.

Both the possibly homogeneous linewidth of the C–Ni mode and the inhomogeneously broadened C≡O mode need, and are receiving, further investigation. The data of Figures 13 and 14, however, are sufficient to demonstrate the power of infrared emission spectroscopy. The extension of high-resolution spectroscopy to the weak vibrational modes in the frequency range of a few hundred cm^{-1} is a special feature of this instrument. The possibility exists of extending the frequency range still further, into the range of substrate phonon frequencies. The modifications necessary, while significant, appear possible.

4.3. Arrays and the Multichannel Advantage

In the spectrometer described here, a single detector is used, and a spectrum is measured by rotating the grating, sweeping the dispersed radiation across the detector. In principle, one could instead place a linear array of n detectors in the focal plane, and measure a range of frequencies simultaneously. Such a multichannel system has an $n^{1/2}$ advantage in signal-to-noise ratio over a single-detector instrument. In addition, the grating need not be moved, so that problems of the reproducibility of the grating position, and the generation of mechanical noise by the grating drive, are eliminated.

Arrays of ~10 discrete detectors are commonly used in infrared astronomy. Recently, however, monolithic arrays of large numbers of photoconductive detectors ($N = 10$–1000), have appeared.[73] These detectors use technology developed for visible light detector arrays, and do not need the cumbersome individual amplifiers of discrete arrays. Because the detectors are arrayed on a single chip, high packing densities ($\sim 100\,cm^{-1}$), can be achieved. It can be expected that in the next few years, such detector arrays will become available for laboratory use, opening the possibility of large gains in sensitivity for infrared surface spectroscopy.

The exploitation of these detector arrays will require a spectrometer with a different optical design. The short focal length off-axis design of the spectrometer we have described here is very compact, but off-axis aberrations make it unsuitable for use with large arrays. As the size of the instrument is increased, the difficulties of cooling it increase as well. Such spectrometers have been successfully built and operated, however, for astronomical measurements.[74]

4.4. Nonequilibrium Emission and Chemiluminescence

The discussion to this point has concerned the use of emission spectroscopy to observe the thermal equilibrium emission from adsorbed molecules. For this application, emission spectroscopy has an advantage over the more conventional technique of reflection–absorption spectroscopy only in its lower background. There is another class of experiments, however, for which emission spectroscopy is uniquely suited; these are experiments involving the detection of nonequilibrium radiation from excited species. Such an experiment has been carried out by Chuang[75] in the near infrared (1250–$4000\,cm^{-1}$) region. The excitation mechanism was the chemical reaction of XeF_2 gas with a Si surface. The large amounts of energy released in this extremely fast, exothermic reaction resulted in intense chemiluminescence. With the more sophisticated spectroscopic technology described here, much weaker signals could potentially be detected with higher spectral resolution opening an area rich in interesting phenomena.

As an example, consider an adsorbed molecule that has been electronically excited by a laser or a chemical reaction. In its electronic excited state, the equilibrium internuclear distance is perturbed, so that an electronic transition into the ground state is likely to leave the molecule vibrationally excited, via the familiar Frank–Condon effect. If infrared emission is emitted as the molecule then relaxes vibrationally, this radiation could potentially be detected with a cooled spectrometer. The problem with such an experiment in systems such as CO on transition metals, is that the nonradiative relaxation rates are apparently many orders of magnitude faster than the radiative rate, with the result that the efficiency of the luminescence process is prohibitively small. The small probability of radiative decay does not affect the equilibrium experiment—the principle of detailed balance ensures that the decay and excitation rates for all of the nonradiative processes are equal—but it makes the measurement of nonequilibrium radiation very difficult. Systems with molecules on insulating substrates with vibrational frequencies much larger than substrate phonon frequencies, such as those studied by Chuang,[75] are more promising. In such a case, the nonradiative decay processes can be slow enough to make the luminescence experiment feasible. Other strategies could include the use of cooled samples or samples with very high surface areas, such as dispersed catalysts.

Acknowledgments

We are grateful to P. Dumas for help in characterizing the bolometric sample, and for permission to quote direct absorption results before publication. Thanks are also due to E. E. Haller and his group for providing some of the thermometer chips used in the direct absorption experiment. Our co-workers R. B. Bailey and S. Chiang had a major role in the development of the direct absorption and emission techniques, respectively. This work was supported by the Director, Office of Energy Research, Office of Basic Energy Sciences, Materials Sciences Division of the U.S. Department of Energy under contract No. DE-AC03-76SF00098.

References

1. F. M. Hoffmann, Infrared reflection-absorption spectroscopy of adsorbed molecules, *Surf. Sci. Rep.* **3**, 107–192 (1983).
2. R. Ryberg, Infrared spectroscopy of adsorbed molecules: Some experimental aspects, *J. Phys. (Paris) Coll.* **44**, CIO-421–426 (1983).

3. R. G. Greenler, Reflection method for obtaining the infrared spectrum of a thin layer on a metal surface, *J. Chem. Phys.* **50,** 1963–1968 (1969).
4. J. F. Blanke and J. Overend, Infrared spectroscopy of surface species; emission spectra from a semi-blackbody, *Spectrochim. Acta* **32A,** 1383–1386 (1976).
5. Y. J. Chabal, Hydrogen vibration on Si(111)7 × 7: Evidence for a unique chemisorption site, *Phys. Rev. Lett.* **50,** 1850–1853 (1983).
6. Y. J. Chabal and A. J. Sievers, High-resolution infrared study of hydrogen (1 × 1) on tungsten (100), *Phys. Rev. Lett.* **44,** 944–947 (1980).
7. D. K. Lambert, Observation of the first order Stark effect of CO on Ni(110), *Phys. Rev. Lett.* **50,** 2106–2109 (1983).
8. F. M. Hoffmann and A. M. Bradshaw, Infrared spectroscopy of CO adsorbed on palladium (100) and (111) surfaces, in: Proceedings Third International Conference on Solid Surfaces, (Vienna, 1977), pp. 1167–1170.
9. W. G. Golden, D. S. Dunn, and J. Overend, A method for measuring infrared reflection-absorption spectra of molecules adsorbed on low-area surfaces at monolayer and submonolayer concentrations, *J. Catal.* **71,** 395–404 (1981).
10. R. Ryberg, Carbon monoxide adsorbed on Cu(100) studied by infrared spectroscopy, *Surf. Sci.* **114,** 627–641 (1982), and references therein.
11. D. P. Woodruff, B. E. Hayden, K. Prince, and A. M. Bradshaw, Dipole coupling and chemical shifts in IRAS of CO adsorbed on Cu(110), *Surf. Sci.* **123,** 397–412 (1982).
12. W. G. Golden, D. D. Saperstein, M. W. Severson, and J. Overend, Infrared reflection-absorption spectroscopy of surface species: a comparison of Fourier transform and dispersion methods, *J. Phys. Chem.* **88,** 574–580 (1984).
13. W. B. Jackson, N. M. Amer, A. C. Boccara, and D. Fournier, Photothermal deflection spectroscopy and detection, *Appl. Opt.* **20,** 1333–1344 (1981).
14. C. K. N. Patel and A. C. Tam, Pulsed photoacoustic spectroscopy of condensed matter, *Rev. Mod. Phys.* **53,** 517–550 (1981).
15. A. Otto, in: *Light Scattering in Solids* (M. Cardona and G. Guntherolt, eds.), Vol. 4, pp. 289–418, Springer, Berlin (1984), and references therein.
16. Y. R. Shen, in: *Novel Materials and Techniques in Condensed Matter* (G. W. Crabtree and P. Vashista, eds.), pp. 193–208, Elsevier (1982), and references therein.
17. A. Mooradian, Tunable infrared lasers, *Rep. Prog. Phys.* **42,** 1533–1564 (1979).
18. C. K. N. Patel, *The Free Electron Laser,* National Academy Press, Washington, D.C. (1982).
19. S. Silver, *Microwave antenna theory and design,* pp. 50–51, McGraw-Hill, New York (1949).
20. Kwang-Je Kim, private communication.
21. W. D. Duncan and G. P. Williams, Infrared synchrotron radiation from electron storage rings, *Appl. Opt.* **22,** 2914–2923 (1983).
22. E. H. Putley, Solid state devices for infra-red detection, *J. Sci. Instrum.* **43,** 857–868 (1966).
23. M. R. Hueschen, P. L. Richards, and E. E. Haller, Performance of Ge:Ga far infrared detectors, Proceedings of the NASA Infrared Detector Technology Workshop (August 1983), p. 3-1.
24. E. L. Dereniak, R. R. Joyce, and R. W. Capps, Low noise preamplifier for photoconductive detectors, *Rev. Sci. Instrum.* **48,** 392–394 (1977).
25. F. J. Low, Integrating amplifiers using cooled JFETS, *Appl. Opt.* **23,** 1309–1310 (1984).
26. P. L. Richards and L. T. Greenberg, in: *Infrared and Millimeter Waves* (K. J. Button, ed.), Vol. 6, pp. 149–207, Academic, New York (1982).

27. N. S. Nishioka, P. L. Richards, and D. P. Woody, Composite bolometers for submillimeter wavelengths, *Appl. Opt.* **17**, 1562–1567 (1978).
28. A. E. Lange, E. Kreysa, S. E. McBride, and P. L. Richards, Improved fabrication techniques for infrared bolometers, *Int. J. Infrared Millimeter Waves* **4**, 689–706 (1983).
29. R. J. Bell, *Introductory Fourier Transform Spectroscopy*, Academic, New York (1972).
30. N. P. Palaio, M. Rodder, E. E. Haller, and E. Kreysa, Neutron transmutation-doped germanium bolometers, *Int. J. Infrared Millimeter Waves* **4**, 933–943 (1983).
31. F. J. Low and A. R. Hoffman, The detectivity of cryogenic bolometers, *Appl. Opt.* **2**, 649–650 (1963).
32. C. Kittel and H. Kroemer, *Thermal Physics*, W. H. Freeman, San Francisco (1980).
33. R. B. Bailey, T. Iri, and P. L. Richards, Infrared spectra of carbon monoxide on evaporated nickel films: A low temperature thermal detection technique, *Surf. Sci.* **180**, 626–646 (1980).
34. J. C. Burgiel and L. C. Hebel, Far infrared spin and combination resonance in bismuth, *Phys. Rev.* **140**, A925–A929 (1965).
35. H. E. Grenga, K. R. Lawless, and L. B. Garmon, Structure and topography of monocrystalline nickel thin films grown by vapor deposition, *J. Appl. Phys.* **42**, 3629–3633 (1971).
36. J. Kleefeld, B. Pratt, and A. A. Hirsch, Epitaxial growth of nickel from the vapour phase, *J. Crystal. Growth* **19**, 141–146 (1973).
37. Stycast 2850 GT, Emerson and Cuming Co., Canton, Massachusetts.
38. P. Dumas, R. G. Tobin, and P. L. Richards, Study of adsorption states and interactions of CO on evaporated noble metal surfaces by infrared absorption spectroscopy, I. Silver, *Surf. Sci.* **171**, 555–578 (1986).
39. P. Dumas, R. G. Tobin, and P. L. Richards, Study of adsorption states and interactions of CO on evaporated noble metal surfaces by infrared absorption spectroscopy, II. Gold and copper, *Surf. Sci.* **171**, 579–599 (1986).
40. S. Klause, C. Mariani, K. C. Prince, and K. Horn, Screening effects in photoemission from weakly bound adsorbates: CO on Ag(110), *Surf. Sci.* **138**, 305–318 (1984).
41. A. Crossley and D. A. King, Adsorbate island dimensions and interaction energies from vibrational spectra: CO on Pt{001} and Pt{111}, *Surf. Sci.* **95**, 131–155 (1980).
42. A. Crossley and D. A. King, Infrared spectra for CO isotopes chemisorbed on Pt{111}: evidence for strong adsorbate coupling interactions, *Surf. Sci.* **68**, 528–538 (1977).
43. R. M. Hammaker, S. A. Francis, and R. P. Eischens, Infrared study of intermolecular interactions for carbon monoxide chemisorbed on platinum, *Spectrochim. Acta* **21**, 1295–1309 (1965).
44. B. N. J. Persson and R. Ryberg, Vibrational interaction between molecules adsorbed on a metal surface: The dipole–dipole interaction, *Phys. Rev. B* **24**, 6954–6970 (1981).
45. M. Moskovits and J. E. Hulse, Frequency shifts in the spectra of molecules adsorbed on metals, with emphasis on the infrared spectrum of adsorbed CO, *Surf. Sci.* **78**, 397–418 (1978).
46. P. Geraghty, M. Wixom, and A. H. Francis, Photocalorimetric spectroscopy and ac calorimetry of thin surface films, *J. Appl. Phys.* **55**, 2780–2785 (1984).
47. S. Chiang, R. G. Tobin, and P. L. Richards, Vibrational spectroscopy of chemisorbed molecules by infrared emission, *J. Vac. Sci. Technol. A* **2**, 1069–1074 (1984).
48. P. R. Griffiths, *Chemical Infrared Fourier Transform Spectroscopy*, Wiley, New York (1975).
49. D. Kember and N. Sheppard, The use of ratio-recording interferometry for the

measurement of infrared emission spectra: applications to oxide films on copper surfaces, *Appl. Spectrosc.* **29,** 496–500 (1975).

50. L. M. Gratton, S. Paglia, F. Scattaglia, and M. Cavallini, Infrared emission spectroscopy applied to the oxidation of molybdenum, *Appl. Spectrosc.* **32,** 310–316 (1978).

51. M. Adachi, K. Kishi, T. Imanaka, and S. Teranishi, Infrared emission spectra of formic acid adsorbed on V_2O_5, *Bull. Chem. Soc. Jpn* **40,** 1290–1292 (1967).

52. O. Koga, T. Onishi, and K. Tamaru, Infrared emission spectra of formic acid adsorbed on alumina, *J. Chem. Soc. Chem. Commun.,* 464 (1974).

53. M. Primet, P. Fouilloux, and B. Imelik, Propene–V_2O_5 interactions studied by infrared emission spectroscopy, *Surf. Sci.* **85,** 457–470 (1979).

54. M. Primet, P. Fouilloux, and B. Imelik, Chemisorptive properties of platinum supported on zeolite Y studied by infrared emission spectroscopy, *J. Catal.* **61,** 553–558 (1980).

55. D. L. Allara, D. Teicher, and J. F. Durana, Fourier transform infrared emission spectrum of a molecular monolayer at 300 K, *Chem. Phys. Lett.* **84,** 20–24 (1981).

56. JFET 00, Infrared Laboratories, Inc., Tucson, Arizona.

57. L. Harris and J. K. Beasley, The infrared properties of gold smoke deposits, *J. Opt. Soc. Am.* **42,** 134–140 (1952).

58. L. Harris, The transmittance and reflectance of gold black deposits in the 15- to 100-micron region, *J. Opt. Soc. Am.* **51,** 80–82 (1961).

59. L-51 Taut band chopper, Bulova Watch Co., Woodside, N.Y.

60. H. Ibach, Comparison of cross sections in high resolution electron energy loss spectroscopy and infrared reflection spectroscopy, *Surf. Sci.* **66,** 56–66 (1977).

61. R. G. Tobin, S. Chiang, P. A. Thiel, and P. L. Richards, The C≡O stretching vibration of CO on Ni(100) by infrared emission spectroscopy, *Surf. Sci.* **140,** 393–399 (1984).

62. S. Chiang, R. G. Tobin, P. L. Richards, and P. A. Thiel, The molecule-substrate vibration of CO on Ni(100) studied by infrared emission spectroscopy, *Phys. Rev. Lett.* **52,** 648–651 (1984).

63. R. G. Tobin and P. L. Richards, An infrared emission study of the molecule-substrate mode of CO on Pt(111), *Surf. Sci.* **179,** 387–403 (1987).

64. J. C. Ariyasu, D. L. Mills, K. G. Lloyd, and J. C. Hemminger, Anharmonic damping of adsorbate vibrational modes, *Phys. Rev. B* **28,** 6123–6126 (1984).

65. M. J. Dignam, in: *Vibrations at Surfaces: Proceedings of an International Conference at Namur, Belgium* (R. Caudano, J. M. Gilles, and A. A. Lucas, eds.), pp. 265–288, Plenum Press, New York (1982).

66. J. C. Tracy, Structural influences on adsorption energy. II. CO on Ni(100), *J. Chem. Phys.* **56,** 2736–2747 (1972).

67. G. E. Mitchell, J. L. Gland, and J. M. White, Vibrational spectra of coadsorbed CO and H on Ni(100), *Surf. Sci.* **131,** 167–178 (1983).

68. S. Andersson, Vibrational excitations and structure of CO adsorbed on Ni(100), *Solid State Commun.* **21,** 75–81 (1977).

69. J. C. Bertolini and B. Tardy, Vibrational EELS studies of CO chemisorption on clean and carbided (111), (100), and (110) nickel surfaces, *Surf. Sci.* **102,** 131–150 (1981).

70. M. Trenary, K. J. Uram, F. Bozso, and J. T. Yates, Jr., Temperature dependence of the vibrational lineshape of CO chemisorbed on the Ni(111) surface, *Surf. Sci.* **146,** 269–280 (1984).

71. B. N. J. Persson and R. Ryberg, Vibrational phase relaxation at surfaces: CO on Ni(111), *Phys. Rev. Lett.* **54,** 2119–2122 (1985).

72. B. E. Hayden and A. M. Bradshaw, The adsorption of CO on Pt(111) studied by infrared reflection-absorption spectroscopy, *Surf. Sci.* **125,** 787–802 (1983).
73. C. R. McCreight, Two-dimensional infrared detector arrays, *Proc. IAU Colloq.* **79,** 585–602 (1984).
74. D. M. Rank, Astronomical applications of IR CID technology, final report, NADA CR 166–584 (1984).
75. T. J. Chuang, Infrared chemiluminescence from XeF_2-silicon-surface reactions, *Phys. Rev. Lett.* **42,** 815–817 (1979).

Index